CSIR	Council of Scientific and Industrial Research
CSISA	Cereal Systems Initiative for South Asia
CSO	Central Statistics Office
CSR	Corporate Social Responsibility
CU	Central University
CWR	Crop Wild Relatives
DAC	Department of Agriculture and Cooperation
DAC&FW	Department of Agriculture Cooperation and Farmers Welfare
DAGAT	Diacylglycerol Acyltransferase
DAHDF	Department of Animal Husbandry, Dairying and Fisheries
DARE	Department of Agricultural Research and Education
DBT	Department of Biotechnology
DES	Directorate of Economics and Statistics
DFI	Development Financial Institution
DGCIS	Directorate General of Commercial Intelligence and Statistics
DGR	Directorate of Groundnut Research
DGWG	Dee-Geo-Woo-Gen
DM	Dry Matter
DNA	Deoxyribonucleic Acid
DOA	Department of Agriculture
DoAC	Department of Agriculture and Cooperation
DoC	Department of Commerce
DoCS	Department of Civil Supplies
DoR	Directorate of Oilseed Research
DoVVF	Directorate of Vanaspati, Vegetable Oil and Fats
DPAP	Drought Prone Area Programme
DREB	Gene Dehydration Responsive Element Binding Gene
DSR	Direct Seeded Rice
DSR	Directorate of Soybean Research
DST	Department of Science and Technology
DU	Deemed-to-be University
EEZ	Exclusive Economic Zone
e-NAM	Electronic National Agriculture Markets
ENSO	El Niño–Southern Oscillation
EPA	Environment Protection Act
ESCAP	Economic and Social Commission for Asia and the Pacific
ET	Evapotranspiration
ETL	Economic Threshold Level
EXIM	Export–Import
FAI	Fertiliser Association of India
FAO	Food and Agriculture Organization
FAOSTAT	Food and Agriculture Organization Statistics
FARA	Forum for Agricultural Research in Africa
FFDA	Fish Farmers Development Agency
FICCI	Federation of Indian Chambers of Commerce and Industry
FIRB	Furrow Irrigated Raised Bed System
FLD	Front-line Demonstration
FMD	Foot and Mouth Disease
FPO	Farm Producer Organization
FRP	Fibreglass Reinforced Plastic
FSHG	Farmers' Self-Help Groups
FSI	Forest Survey of India

FSII	Federation of Seed Industries of India
FSR	Farming Systems Research
FTEP	Farmers Training and Education Programme
FY	Financial Year
FYM	Farmyard Manure
GAP	Good Agricultural Practice
GAV	Gross Value Added
GB	Grameen Bank
GBPUA&T	GB Pant University of Agriculture and Technology
GCARD	Global Conference on Agricultural Research for Development
GCDT	Global Crop Diversity Trust
GCM	Global Climate Model
GCWA	Global Conference on Women in Agriculture
GDP	Gross Domestic Product
GEAC	Genetic Engineering Approval Committee
GEM	Genotype × Environment × Management
GFAR	Global Forum on Agricultural Research
GHG	Greenhouse Gas
GHI	Global Hunger Index
GIPB	Global Initiative for Plant Breeding
GIS	Geographic Information System
GLF	Global Leadership Forum
GM	Genetically Modified
GNP	Gross National Product
GnRH	Gonadotropin Releasing Hormone
GNV	Gross Net Value
GODAN	Global Open Data for Agriculture and Nutrition
GoI	Government of India
GPA	Global Plan of Action
GR	Genetic Resource
GRFA	Genetic Resources for Food and Agriculture
GRM	Genetic Resource Management
GSDP	Gross State Domestic Product
GVA	Gross Value Added
GWS	Genome-wide Selection
HCF	Human Chorionic Gonadotropin
HIS	High Income States
HIV/AIDS	Human Immunodeficiency Virus
HRD	Human Resource Development
HVOC	Hindustan Vegetable Oils Corporation
HYV	High Yielding Varieties
HYVP	High Yielding Varieties Programme
IAA-IPB	Incubator for Agribusiness and Agroindustry at Bagor Agriculture University, Indonesia
IAAP	Intensive Agriculture Area Programme
IAC	International Agrobiodiversity Congress
IADP	Integrated Agriculture Development Programme
IAE	Impact Assessment and Evaluation
IARC	International Agricultural Research Centres
IARI	Indian Agricultural Research Institute
IASRI	Indian Agricultural Statistics Research Institute
IAUA	Indian Agricultural University Association
IBPGR	International Board for Plant Genetic Resources

17 goals, considering the current levels of poverty and hunger that exist in India (NITI Aayog, 2015).

Meeting the Targets of Sustainable Development Goals

In order to meet the SDG targets, India will have to: (i) double its agricultural income by 2030 from small-scale food producers, particularly women, family farmers, pastoralists and fishermen, through secure and equal access to land, other productive resources, inputs, knowledge, financial services, markets, and opportunities for value addition, as well as non-farm employment; (ii) maintain, by 2020, available genetic diversity of seeds, cultivated plants and domesticated animals and their related wild species, and promote access to and fair and equitable sharing of benefits arising from their use; (iii) increase investment in rural infrastructure, agricultural research, technology development and extension services; (iv) correct trade restrictions and distortions in world agricultural markets, including possible elimination of agricultural subsidies; and (v) adopt measures to ensure proper functioning of food commodity markets and their derivatives, and facilitate timely access to market information, including on food reserves, in order to help limit extreme food price volatility. The goal of SDG1 relates to elimination of poverty and SDG2 calls specifically to 'end hunger, achieve food security and improved nutrition through sustainable agriculture'. Sustainability means using fewer natural resources to produce food and reducing food waste and loss. Improved nutrition means reducing both hunger and obesity through improved education, and access to and availability of good-quality foods (Farming First, 2015; Paroda, 2017).

SDG1 and SDG2 resonate strongly with the Indian development agenda since elimination of poverty and hunger continues to be a major goal in the future. Fortunately, the database for poverty indicators is robust and India has adopted some of the elements of a social protection network. Food Security Act India is justly proud of its success at the food front but this has not taken care of existing hunger. If India succeeds in its goal of poverty reduction, it will contribute substantially to the elimination of hunger. Indian policy, however, has placed too much emphasis on hunger measured in terms of low dietary energy intake. The country faces a serious problem of poor nutrition. Many of its children are stunted and weigh less than the children in many other countries in the region. This could be partly due to the young age at which girls marry and their poor nutritional status. This is a principal challenge today, and if we can address this, it would take us a long way to meeting the SDGs. Interestingly, SDGs concerning hunger, decline in poverty and average per capita calorie intake seem to have been addressed well in recent years. For the country as a whole, rural poverty declined from 45.61% in 1983 to 28.30% in 2004–05, and urban poverty declined from 42.15% to 25.70% in a similar period. During the intervening period, the average calorie intake per capita declined from 2220 to 2040 and from 2089 to 2020 kcal in the rural and urban sectors, respectively. In fact, as regards calorie deprivation, its extent has increased from 69% to 85% in rural India and from 60% to 65% in urban India (NITI Aayog, 2015).

Urbanization comes with challenges to agriculture and nutrition. Higher urban incomes are associated with a dietary transition to more fruits and vegetables, animal-sourced food, fats and oil, and refined grains, which require more intensive use of natural resources. Urban lifestyles tend to increase consumption of processed foods and the urban poor are often limited to cheap, unhealthy foods. At the same time, as the urban population grows, hunger and malnutrition will increase. In addition to access to healthy and nutritious foods, access to clean water, toilets and sanitation will also present challenges. Yet rapid urbanization brings opportunities, as the rise in demand for increased and diversified food production in rural areas can contribute to improved farmers' livelihoods. To take advantage of these opportunities, strong rural–urban links are needed. Where links are strong, rural farmers can sell larger shares of produce in urban markets, and labourers can migrate or commute to nearby towns for seasonal work and have better options for their livelihood.

The agriculture sector in India is currently facing numerous challenges such as: decline in the size of land holdings, natural resources (especially soil and water), adverse impact of climate change, factor productivity decline, costly inputs, fluctuating markets and decline in income. In the country there is a huge gap between the actual yields and the potential yields, and this yield gap is more in the case of pulses, oilseeds and other neglected crops. At the same time, we have to look into the public distribution system (PDS) by plugging the leakages and the diversions. There are various factors for low yields of crops in the country as compared to most of the developed countries. With the passing of time there is a decline in yield and these varieties become susceptible to diseases and pests. There is also the problem of low seed replacement rate (SRR) in the country, mostly of pulses and oilseeds. As such, greater emphasis is needed on increasing the seed replacement rate using high-yielding varieties and hybrids. The main questions before us now are:

- How can agriculture contribute towards achieving SDGs?
- What should be the strategy to promote agriculture for achieving SDGs?
- What lessons can other developing countries, especially in south Asia, learn from India, or vice versa?

We have to achieve the SDGs with limited and shrinking resources, and with a changing climate scenario. We have considerably harmed our agro-ecology and lost considerable diversity of our flora and fauna; many insect and weed species have become resistant to various antibiotics; many new weed species have emerged; many new diseases are taking their toll and soils have become sick and degraded. Thus, to achieve SDGs, we have to mainly focus now on climate-smart agriculture, like zero- or no-till cultivation, rainwater harvesting, practices that make best possible use of available resources with minimum loss of natural resources, and, above all, the loss of agrobiodiversity. The role of improved varieties/hybrids and management practices have immense potential in achieving the SDGs. It is encouraging that the National Agricultural Research System (NARS) has developed several technologies that promise to increase income, reduce production costs, conserve natural resources, improve food quality and nutrition, and minimize various risks. The need now is to create an enabling environment to scale out useful and efficient innovations like conservation agriculture (CA) for greater adoption and large-scale impact on the income of our smallholder farmers.

Farm mechanization also saves a lot of energy and labour. Our policies and institutions should support the marginal and small farmers to adopt farm mechanization. The financial institutions must provide better credit at lower interest rates. Similarly, more farmers and more crops should be brought under insurance cover. Mobilization of farmers by organizing them into farmers' cooperatives, producer companies or commodity interest groups should now be the major aim of all developmental institutions. These groups could then be linked to the markets to increase their income substantially.

The Indian Council of Agricultural Research (ICAR) coordinates research and education conducted by 107 specialized institutes/research centres and 67 agricultural universities across the country. Technological innovations are the backbone of productive and resilient farms, fisheries and livestock operations, and a safe, wholesome food supply. They contribute to improvements in the quality of seeds, animal stock and inputs, labour-saving devices, effective production and conservation practices, reduction of post-harvest losses, efficient price discovery mechanisms, and control of pests, diseases and contamination. Access to these innovations will be essential if farmers and producers along the value chain are able to meet the rising global demands of climate change. Climate change, resource constraints, and storage and distribution of food are some concerns that threaten India's food security. With increasing population and socioeconomic development needs, access and availability of resources for food production can be seen as a critical constraint in ensuring food security. Agriculture is undeniably a resource-intensive sector, and this fact comes along with a need for efficient and effective management of finite resources in order to ensure long-term sustainability of agriculture and food security for all.

Recent Government Initiatives

The Indian government is giving high priority to the agriculture sector to make it more efficient, competitive, sustainable and resilient. Doubling farmers' income by 2022 is a recent policy initiative of the government. In this context there are several programmes that aim to increase farmers' income, conserve soil and water resources, improve resilience and reduce climatic risks. These programmes include: the Prime Minister Irrigation Programme, the Prime Minister Agricultural Insurance Scheme, the National Food Security Mission, the National Horticulture Mission, the National Mission on Sustainable Agriculture, the National Agricultural Development Plans, the National Livestock Mission, the Midday Meal Scheme, and the Anganwadi Centres, contributing to tackling food and nutrition insecurity. To strengthen value chains of agricultural commodities and improve market efficiency, a provision has been made to develop e-NAM (One Nation, One Market). However, to establish efficient and inclusive rural-urban value chains, institutional arrangements that support the participation of marginal and smallholder farmers, who often have little marketable surplus, are needed. Production in urban and peri-urban areas is shifting towards resource-intensive foods such as vegetables, dairy, meat and poultry to meet the rapidly growing demand. To veer production to rural areas, thereby reducing pressure on increasingly scarce urban and peri-urban lands, rural agri-infrastructure such as cold chains, cold storage and processing facilities are necessary. Leveraging towns and intermediate cities to facilitate economic and social connections between rural and urban areas, and improving rural infrastructure, is therefore crucial. All these efforts demonstrate India's commitment to accomplish the SDGs that relate to agriculture. There is, however, an urgent need to ensure reorientation of ongoing efforts towards higher efficiency and effectiveness of initiatives by developing a road map by which we are able to achieve the goals well before 2030. To end hunger and malnutrition in India and beyond, we must find solutions that take account of the ongoing trend of urbanization. Doing so is key in India where, despite progress, 20% are still hungry and around 39% of children are stunted. Improving links between rural and urban areas is therefore a critical start.

Indicators of Achieving Sustainable Development Goals

Major dimensions of hunger include calorie deprivation and protein hunger (including hidden hunger). Some specific policies to achieve sustainability include: focus on hunger (including hidden hunger) and malnutrition, taking a 'zero hunger' by 2025 challenge; links between agriculture and nutrition; increased investment; raising the productivity of small farmers; assessing climate change and thereby improving productivity and resilience in agriculture; and gender-sensitive policies in agriculture and health. The time has come to focus on small farmers, rainfed agriculture, the plight of women farmers and youth, and also on biofortified crops for nutritional security. It has also been observed that there was intense desertification through the warming of cold desert areas and land degradation in the eastern region between 1975 and 2006. Due to this, agriculture is becoming distressed due to crop failures. Also, in the southern region, the coconut-based farming system has become uneconomical. Due to land degradation there is an increase in arsenic and fluoride contamination, a shift in rivers, a shift in the Sundarban delta, and increased aridity and incidence of drought, floods and cyclones, which aggravate the situation further. There is a need to develop site-specific information through land resource inventory (LRI) on a 1:10,000 scale, along with the use of balanced fertilizers, boosting rainfed agriculture, and land management in hills. Land use plans need to be developed for plateaux, the drought-hit area of central India, the coastal region, the flood plains and areas with potential for carbon sequestration and geoportal or mobile apps.

The impact of climate change is clearly visible across the globe and tropical countries like India are most vulnerable. In the past 15 years the country has observed simultaneous occurrence of drought and floods affecting agriculture, food and nutrition, and the livelihoods and sustainability of smallholder farmers. Setting up integrated farming systems (IFS) models for households, use of community participation, zero tillage, stopping burning of crop residues, and expanding climate-resilient villages could be major solutions for climate risk reduction. Contingency plans are required to be in place, such as water-saving cultivars, crop diversification, rainwater harvesting and conservation, building large farm ponds, sustainable vegetables and horticulture systems, and increased production of pulses and fodder, so as to increase household farm income.

New Technologies and Innovations

There is a need to accelerate the breeding of self-pollinating crops with a wider gene pool, to develop and deploy high-yielding, nutrient-rich hybrids in both field and horticulture crops, especially vegetable crops, and to promote biofortified crops and the use of genome engineering/gene editing to gain more yield and to resist drought and disease more effectively. Crop intensification, rainwater harvesting, recycling of wastewater, managing blue water, mechanization and value chain/crop cycle (from tillage and seedbed preparation to post-harvesting) to enhance crop productivity also need to be addressed. Scaling-up farm mechanization by promoting both pre- and post-harvest machineries brings efficiency in the food value chain by improving cropping intensity, reducing the cost of production and drudgery, enhancing farm power supply and maintaining a socially desirable mix of human labour, animal power and mechanical power. IT-based skill development programmes for extension workers, decision support systems, appropriate technologies for mechanizing horticultural crops, especially in hilly areas, and cost-effective technologies like smart tractors, unmanned aerial vehicles and wireless technology are some areas that need attention. Also, a pluralistic extension approach needs to be promoted along with empowerment models like commodity groups, farmers' organizations and producer companies to strengthen market links. Programme delivery mechanisms in disadvantaged areas need to be streamlined with emphasis on socioeconomic mapping. Extension services in allied sectors like horticulture, animal husbandry, fisheries, poultry, sericulture etc. need to be strengthened. The competency of extension agencies, especially youth as 'technology agents', needs to be improved by systematic training and capacity-building programmes, enabling them to respond to emerging issues like climate change adaptation, use of ICT, input-use efficiency, integrated nutrient management (INM) and integrated pest management (IPM) technologies. Agricultural extension planning at block or cluster level needs to be addressed jointly by agricultural technology management agencies (ATMA), Krishi Vigyan Kendras (KVK), non-governmental organizations and the private sector, at micro level, keeping in view the specific requirements to meet the SDGs.

India must again strengthen conventional plant breeding (including pre-breeding) and pursue the adoption of GM technology both in field and horticultural crops, for which policy support is badly needed. Availability of good-quality seed, including hybrids and planting material, is the pressing need. Research on pre- and post-harvest losses also needs to be strengthened. Besides characterization of bioresources, a multidisciplinary/multifunctional approach will have to be followed in natural resource management in a way that enables farmers and scientists to work in unison on a long-term basis. In the livestock sector, India may expand successful models like Amul Dairy with still better efficiency and investigate the reasons for not scaling up this model in other states. Also, there is a need to reduce the number of non-productive animals, to conserve and improve indigenous breeds, to reduce methane emissions through better housing and feeding of large animals, to promote backyard poultry and to enhance feed resources that can be produced locally,

Role of Public Policies

Changing goals and approaches have invariably led to the failure of policies to reduce poverty and inequality. Many times, administrative incapacity, and uncoordinated and duplicate efforts have resulted in not achieving the targets. There is a need to bring in socioeconomic reforms to insulate the poor from adverse shocks. The strengthening of institutions for the effective implementation of policies is required. A different mindset is necessary to set targets commensurate with the right policies. We know that agricultural spending is still low in India (0.4%, to be raised to a minimum of 1% of AgGDP). Also, more capital investment in agriculture-related activities is necessary in high-income states, middle-income states and low-income states. High-income states need investment in agricultural R&D, health and education, with greater focus on non-farm employment opportunities; whereas rural infrastructure development is required in low-income states. Rationalization of subsidies/reduction in input subsidy and technology interventions are also required to improve the efficiency of public spending. To meet the target of doubling farmers' incomes by 2022, an innovative strategy is required for increasing the livelihood of resource-poor marginal farmers through diversification towards sub-sectors of agriculture like livestock, horticulture and fisheries, and to move towards secondary and speciality agriculture with a focus on marketing reforms, including price management. Also, there is a need to put in place policies to promote low-volume, high-value crops, through market links, and for exports and value addition.

Climate Change-related Policies

India faces many climatic challenges, such as serious droughts in one region and dangerous floods in another. The reason it is so vulnerable is because it is a large country with many citizens living in poverty, inadequate infrastructure and lack of government planning to deal with complex weather systems. Recently, a World Bank report emphasized how India will be subject to irregular monsoons, flooding, rising sea levels and higher temperatures. The monsoon season is vital to the Indian economy. Preparation for weather irregularities is thus essential in order to protect the lives of Indian people and the growth of the Indian economy.

Climate change can have a dramatic impact on natural resources, economic activities, food security, health and physical infrastructure. The threat is especially severe in places where people's livelihoods depend on natural resources. In such areas, climate adaptation measures take on a special significance for safeguarding rural livelihoods and ensuring sustainable development. The Indian government launched the country's first National Action Plan on Climate Change (NAPCC) in 2008, with the main themes of: (i) further expansion of solar power generation; (ii) further increases in energy efficiency; (iii) measures to sustain India's environmental and water assets; (iv) further expansion of forests for carbon sink purposes; (v) sustainable agriculture; and (vi) developing a knowledge base for dealing with climate change issues. India's NAPCC recommended that the country should generate 10% of its power from renewable sources by 2015 and 15% by 2020. There are

three main areas of policy, focused on targeting, mitigating and adapting to climate change. First, energy access is a priority. Providing energy to 400 million people who do not have access to electricity is a necessity; using off-grid solutions such as solar energy is key to reaching these people and providing sustainable, clean energy sources. Secondly, India has adopted an NAPCC, and many of its smaller states are developing state action plans (SAP) that include climate change adaptation. Many of the policies are already being implemented as part of the centralized economic plan drawn up by India's Planning Commission (now NITI Aayog). Thirdly, India is keen to further develop its economy and to continue its policies aimed at poverty alleviation, and it appears determined to pursue these goals in addition to policies aimed at reducing greenhouse gas (GHG) emissions.

Under the Paris Convention, countries responsible for more than 80% of global greenhouse gas emissions made specific commitments to reduce their emissions by 2020. The Paris agreement also includes commitments going beyond 2020, and this reflects a greater level of ambition than was seen in previous agreements. Countries' emissions reduction commitments reflect their various levels of development and capability. The Indian government has voluntarily agreed to reduce the emissions intensity of its GDP by 20–25%, from 2005 levels, by 2020. It also has agreed that its GHG emissions from one unit of GDP will reduce by one third by 2030, from what they were in 2005. India intends to produce about 40% of its electricity from non-fossil fuel-based sources, like solar, wind and hydropower, by 2030. These promises have been made in an action plan that India submitted to the UN's climate body, the United Nations Framework Convention on Climate Change (UNFCCC), outlining the steps it wants to take, up to 2030, in the global fight against climate change. India has sought international help of at least US$2.5 trillion, at current prices, in order to implement these plans.

India is the nation that made the key efforts to impress the international community with its intent to shift to a sustainable, low-carbon path that will confront climate change, improve human health and foster prosperity for all. In India, climate change-related action seems to be the most successful since it is integrated with efforts to tackle existing challenges of energy access, water security, agricultural productivity, disaster resilience and broader economic development goals. India is now better prepared to deal with the multifaceted nature of climate change. However, the current challenge is to develop a cross-sectoral, integrated approach. In common with other developing countries, India considers that the solution to the world's climate change problems are primarily the responsibility of the developed, industrialized world. It has resisted calls for a limit to be placed on its own GHG emissions. It is concerned to further develop its economy and continue its policies aimed at alleviating poverty, and it appears determined to pursue these goals in addition to policies aimed at reducing GHG emissions. India is the world's fourth-largest producer of greenhouse gases.

Addressing climate change effectively will be the key to achieving the SDGs. Many investments in mitigation and adaptation, such as low-carbon energy plants or climate-resilient infrastructure, are operationally indistinguishable from investments in 'development', and the two must be structured and executed together. Some of the policy action points related to climate risk management are: to invest in climate-smart technologies and capacity-building with a synergy of food security and integrated/scientific land-use policy. Also, there is a need to review the SAP for climate change. Emphasis needs to be placed on proper analysis and effective adoption of soil health cards (SHC) and soil-testing laboratories (STL), at least at the block level, to enhance risk-coping abilities of resource-poor farmers to deal with weather and market fluctuations. Capacity-building and the adoption of efficient irrigation systems like drip and sprinkler irrigation to reduce excess water use and increase productivity are vital. The rationale for climate-smart agriculture (CSA) has to be appreciated by decision makers and scaled to benefit smallholder farmers.

Towards 'No Poverty' and 'No Hunger'

Income inequalities continue to grow and poverty remains largely a socioeconomic problem. Approximately three quarters of the world's poor live in rural areas, with the share even bigger in low-income countries. In addition, certain groups are disproportionately represented among the poor: women, the disabled, children and people living in tribal areas. The degradation of the productive assets of the poor, exacerbated by lack of access to modern infrastructure and amenities, creates a poverty trap that reinforces further degradation and worsening of poverty. While its extreme manifestations are in low-income countries, developed countries also need to address problems of poverty and malnutrition. Reducing by half the number of poor people, as defined nationally, and ending all forms of malnutrition requires developing countries like India to initiate focused action, including addressing the structural causes of poverty, hunger and malnutrition. Feeding the growing world's population, expected to exceed 9 billion by 2050, will require food production to increase by 70% at a time when agriculture is already facing unprecedented pressures from a degraded natural resource base, coupled with the effects of climate change. What is more, the investment gaps in agriculture and the social sector are substantial.

'The Future We Want' has set out SDGs to end poverty in all its forms, everywhere, to end hunger, to achieve food security and improved nutrition, and to promote sustainable agriculture, ensuring healthy lives for all, at all ages. It aims to ensure sustainable consumption and production patterns and to take urgent action to combat climate change and its impact. Hence, alleviation of poverty and hunger is of the utmost importance. Hunger can be removed only when households have a continuous flow of income. For this to happen, there is an urgent need for agricultural diversification. Skills development of young people can help them to get jobs and provide regular income for their families. Hence, vocational training and entrepreneurship are urgently needed; India needs young entrepreneurs who are job creators rather than job seekers.

The Indian government passed the National Food Security Act (also known as the Right to Food Act) on 10 September 2013 with the objective to provide food and nutritional security by ensuring access to adequate quantities of quality food at affordable prices so that people may live a life with dignity. The Act provides for up to 75% of the rural population and 50% of the urban population to receive subsidized foodgrains under a targeted public distribution system

(TPDS). This represents about two thirds of the total population. Eligible persons will be entitled to receive 5 kg of foodgrains/person/month at subsidized prices of Rs 3, 2, and 1/kg for rice, wheat and coarse grains, respectively. The existing Antyodaya Anna Yojana (AAY) households, which constitute the poorest of the poor, will continue to receive 35 kg of foodgrains/household/month. The Act also focuses on nutritional support for women and children. In addition to meals for pregnant women and lactating mothers, during pregnancy and for six months after childbirth, women will be entitled to receive maternity benefit of not less than Rs 6000. Children up to 14 years of age will be entitled to nutritious meals as per the prescribed nutritional standards. In the case of non-supply of entitled foodgrains or meals, the beneficiaries will receive a food security allowance. The Act also contains provisions for setting up a grievance redress mechanism at district and state levels. Separate provisions have also been made in the Act for ensuring transparency and accountability. At present, 32 states/union territories (UTs) are implementing the Act – Andhra Pradesh, Assam, Bihar, Chandigarh, Chhattisgarh, Daman & Diu, Delhi, Goa, Haryana, Himachal Pradesh, Jharkhand, Karnataka, Lakshadweep, Madhya Pradesh, Maharashtra, Odisha, Puducherry, Punjab, Rajasthan, Sikkim, Telangana, Tripura, Uttarakhand, West Bengal, Uttar Pradesh, Meghalaya, Jammu & Kashmir, Andaman & Nicobar, Mizoram, Dadra & Nagar Haveli, Gujarat and Arunachal Pradesh.

To achieve 'no hunger' status by 2030, the government must build on the approaches that have proved to be effective. These feature three important elements:

- promoting immediate access to food and nutrition-related services for hungry people through a social protection net;
- creating opportunities for the poor and hungry to improve their livelihoods by promoting decent labour conditions, and increasing investment to improve farm productivity, rural infrastructure and better market access; and
- increasing the sustainability of food production and consumption systems by conserving natural resources, adopting sustainable agricultural practices, reducing food losses, diversifying dietary preferences, reducing

levels of food waste, and reducing emissions of GHGs from agriculture and other sectors so as to slow down the pace of climate change and ensure better food security for future generations.

Investing in agriculture is the best way to increase the productivity of agricultural labour and the land. Productivity increases enable better remuneration, thus contributing to raising the living conditions of food-insecure people while helping to reduce pressure on scarce natural resources. Public investment in institution-building, productivity-enhancing research, rural transport, markets, health, education and social protection is needed to ensure food security, nutrition and inclusive growth as well as sustainable development.

Role of Institutions

There is a need to empower farmers with the right information to enable them to improve agricultural productivity as well as efficiency. Input providers should be competitive. The government should make increased investment in ARI4D and in strengthening rural institutions and farm services, integrating approaches to germplasm improvement, building capacity for knowledge integration and dissemination, and promoting competitiveness of technology-based input markets for access by small farmers to improved technology and reforms in land, market and trade, to realize desired outcomes. There is an urgent need to address the following areas:

- strengthening ICAR as an apex organization by tripling its budget;
- achieving autonomy of state agricultural universities;
- promoting the Institution Village Linkage Programme (IVLP) through farmers' participatory approach; and
- fostering vocational training/informal education.

The role of agri-markets is essential to increase farmers' income through price realization and crop diversification. A policy needs to be put in place to denotify fruit and vegetable crops from the Agricultural Produce Marketing Committee (APMC) Act, to promote perishable produce markets, to focus on soil health and water management and to develop more direct links with farmers. Institutions need to be more vigilant in the implementation of policies through effective monitoring and oversight and for coordination and convergence of ongoing programmes and activities.

Public–Private Partnership

There is a need for the private sector to start focusing on R&D to deliver better services and products and to involve themselves thoroughly in research, development and policy planning. The agricultural sector – dairy, animal husbandry, poultry etc. – managed by private enterprise contributes 32% of domestic GNP and provides employment for 67% of the working population. Over the years, the public sector has played a key role in agriculture in India in setting up guiding policies and providing goods and services such as fertilizers, extension and marketing. The National Agricultural Policy (NAP) 2000 also envisaged promoting private sector participation in agriculture through contract farming, land-leasing arrangements, direct marketing and setting up of private markets to allow accelerated technology transfer, capital inflow and assured markets for crop production. The private sector can offer their services through various ways throughout the agricultural value chain. Conducting research, introducing improved technologies, provision of credit through cooperatives and self-help groups, creating infrastructure (for seeds, fertilizers, pesticides, transportation and processing), helping with extension services, passing on accurate and timely information, and diffusing crop insurance are key areas where the private sector can further enhance their engagement.

India is now one of the fastest-growing economies with a target annual growth rate of over 8%. For the economy to grow at this rate there is a need to upgrade the country's infrastructure. Public–private partnership (PPP) has been recognized as one of the most effective mechanisms to achieve this. There is scope to leverage PPP as a relevant vehicle in the agriculture sector. Enhanced yield and productivity is needed, with India still battling food insecurity and poverty. Improved technology, better inputs

and improved farming practices can make this possible. Over the past 65 years, Indian agriculture has recorded an average growth rate of 2.7% p.a., making it the slowest-growing sector. The failure to consistently hit 4% growth, as targeted in the five-year plans, indicates the challenges that are faced in agriculture. Agriculture is a key sector for research, investment and development. There is an urgent need to innovate via PPP and between farmers and the government to meet India's agricultural needs through new technology and intervention models. Several partnerships have already been developed between the public and private sectors with the objective of achieving these goals. Monsanto India Limited (MIL) is an important stakeholder in the agricultural PPP space through its multiple partnerships with state governments. India has reached out to more than 900, 000 farmers through PPP alone and has helped improve yields and rural incomes significantly in the areas where these partnerships have been implemented.

Corporate Social Responsibility

India is the first country in the world to make corporate social responsibility (CSR) mandatory, following an amendment to the Company Act 2013 in April 2014. Businesses can invest their profits in areas such as education, poverty, gender equality and hunger. The Act advocates that those companies with a net worth of Rs 4.96 billion or more, or an annual turnover of Rs 9.92 billion or more, or a net profit of Rs 50 million or more, earmark 2% of their average net profit over three years for CSR. The agriculture sector can benefit from CSR to a great extent.

The Way Forward

SDGs present a unique opportunity for the entire agricultural sector to become aligned to achieve a better tomorrow. If India can accelerate its pace to achieve the SDGs, then globally we could soon eliminate hunger, achieve food security and improve household nutritional security. At the same time it is imperative that policy makers give high priority to ARI4D, ensure enhanced allocations (a minimum of 1% of agricultural

GDP) to NARS and strengthen physical and economic access to food for resource-poor people residing in rural and urban areas. In fact, the agricultural sector can be seen as an important sector for achieving the goals of eliminating poverty and hunger as well as ensuring nutrition, environmental security and protection of fast-degrading natural resources. However, success in achieving the SDGs will require a 'mission-mode' approach in implementing and effectively monitoring the progress towards the defined goals. Strategies to accomplish SDGs must therefore address the following:

- Despite witnessing Green, White and Blue Revolutions, and having attained impressive food production of 277.49 million t, milk production of 165 million t and inland and marine fish production of 11.4 million t, India ranks 100 out of 113 countries on the global hunger index (GHI), and the prevalence of poverty is around 20%. Despite physical access, its major aim should now be to provide economic access to food through effective implementation of the National Food Security Act and other safety-net initiatives, especially in the regions/states where maximum poverty and hunger still persist.
- Ensure meaningful engagement of all stakeholders in the formulation of national strategies, implementation plans and monitoring of progress towards achieving SDGs, using baseline data for defined goals.
- The functioning of NARS, involving ICAR institutes and the SAUs, must involve other stakeholders such as NGOs, farmer-producers organizations (FPOs), private-sector institutions, farmers and agribusiness entrepreneurs.
- Continuous prioritization as well as re-prioritization of the development research portfolio is needed in tune with fast-changing global, regional and national needs. The 'top-down' approach adopted in the past will have to be changed to a 'bottom-up' approach. A shift from project to programme mode, and also from commodity/crop to farming system mode is urgently needed. In this context, the focus on crop diversification, hybrid seeds/high-value crops, biotechnology, ICT, GIS and good agronomic practice (GAP) would help double farmers' incomes and obtain

resilience in agriculture with efficient inputs (water, fertilizers, chemicals for pesticides).
- Adopting ecofriendly and climate-resilient technologies, with emphasis on efficient farming systems in different ecoregions, strengthening of activities for improving soil health through organic matter recycling, conservation agriculture, efficient and needs-based use of nutrients, using decision support systems and soil test results, improved water-use efficiency using micro-irrigation techniques etc., would foster resilience in agriculture.
- Make best use of available knowledge and technologies through: (i) defining recommendation domains (technology targeting); (ii) increased investment (doubling) in managing land and water resources efficiently; and (iii) strengthening input delivery as well as market linkage mechanisms.
- The National Livestock Mission should focus on: quality feed and fodder; improved risk coverage including animal insurance; conservation and improvement of indigenous breeds; higher productivity and production; value addition; enhanced livelihood opportunities; increased awareness; and better availability of quality animal products for consumers at affordable prices.
- There is a need to develop new agri-food systems for pre- and post-production management through processing and value addition and by ensuring minimum wastage of food during storage, transportation and consumption.
- Knowledge update for farmers concerning new technologies, practices and recent advancements is a must, rather than merely providing subsidies. Building multilateral and multisectoral technology-transfer mechanisms for linking science to society, with greater emphasis on attracting and retaining youth in agriculture, especially through diversification, secondary and speciality agriculture, needs to be pursued in order to empower farmers.
- Dissemination of available high-value technologies; market linkages through e-NAM; revision of APMC; provision of pledged storage; developing and providing need-based technologies for immediate use and also for anticipatory long-term needs of

farmers/industries/consumers is needed. We need to remain competitive in order to take full advantage of the globalization of agriculture and to be prepared for the emerging new WTO regime.

- India must increase its capital investment in creating much-needed infrastructure, by involving the public and private sectors, especially in the eastern and north-eastern regions, so as to capitalize on rich natural resources that we have the potential to hasten agricultural growth and the 'evergreen revolution' (MoE and CC, 2015).
- SDGs have several interconnected goals, and thus require effective coordination and convergence mechanisms at all levels through an interdisciplinary and inter-institutional/departmental approach, to draw collective strength for desired impact. Such coordination mechanisms have to be top-down for effective monitoring and evaluation.
- Widening the policy space with much-needed faith in agricultural science and new technology, without fear and with a human face, is greatly needed to accelerate growth. Therefore an aggressive approach on policy advocacy and reform is warranted to scale innovation to achieve the SDGs in the given timeframe, i.e. 2030.

References

Farming First (2015) The Story of Agriculture and the Sustainable Development Goals. Available at: https://farmingfirst.org/sdg (accessed 4 May 2018).

IFPRI (2017) *Sustainable Development Goals: Preparedness and Role of Indian Agriculture*. International meeting organized by the International Food Policy Research Institute and TAAS, May.

MoE and CC (2015) *Achieving the Sustainable Development Goals in India: A Study of Financial Requirements and Gaps*. Ministry of Environment, Forest and Climate Change, India.

NITI Aayog (2015) *An Overview of the Sustainable Development Goals*. Technical Support by NIC and content owned and provided by National Institution for Transforming India, Government of India.

Paroda, R S (2017) Indian Agriculture for Achieving Sustainable Development Goals. Strategy paper. Trust for Advancement of Agricultural Sciences, New Delhi.

3

Fifty Years of the Green Revolution and Beyond

Preamble

During his speech delivered on the occasion of receiving the Nobel Peace Prize in 1970, Dr Norman Borlaug prophetically said:

> The Green Revolution has won a temporary success in man's war against hunger and deprivation; it has given man a breathing space. If fully implemented, the revolution can provide sufficient food for sustenance during the next three decades. But the frightening power of human reproduction must also be curbed; otherwise the success of the Green Revolution will be ephemeral only.

While delivering a special 30th anniversary lecture at the Norwegian Nobel Institute, Oslo, in 2000, he reviewed his prophecy and said:

> The world has the technology – either available or well advanced in the research pipeline – to feed on a sustainable basis a population of 10 billion people. The more pertinent question today is whether farmers and ranchers will be permitted to use this new technology. While the affluent nations can certainly afford to adopt ultra-low-risk positions, and pay more for food produced by the so-called 'organic' methods, the one billion chronically undernourished people of the low-income, food-deficit nations cannot. It took some 10,000 years to expand food production to the current level of about 5 billion tonnes per year. By 2025, we will have to nearly double current production. This cannot be done unless farmers across the world have access to

current high-yielding crop-production methods as well as new biotechnological breakthroughs that can increase the yields, dependability and nutritional quality of our basic food crops.

(Borlaug, 2000a, b)

Dr Borlaug's foresight and realization of the UN's 2030 agenda for Sustainable Development Goals (SDGs) to end poverty while protecting the planet can only happen if agricultural development and world demographic projections are kept in mind by all stakeholders, especially policy makers.

With reference to India, science and technology-based Green, White and Blue revolutions have significantly altered agricultural production and the agrarian economy in the last six decades. Between 1951 and 2017, foodgrain production increased fivefold, from 51 million to 276 million t; horticultural production swelled to 305.4 million t (the second-largest in the world); milk production grew ninefold, from 17 million to 155 million t (the highest in the world), and fish production increased 15-fold, from 0.75 million to 11.4 million t (the second-largest in the world). These unprecedented production gains, coupled with efficacious policies and actions, have resulted in more than halving the number of hungry, undernourished and ultra-poor. The years 2015–2016 marked the golden jubilee of the Green Revolution in India (Alagh, 2015). A review of the history of the Green Revolution, the lessons learnt and the policies pursued for achieving the SDGs in a contemporary context are discussed below.

The Green Revolution – A Snapshot

'In the early 1900s, the projections showed that there would be 6 billion people to feed by the turn of the century. With most of the population inhabiting the less-developed world, there was increasing fear of a world famine' (Paddock and Paddock, 1967). The narrative of the Green Revolution dates back to early 1941, when US Vice-President Henry Wallace toured Mexico as a special ambassador and was appalled by the poor state of Mexican agriculture. Thereafter, he urged the Rockefeller Foundation to look at ways of helping the Mexicans. The Rockefeller Foundation developed the Mexican Agricultural Programme (MAP) to boost Mexican agriculture with a team of four dedicated scientists. The team was headed by J. George Harrar and the other members were John Niederhauser (in charge of potato improvement), Edwin Wellhausen (maize improvement) and a young biologist from Iowa, Norman Ernest Borlaug, in charge of the wheat improvement programme (Borlaug and Dowswell, 2003). As a result of his dedication, ingenuity and use of germplasm from far and wide, Dr Borlaug developed the semi-dwarf-statured, fertilizer-responding 'miracle wheat' in 1954, which was spread by the Rockefeller and Ford Foundations throughout the world in the 1950s and 1960s, including India. Breeders incorporated dwarfing genes that allowed the development of shorter, stiff-strawed varieties of wheat. These varieties devoted much of their energy towards producing grain and relatively little towards producing straw or leaf material. They also responded better to fertilizer application than did traditional varieties. Farmers adopted the new semi-dwarf modern varieties rapidly in some areas, chiefly those with access to irrigation or reliable rainfall, and the new varieties yielded substantially more grain than previous ones (Evenson and Gollin, 2003). The highlights of the Green Revolution in India are given in Box 3.1.

The term Green Revolution was first coined in 1968 by William S. Gaud, an administrator of the US Agency for International Development (USAID), at a meeting of the Society for International Development, to describe the remarkable increases in cereal crop yields achieved in developing countries during the 1960s. The keys to this revolution were new plant varieties that fully utilized improved fertilizers and other new agrochemicals that had become available during the period. When planted using improved irrigation and crop management techniques, these new varieties gave dramatic increases in yield. The success of the Mexican programme prompted the setting up of a similar programme for rice at IRRI in the Philippines, funded jointly by the Rockefeller and Ford Foundations.

Prior to the 1960s, India struggled with feeding its increasing population, and famine was a regular feature (e.g. great Bengal famine of 1942–43), resulting from meagre food production, poor distribution, droughts and floods. Two consecutive droughts in the mid-1960s led to a famine situation, which was averted by substantial foodgrain imports from the USA under its Title I Public Law 480 (PL 480) scheme. Foodgrain import steadily rose from 1.5 million t in 1946 to 4.8 million t in 1950, peaking at 10.4 million t in 1966. While branding India with epithets like the 'begging bowl' and 'ship-to-mouth', it was predicted that Indians would die in their millions by 1975 and that no food aid could save them (Paddock and Paddock, 1967).

During the early days of independence, food crop production and productivity were very low, and coverage of high-yielding varieties was less than 5% in all crops except sugarcane, cotton and jute. Traditional wheat varieties grown in India at that time were low-yielding, albeit tall and fairly resistant to several races of rusts (Swaminathan, 1993). The introduction of high-yielding technology to India is attributed to the initiative of the political leader C. Subrahmaniam and the civil servant B. Sivaraman, who took a bold decision to import large quantities (18,000 t) of seeds of the variety Lerma Rajo 64A and Sanora 64 from Mexico in 1966. Prior to that, in 1960, Professor M.S. Swaminathan procured seeds of the semi-dwarf wheat variety 'Gaines', keeping in view its high productivity. In 1960–61, from the wheat-breeding lines received from USDA by the Indian scientists under the International Wheat Rust Nursery, Professor Swaminathan and Dr M.V. Rao identified semi-dwarf lines with long panicle and high yield potential. These lines were traced to the wheat-breeding programme of Dr Borlaug who had incorporated the Norin-10 dwarfing gene to spring wheat.

Box 3.1. The story of the Green Revolution. (Excerpts from Swaminathan, 2016)

Indian independence was born against the backdrop of the great Bengal famine where nearly 3 million men, women and children died. The Bengal famine was partly due to World War II, when Myanmar was under the occupation of Japan. Myanmar used to be a major supplier of rice to India, and that source was cut off during the war. Whatever the cause, there was a great deal of awareness among political leaders of India that agriculture would have to receive priority. In fact Mahatma Gandhi said at Noakhali, 'to those who are hungry, bread is God' and it should be available to every individual in India. Therefore, when India became independent in 1947, Jawaharlal Nehru said, 'Everything else can wait, but not agriculture.' Hence, agricultural progress received attention. The Green Revolution can be divided into three distinct periods.

The first period, between 1945 and 1955, was related to the search for methods of improving the production and productivity of major crops. In the case of rice, this was accomplished by identifying japonica strains of rice from Japan, which could respond to fertilizer and irrigation water more effectively. The indica-japonica hybridization programme was first started at the ICAR-National (earlier Central) Rice Research Institute (NARI), Cuttack, in the early 1950s, which could not lead to the desired results largely because of sterility in the hybrids. Although there were some varieties like ADT 27, identified at Aduthurai and Mashuri in Malaysia, which gave higher yields compared to earlier ones, there was a need to search for new genes for a kind of plant architecture that could help plants not to lodge or fall down even when there is good soil fertility. NARI was the starting point in the search for new genes, new plant architecture, new physiological rhythm and photo-insensitivity. This was followed by importing the Dee-Geo-Woo-Gen dwarfing gene from Taiwan and the International Rice Research Institute (IRRI).

The second phase consisted of introducing genes for a new kind of plant type and plant architecture – dwarf varieties and semi-dwarf varieties that could respond to irrigation and fertilizer well. This was accomplished in the case of wheat by introducing the norin dwarfing genes from Japan through Norman Borlaug in Mexico. Inputs are needed for output, and unless the plant is capable of utilizing more inputs there will be difficulty in producing more. So the second phase consisted of genes for dwarfing. At ICAR-IARI, work was started, and in the very first year it was realized that these new plant types could increase the yield potential of the crop substantially. This created a lot of interest among farmers.

The third phase consisted of appropriate government policies to support technology. The interaction of technology and public policy is now related to input–output and procurement pricing as well as storage and public distribution. Furthermore, it is only assured and remunerative marketing that can help farmers to take an interest in technology. However good the technology is, if the net income is not high, farmers will not take to it. In this third phase many changes took place – much more interest in technology, seed production and the distribution of inputs like fertilizers, seeds etc. As a result, in 1968, wheat production went up to 17 million t from about 7 million in 1947. Between 1964 and 1968 more wheat was added and the wheat revolution was now underway. Since then, the country has never looked back; wheat production is now 97.11 million t (2017–2018). Farmers have tried to do their best under difficult circumstances. Scientists also have been continuously producing new varieties with more resistance to pests and diseases, particularly the three rusts – stem, stripe and leaf. All three are now under control, but stem rust could become a threat again because of climate change. The story of the Green Revolution is condensed in the period 1950–1970 largely because of a new kind of plant architecture. The Green Revolution may be defined as an increase in production through productivity improvements. Hybrid corn of the USA would qualify as the starting point of the Green Revolution, as it was the exploitation of hybrid vigour in corn that started the high-yield movement. But whatever the factors that caused increased production, the Indian Green Revolution was unparalleled. The Green Revolution was the beginning of a new era in agriculture. The reasons for its success were many – technology, public policy, farmers' enthusiasm and assured and remunerative marketing. When all these came together, India made significant progress. Now we talk of the 'Evergreen Revolution', i.e. increase in productivity in perpetuity without ecological harm. We need to see a hunger-free India, an India that will not go with a 'begging bowl' or exist hand-to-mouth.

Later, in 1963, Dr Borlaug visited India. On confirmation that many of the semi-dwarf wheat lines were doing well in the Punjab province of Pakistan, India initiated a major programme to import large quantities of seeds of two semi-dwarf varieties, Sonara 64 and Lerma Rojo 64-A, from Mexico and seeds of large segregating populations from the breeding nursery of Dr Borlaug.

India also developed a five-year road map (1963–1968) for transforming wheat productivity using semi-dwarf spring wheat varieties bred at the International Maize and Wheat Improvement Center (CIMMYT), Mexico (Swaminathan, 1993).

The Indian breeders subsequently bred semi-dwarf wheat varieties using the genetic background of Indian wheat germplasm and released high-yielding varieties with superior yield, quality and resistance to major diseases. The Indian wheat revolution gave rise to record harvests of 16 million t, from 11 million t, during 1967/68. A change of this magnitude became a reality not as a result of high-yielding semi-dwarf seeds alone, but also by the imaginative use of complementary enabling institutional mechanisms, including the organization of demonstrations on the small farms of resource-poor farmers and a ready supply of quality seeds, produced with farmers' active participation.

As in the case of wheat, the Green Revolution story was also unfolding with rice. In 1966, IR 8, a new high-yielding variety (with dwarfing genes sourced from the Taiwanese variety Dee-Geo-Woo-Gen), then described as the 'miracle rice', reached India from the IRRI in the Philippines. This variety, with a unique genotype possessing photoperiod insensitivity, semi-dwarf stature, high fertilizer responsiveness-linked high-yield potential and medium maturity duration offered an unprecedented opportunity for a rice

revolution in the country. Subsequently, Indian rice breeders developed several high-yielding varieties, some of them surpassing IR 8 in yield and quality. Interestingly, the rice production revolution came from an unconventional rice area, the irrigated Indo-Gangetic wheat belt during the late 1960s, triggered by the medium maturing, Basmati-derived, high-yielding fine rice varieties bred at the Indian Agricultural Research Institute (IARI) (Swaminathan, 1993; ICAR, 2015).

Undoubtedly, the Green Revolution is one of the great technological achievements of the 20th century, enabling humanity to defy the Malthusian catastrophe, with food production outpacing population growth (Sharma, 2016). The high-yielding varieties of wheat and rice, bred scientifically to respond to the application of fertilizers and irrigation, heralded the Green Revolution in India, propelled primarily by the public research and extension system. Concomitantly, political will, suitable input support, appropriate pricing policies and progressive farmers' participation resulted in the transformation of Indian agriculture from a 'ship-to-mouth' status in the 1960s to a 'right to food' (under the National Food Security Act 2013) situation. It transformed Indian agriculture, while providing a foundation for subsequent strides in overall agricultural production (Fig. 3.1). It helped India triple its foodgrain production between 1968 and 2000 and halve the percentage of food insecurity and poverty (even

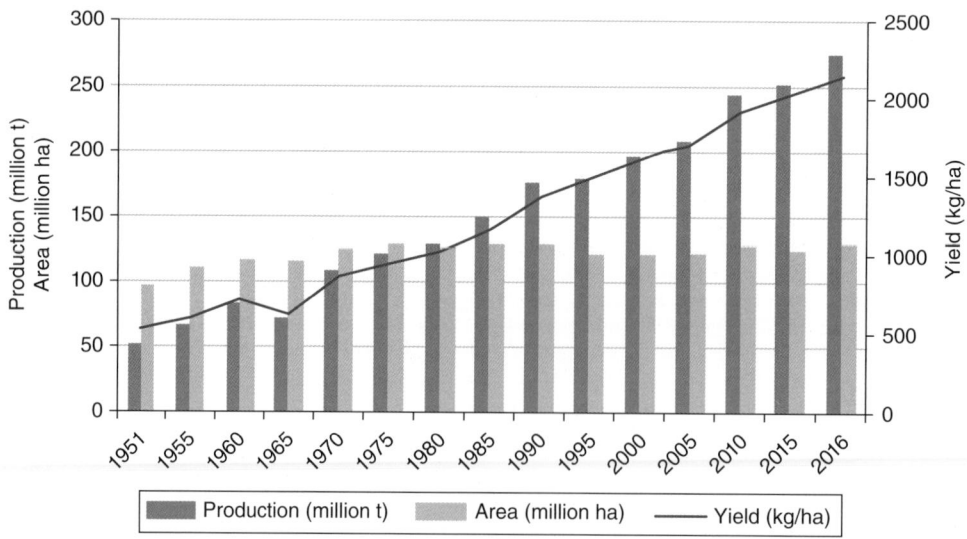

Fig. 3.1. Area, production and yield of foodgrains in India since 1951.

though the population had almost doubled during the same period), thus rendering India a self-sufficient nation at the macro level (Singh, 2014). Green Revolution technologies addressed two important impact indicators – total factor productivity (TFP) and poverty alleviation. The TFP index for crop-livestock reached 290% in 1991/92 from the base year (1964/65 = 100), of which research contributed 48%. The poverty ratio declined from 55% in 1973/74 to 36% in 1993/94, proving a strong positive correlation between research outputs and poverty alleviation (Joshi *et al.*, 2005). This transition was achieved not only through pioneering agricultural research to increase food production but also through political and administrative support of building grain reserves, operating an extensive public distribution system (PDS), protective social security measures like food for work, mid-day meals for children in schools, employment guarantee, land reforms and asset creation measures.

Constraints of Technologies and their Implementation

The Green Revolution strategy for food crop productivity growth was overtly based on the hypothesis that, given appropriate institutional mechanisms, technology spillovers across political and agroclimatic boundaries could be captured (Pingali, 2012); and although Green Revolution technologies averted a famine-like scenario in India, avoiding the conversion of thousands of hectares of land for agricultural cultivation, they also spurred unintended negative consequences. The way in which Green Revolution technologies were applied gave rise to many challenges, not because of the technologies themselves but because of the policies that were used to promote them. A few of these are described below.

Impact of Technologies in Selected Regions

One downside of the Green Revolution was selectivity in impact, remaining confined to the well-endowed, well-irrigated and high-rainfall areas. This was in spite of the fact that international breeding programmes had aimed to provide broadly adaptable germplasm for cultivation in a wide range of geographies; but adoption was highest in favoured areas. Many of the agricultural technologies were not suitable for, or adopted by, small and marginal farmers, especially those in rainfed ecologies. For example, high-yielding varieties (HYV) of wheat provided yield gains of 40% in irrigated areas with modest use of fertilizers, while in dry areas the gains were often no more than 10%, and the technology adoption was strongly correlated with water supply (Pingali, 2012). Inter-regional disparities widened, as did the gap between rich and poor farmers. Technologies in the Green Revolution period did not focus on the constraints to production in more marginal environments, especially tolerance to stresses such as droughts or floods.

Natural Resource Imbalance

The technology application invariably led to loss of traditional varieties to the HYVs, loss of soil fertility under indiscriminate application of chemical fertilizers, over-exploitation of underground water, soil health degradation (saline or alkaline) with overuse of surface or underground water, and increasing environmental toxicity with reckless use of chemical pesticides. The high response of HYVs to nitrogen encouraged farmers to use urea indiscriminately, completely ignoring the recommendations of organic manure application along with chemical fertilizers. Such practices resulted in soil health loss, including decline in organic matter content, in two to three decades, while promoting incidence of pests, which in turn led to increasing indiscriminate use of pesticides and toxic contamination of food, feed, water and the environment. Unregulated use of irrigation water and unchecked extraction of underground water, aided by subsidized or free electricity together with poor drainage, led to degradation of soil and decline in factor productivity (Swaminathan, 1993).

Incomplete Reduction of Poverty and Food Insecurity

Food security exists when all people, at all times, have physical and economic access to sufficient safe and nutritious food to meet their dietary needs and food preferences for an active and healthy life (Mishra, 2017). The Green Revolution fuelled an increase of food availability per capita with a fall in prices, but poverty continued to limit access to food, leaving hundreds of millions of people undernourished in developing countries (Serageldin, 1999). The 2016 *Global Hunger Index Report* ranked India 97th among 118 countries (GHI, 2016). In 2011, 21.9% of the Indian population lived below the poverty line, while according to the international poverty line of US$1.90/day, India has 224 million people living under poverty (World Bank, 2016; ADB, 2017); and this is in spite of the fact that India has witnessed more than a fivefold increase in its staple cereal production since 1950 due mainly to the gains from the Green Revolution (Fig. 3.1). Thus the Green Revolution alone was not the panacea for solving the myriad complexities of poverty, food security and nutritional problems, because the amount and quality of food available globally, nationally and locally can be affected, temporarily or for long periods, by many factors including climate, disasters, wars, civil upset, population size and growth, agricultural practices, the environment, social status, per capita income and agricultural trade.

Technology Fatigue

Broadly speaking, technologies reached a plateau by the end of the 1980s for many crops and regions. Technology fatigue has been identified as one of the two main reasons for the problems with India's agriculture, which has been going through a difficult phase in recent years with declining rates of growth in agricultural productivity and profitability. Intensive agriculture led to loss of soil fertility due to excessive mining of nutrients, groundwater depletion from over-exploitation of underground aquifers, increased soil salinity from poor drainage, reduction of diversity due to monocropping, and threat to environmental and human health due to excessive use of agrochemicals. These effects had been most severe in the intensively cultivated areas of Punjab, Haryana and western Uttar Pradesh, where the technologies were first adopted. Some experts opine that not the technologies per se

but injudicious implementation of policies like highly subsidized electricity and water are to be blamed (Sharma, 2016). Agricultural research was also focused on a few selected crops, and participatory breeding (especially involvement of farmers) was inadequate. The overall performance of the agriculture sector in terms of growth suffered a setback for some years, especially after the mid-1990s and this slow-down had several consequences including widespread agrarian distress. The first decade of the new millennium witnessed a growth in production of 2.7% p.a., compared to 3.4% p.a. during the 1990s and 4.7% p.a. during the 1980s. It was a great challenge and a formidable task to arrest the decline and reverse the slowing growth of the agriculture sector. Several initiatives were taken by the central and state governments to reverse the slow-down (Chand, 2014). Since 2004/05, a revival of the growth rate to 3.75% has been achieved. Thus there is an urgent need to rectify the fatigued technologies, remove regional disparities and rejuvenate the agricultural sector afresh to address the issue of food, nutrition and environmental security.

Global Population, SDGs and Food Security

In 1901, the world population was 1.6 billion. It increased rapidly to 3 billion by 1960, 5 billion by 1987 and 6 billion by 1999. By mid-2017 it had crossed 7.6 billion, an addition of nearly a billion during the last 12 years (UNDESA, 2017). Of greater concern is the fact that 60% of the world's people live in Asia (4.5 billion), with China (1.4 billion) and India (1.3 billion) continuing to be the two most populous countries (Fig. 3.2). Within the next seven years, India is projected to overtake China as the world's most populous country and will continue to grow until c.2060; its population only starting to decline when it has reached approximately 1.68 billion.

The Green Revolution helped India to achieve some of the MDGs, including the target for reducing poverty and hunger by half (Goal 1). However, progress was uneven, as poverty became increasingly concentrated in poorer states. Since the mid-2000s, the economic growth, including in agriculture, as well as increased social spending on interventions such as the

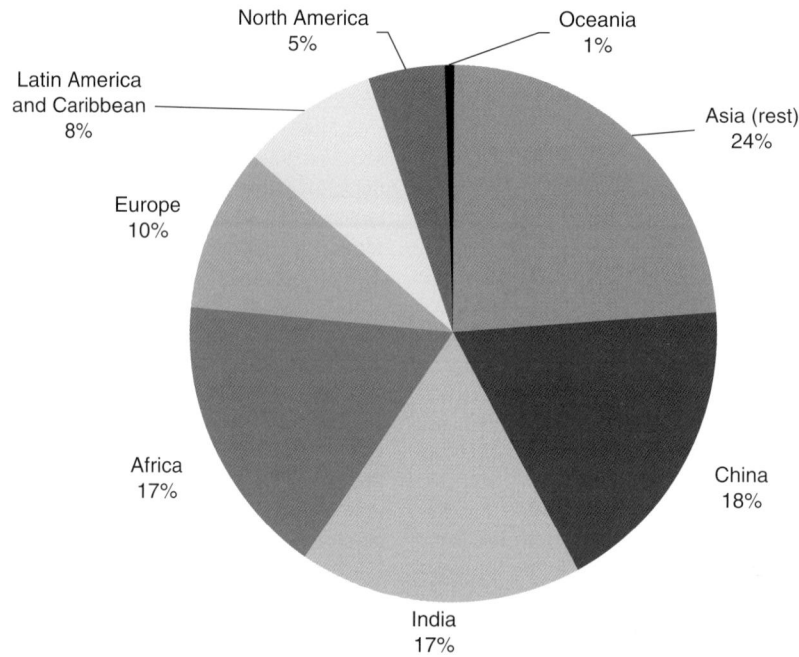

Fig. 3.2. Distribution of the world's population (2017).

Mahatma Gandhi Rural Employment Guarantee Act (MGNREGA) and the National Rural Health Mission (NRHM) facilitated India to halve the incidence of poverty from the 1990 level. However, it is apparent that the tasks of meeting the consumption needs of the projected population are going to be more difficult, given that previous strategies of generating and promoting technologies have contributed to serious and widespread problems of environmental and natural resource degradation. This implies that in future the technologies that are developed and promoted must result not only in increased agricultural productivity levels but also ensure that the quality of the natural resource base is preserved and enhanced. In short, they lead to sustainable improvements in agricultural production.

The 17 SDGs of the UN, also known as the Global Goals, are an inclusive agenda to make the right choices now to improve life, in a sustainable way, for future generations (UN, 2017). They build on the successes of the MDGs while including new areas such as climate change, economic inequality, innovation, sustainable consumption, peace and justice, among others.

They provide clear guidelines and targets for all countries to adopt in accordance with their own priorities and the environmental challenges of the world at large. The SDGs offer a unique opportunity to put the whole world on a more prosperous and sustainable development path.

The 2017 edition of *State of Food Security and Nutrition in the World* is an important benchmark to gauge hunger and malnutrition prevalent today and actions required to attain the targets of the SDGs, specifically those of ending hunger (Target 2.1) and all forms of malnutrition (Target 2.2) (FAO *et al.*, 2017). According to the report, in 2016 the number of chronically undernourished people in the world is estimated to be 815 million (10.7% of the world's population), up from 777 million in 2015, though down from 900 million in 2000. This increase is alarming, and is ascribed to situations of conflict combined with droughts, floods and climate change, particularly in parts of sub-Saharan Africa, south-east Asia and western Asia.

With reference to India, endemic hunger continues to be a challenge, with over 200 million men, women and children estimated to be undernourished, largely due to inadequate purchasing

power arising from inadequate opportunities for skilled employment. The population of India is growing at a rate of 1.2% per year. If this trend continues, the population will be doubled in less than 40 years. The decade between 1990 and 2000 was characterized by globalization, macroeconomic reforms and trade liberalization in India, which impacted the hitherto protected Indian agriculture. Unfortunately, economic reforms were minimal in agriculture, impacting exports considerably. A strong need was voiced for a second Green Revolution. By the turn of the 20th century, high price volatility, shortages of food and increasing rates of farmer suicides had become prominent issues in Indian agriculture. To address the agrarian crisis a National Policy for Farmers (NPF) was formulated to increase the net income of farmers. Overall, there has been national-level (macro) food security and a boost in overall GDP growth in India, but agricultural production and rural income growths have slowed down considerably, outstripped by the population growth rate (Singh, 2014). There is still a widespread mismatch between production and post-harvest technologies. In perishable commodities, such as fruits, vegetables, flowers, meat and other animal products, this mismatch is often severe, affecting the interests of both producers and consumers. Out of every 3 ha of cultivated land in India, almost 2 ha are under rainfed agriculture. With little reduction in the number of undernourished and poor people, the country has to accelerate the pace of reforms for achieving the SDGs. The *State of Food Security and Nutrition in the World 2017* report sends a clear signal that eradicating hunger and malnutrition by 2030 is a challenging task requiring renewed efforts through new ways of working; it advocates that only if agriculture and food systems become sustainable can the issue of food insecurity be addressed adequately (FAO *et al.*, 2017).

Indian Agricultural Research and Development beyond the Green Revolution

The welfare of farmers and farm workers not benefitted by the Green Revolution depends on extending the boundaries of the Green Revolution. A major accomplishment of Independent India is the development of a dynamic National Agricultural Research, Education and Extension System (NARES). A well-established network of SAUs, national research institutions, all-India coordinated research projects and Krishi Vigyan Kendras (KVKs) is supported by ICAR. Productivity gains during the Green Revolution were largely confined to relatively well-endowed areas. Thus, during the 1970–1990s, research priorities and agendas shifted towards conservation and improvement of genetic resources to raise productivity, development of HYV seeds for more crops like pulses and oilseeds, sustainable natural resource management, diversification, post-harvest management, human resource development and infrastructure strengthening. During this period, NARES contributed to usher in a milk, egg, fruit, vegetable, fish and oilseed revolution. Subsequently, in the decade 1990–2000, research agendas and priorities included dryland horticulture, ideal cropping systems, PPPs, HRD and strengthening front-line extension activities through further expansion of KVKs. Emphasis was also laid on tapping new technologies in the field of biotechnology and genetic engineering, molecular biology, remote sensing etc. During the years of the 21st century, NARES's policy agenda and research priorities included the development and diffusion of agricultural technologies, more efforts in biotechnology, strengthening research in natural resource management and climate change, PPP and more international collaboration.

The Green Revolution benefits in economic terms notwithstanding, Indian agriculture still suffers from low productivity, low quality awareness and rising imports. Agriculture imports have increased six times faster than exports in the past 20 years. Large imports in 2016–2017 have been edible oil (US$10.9 billion), pulses (US$ 4.2 billion) and apples, kiwi fruit, almonds and cashews (US$3 billion). These three groups account for 73% of India's agriculture imports, although it has the required soil and climatic conditions to cultivate them indigenously. India accounts for 4% of global production of grapes but its share in global exports is only 1.6%. The case of bananas is even more abysmal; India's share of global production is 30% but its share of exports is less than 0.4%. It is apparent that India needs to focus on continued agricultural growth through strong national and

international research programmes to increase agricultural productivity, improve input-use efficiency, high-value agricultural products and adoption of a package of technologies that would enhance farm income.

Looking Ahead

Changes are needed urgently to respond to the new demands of agricultural technologies. These include increasing pressure to maintain and enhance the integrity of degrading natural resources, changes in demand and opportunities arising from economic liberalization, unprecedented opportunities arising from advances in biotechnology, the information revolution and the urgency to reach the poor and disadvantaged who could not be insulated by the Green Revolution technologies. The gap between potential and actual yields is high in a majority of crops under different farming systems. Further, in view of deriving benefits under the WTO regime, a targeted approach, which accords adequate attention to export commodities and frontier sciences, so as to reduce cost and improve quality, is needed. Future growth needs to be more rapid, more widely distributed and better targeted. The new-generation technologies will have to be much more site-specific, based on high-quality science and an increased opportunity for end-user participation. These must be not only aimed at increasing farmers' technical knowledge and understanding of science-based agriculture but also must take advantage of opportunities for full integration with indigenous knowledge. They will also need to take on the challenges of incorporating the socioeconomic context and the role of markets. Hence, the following initiatives are proposed with a view to transform the Indian agricultural sector and make it resilient, sustainable and profitable for farmers and other stakeholders.

Research

- A new paradigm of regionalization of research, based on well-defined agroclimatic regions, application of frontier sciences, participatory and proprietary approaches

in research and strengthening research–extension links is urgently needed.

- Public agricultural research systems need to be shaped into an innovative system structure that is well-organized, efficient and results-oriented.
- An international network of scientists in both the public and private sectors must work together to provide seeds and plants to farmers and commercial plant breeders for further crossing and testing in different environments. The research community, therefore, must pay specific attention to the development of locally adapted varieties that meet the needs of the world's poorest farmers.
- There is a need for researchers, farmers and extension scientists to come together for location-specific testing and verification of technologies that are scientifically sound, socially appropriate and environmentally relevant. Results from genomics and agronomic research must be connected to the communities that are responsible for evolving new varieties of crops.
- The bureaucratic system needs to be made a more flexible and liberal system of administration. Scientists can be educated in business skills and other knowledge. There is a dire need to bring in an assessment culture in agricultural innovation systems.
- Basic science has generated enormous advances in our understanding of crop/animal biology, stress tolerance, pathogen resistance and many other fields of science. This understanding should lead in due course to improvements in agricultural technologies. Development agencies, faced with public suspicions of new agricultural technologies, and perhaps eager to find shortcuts to development, have tended to shift funding away from agricultural research towards other priorities. This trend needs to be reversed.
- In most of the crops, the present average yield is just one third of the achievable yield. Therefore a massive effort is required to launch a new revolution in farming through cost-saving and efficient input-use technologies both for production and post-harvest management.
- It is necessary to develop and introduce appropriate technologies coupled with sound delivery systems that ensure economic and

ecological sustainability and optimum use of local resources emphasizing capacity-building and technological empowerment, particularly of small and marginal farmers. Top priority should be given to improving the productivity and stability of rainfed agriculture, and more efficient and sustainable use of increasingly scarce land, water and germplasm resources.

Education

- Convert the top ten SAUs into 'centres of excellence'. They will make region-specific strategies to raise crop yields, advise on the creation of integrated supply chains and prepare a plan to promote exports and cut imports.
- ICT and cutting-edge technological tools need to be accessible to small and marginal farmers. They create capacity among community groups and farmers to produce videos on topics that are relevant to local farmers, featuring farmers as the main contributors.

Development

- There is an urgent need to adopt integrated natural resource management so that present production does not erode future prospects. This is particularly required in the traditional Green Revolution areas to defend the gains already made, and to extend the gains to areas bypassed with respect to improved technologies.
- The need for making new agricultural gains is urgent. This can be achieved by utilizing the available agroclimatic/soil maps, watershed/wasteland atlases, GIS mapping and remote-sensing capabilities for developing improved and integrated crop-livestock-fish farming systems, and for developing infrastructure for value addition to farm products at the village level. These changes will provide opportunities for off-farm employment and income generation.
- There is a need to establish at least 2000 farmer centres, one in each sub-district. These agri-clinic centres should be a 'one-stop shop' for all farmers' needs, such as meeting representatives from banks, insurance companies, seed and equipment suppliers and buyers, and input/technology providers. Farmer centres would integrate with the electronic national agriculture markets (e-NAM) to help farmers sell directly to the consumer. Each centre will also have free water-, soil- and nutrient-testing laboratories.
- Ensure active monitoring of government schemes; for example, many of the 35 million farmers who opted for the Pradhan Mantri Fasal Bima Yojna (Prime Minister Crop Insurance Scheme) in the last *kharif* season got their compensation late, as more than half the states did not pay the premium on time. The e-NAM, another useful initiative, needs to check wrong reporting. Many mandis show normal sales as e-NAM sales.
- Institutional and infrastructural support is essential for higher agricultural production. There is an urgent need to provide efficient irrigation, power supply, rural roads, cold storages, godowns and food-processing units, especially in the eastern and north-eastern regions, supported by assured and remunerative marketing.
- Instead of flood irrigation, efficient micro-irrigation practices need to be promoted to increase the area under irrigation through increased productivity and higher water-use efficiency.

Conclusion

The time is ripe to initiate action to ensure household food security and eliminate hunger. The opportunity for a productive and healthy life for every individual depends on the success of achieving an 'evergreen revolution'. The FAO has projected that by 2030 most developing countries will be dependent on imports from developed countries for their food requirements; hence, greater efforts to sustain the gains of the Green Revolution through resource-use-efficient technologies are needed in a 'mission mode' approach. The second-generation problems of the Green Revolution are a 'ticking timebomb' requiring transformation of Indian agriculture. Doubling farmers' incomes in the next five years,

though a big challenge, can be an apt metaphor and goal for much-needed agricultural transformation. A farming system that produces high yields, makes a good living for farm families, protects natural resources and improves the environment, whilst still producing good, affordable food, is what is required. In return, a rising agricultural sector will improve the lives of millions of people who live on the margin. The results will transform the entire fabric of the nation.

References

ADB (2017) Basic 2017 statistics, Asian Development Bank. Available at: https://www.adb.org/countries/india/poverty (accessed 11 December 2017).

Alagh, Y.K. (2015) *Structure of Indian Agriculture – Growth and Policy Epochs*. NAAS Foundation Day Silver Jubilee Lecture, NASC, New Delhi, p. 16.

Borlaug, N.E. (2000a) Ending world hunger: the promise of biotechnology and the threat of antiscience zealotry. *Plant Physiology* 124(2), 487–490.

Borlaug, N.E. (2000b) The Green Revolution Revisited and the Road Ahead. Available at: https://www.nobelprize.org/nobel_prizes/peace/laureates/1970/borlaug-lecture.pdf (accessed 20 November 2017).

Borlaug, N.E. and Dowswell, C.R. (2003) Feeding a world of 10 billion people: a 21st century challenge. Paper presented at the International Congress *In the Wake of the Double Helix: From the Green Revolution to the Gene Revolution*, Bologna, Italy, 27–31 May.

Chand, R. (2014) From Slowdown to Fast Track: Indian Agriculture Since 1995. National Centre for Agricultural Economics and Policy Research Working Paper, 1/2014.

Evenson, R.E. and Gollin, D. (2003) Assessing the impact of the Green Revolution, 1960 to 2000. *Science* 300(5620), 758–762.

FAO, IFAD, UNICEF, WFP and WHO (2017) *The State of Food Security and Nutrition in the World 2017: Building Resilience for Peace and Food Security*. FAO, Rome.

GHI (Global Hunger Index) (2016) Available at: https://reliefweb.int/sites/reliefweb.int/files/resources/130918.pdf (accessed 21 May 2018).

ICAR (2015) *Golden Jubilee of Green Revolution: Reminiscences*. Golden Jubilee Green Revolution 2015, organized by the ICAR-IARI, ICAR and NAAS and published by the ICAR Directorate of Knowledge Management in Agriculture, New Delhi.

Joshi, P.K., Pal, S., Birthal, P.S. and Bantilan, M.C.S. (2005) Impact of agricultural research: an overview. In: Joshi, P.K., Pal, S., Birthal, P.S. and Bantilan, M.C.S. (eds) *Impact of Agricultural Research: Post-Green Revolution Evidence from India*. National Centre for Agricultural Economics and Policy Research (NCAP), New Delhi, pp. 1–8.

Mishra, C.P. (2017) Malnutrition-free India: dream or reality. *Indian Journal of Public Health* 61, 155–162.

Paddock, W. and Paddock, P. (1967) *Famine-1975!* Weidenfeld and Nicholson, London.

Pingali, P.L. (2012) Green revolution: impacts, limits, and the path ahead. *Proceedings of the National Academy of Sciences* 109(31), 12302–12308.

Serageldin, I. (1999) Biotechnology and food security in the 21st century. *Science* 285(5426), 387–389.

Sharma, R. (2016) Green Revolution at 50, looking back and ahead. *The Tribune*, 8 January.

Singh, R.B. (2014) Towards a Food Secure India and South Asia: Making Hunger History. Asia-Pacific Association of Agricultural Research Institutions (APAARI). Available at http://www.apaari.org/wp-content/uploads/2009/08/towards-a-food-secure-india-making-hunger-history.pdf (accessed 21 May 2018).

Swaminathan, M.S. (1993) *Wheat Revolution: A Dialogue*. Macmillan India Ltd, Madras, India.

Swaminathan. M.S. (2016) *50 Years of Green Revolution: An Anthology of Research Papers. The Quest for a World without Hunger*. Vol. 1. World Scientific Publishing Co. Pte Ltd, Singapore.

UN (2017) Sustainable Development Goals. Available at: http://www.un.org/sustainabledevelopment/sustainable-development-goals/ (accessed 21 May 2018).

UNDESA (2017) World Population Prospects: The 2017 Revision, Key Findings and Advance Tables. *United Nations, Department of Economic and Social Affairs, Population Division ESA/P/WP/248*.

World Bank (2016) *Poverty and Shared Prosperity 2016: Taking on Inequality*. World Bank, Washington, DC. DOI: 10.1596/978-1-4648-0958-3.

4

Intensive Efforts for Food and Nutrition Security

_____

After the Green Revolution, self-sufficiency in foodgrain production was achieved, and the problems of food security were resolved, but in the process, soil texture and useful micro-organisms in the soil were depleted due to nutrient imbalance and excessive use of fertilizers. However, now the scenario has changed, and there are many challenges and concerns that require immediate attention.

The Vision Statement adopted by all the science academies in India was released by Prime Minister Atal Bihari Vajpayee during the Indian Science Congress, in January 2001, where the theme was food, nutrition and environmental security. In India, in fact, the ever-increasing population nullifies all efforts made in this direction. Every year, the population of India grows by the size of Australia's population, and it needs an additional 4–5 million t of foodgrains. Many of the other countries do not face such a challenge, even China. A total of 16% of India's population is sustained by only 2.8% of the land. It is anticipated that India will surpass the population of China by 2020. India also has a livestock population of almost half the human population. Nowhere else in the world is this type of pressure being faced (Sen, 2001; *The Hindu*, 2001).

Over the years, the GDP percentage of the agricultural sector is declining. This is a good sign, since industrial growth is showing an upward trend. However, it is well established that unless India achieves a 4% growth rate from agriculture, the expected 8% industrial growth will not be possible. In rural India, almost 60% of people are dependent on agriculture. Dr M.S. Swaminathan, the father of the Green Revolution, has often highlighted the importance of agriculture for national food security.

In the mid-1960s, India was considered a 'begging bowl' country. Since then, it has progressed considerably thanks to the science-based revolution (such as changing plant-type concept) that made plants respond better to higher inputs, which led to higher productivity. The holy alliance with NARS, supported by policy makers and IARCs, such as the International Maize and Wheat Improvement Center (CIMMYT) and the IRRI, as well as highly intelligent, enthusiastic and hardworking farmers, is well acknowledged.

The Green Revolution enabled India to feed its existing population, which was increasing at a rate of 1.6% per annum. At one time, it was importing 10 million t of foodgrains under the PL-480 scheme. During the past 50 years, it witnessed unprecedented progress and increased agricultural production at a growth rate of 4.5%. Yet the concern is for economic and ecological access to food. Unfortunately, buying power is the limiting factor, which is why the issue of poverty is of great concern.

Over the past five decades, there has been a steady rise in the price of most industrial products. In contrast, prices of foodgrains have followed a declining trend, which has made life much easier for people. Since India became independent, it has been able to reduce poverty by 40% and life expectancy has doubled from 32 to 64.

The MDGs have drawn our attention towards the eradication of poverty and the sustainability of the environment.

Unfortunately, in this part of the world, i.e. south Asia, hunger problems are at their highest. Of the 276 million hungry people in south Asia, 200 million (about 70%) are in India. We need to ensure a better income for people and see that they are above the poverty line and have easy access to food (UNCSD, 2012).

Despite the striking transformation brought about by the Green Revolution, the country is still home to almost a quarter of the world's hungry and poor. The present global concerns are for the 180 million children who are severely underweight for their age; over 800 million chronically undernourished children; 400 million women of child-bearing age are anaemic; and over 200 million children are vitamin A-deficient. Thus, nutritional security is a major concern that needs to be addressed as a priority. The poverty concentration is most acute in south Asia, yet donor organizations focus on Africa. Asia is not given serious attention because it witnessed a Green Revolution. Though around 200 million people are still below the poverty line (earning less than one dollar/day), Asia's per capita calorie consumption is higher than many countries in Africa and parts of Latin America. At the same time, it needs to move from the present 2000 kcal/person intake to 2500–2800 kcal/person. This demands an expansion of the food basket to reduce dependence on cereals. We are experiencing a factor productivity decline on account of second-generation problems associated with the Green Revolution, such as salinity, lowering of the water table and increased incidence of pests and disease.

Lately, owing to policy changes, buffer stocks got depleted, from over 58 million tonnes in 2002 to 17.5 million tonnes in 2007, and again reached all-time high of 66.69 million t in 2013.

Foodgrains are also being diverted as feed, thus making their availability more difficult. India is fortunate to be mostly vegetarian. Its protein demand is mainly met through pulses, vegetables and fruits and not through meat. The USA is diverting its maize production (33%) towards biofuel production, which is ethically wrong (Nandkumar *et al.*, 2010).

Climate change also has had an effect, from 1920 to 2000; average global temperature rose by almost 1° in that period; and it is expected to rise by another 4° if corrective measures are not undertaken. Imagine what will happen if temperatures continue to rise unabated. The impact of climate change is now visible. Emission of GHGs leading to global warming, more intense tropical cyclones, faster wind speeds and heavier precipitation are all a reality. Contraction of Himalayan glaciers by almost 17 km in the last ten years is another reality. As stated earlier, the world's cereal production has been affected adversely in the recent past owing to drought in Australia, Canada, the USA and other countries.

In India, prices are escalating, buffer stocks are depleting and imports from the developed world are rising. We must, therefore, think seriously how to remain self-sufficient. In the years 2007–2008, we had to import wheat for the first time since the Green Revolution. Around 4–5 million t of wheat every year were produced for over a decade until 2002; but in the last 6–7 years, production has remained stagnant. Fortunately, the minimum support price (MSP) for wheat increased from Rs 750/t to Rs 1000/t in 2007, which increased wheat production by almost 3 million t in just one year. So the issue of sufficient production and self-sufficiency depends on the right policies. In recent years, the demand for other commodities, especially horticultural crops, has grown much faster than demand for cereals, which is an encouraging sign.

We need to reorientate our research and development strategy to a twin-pillar approach. This requires a paradigm shift of having not only germplasm improvement (good varieties and hybrids) but also improved natural resource management. We need to also consider socioeconomic aspects and policies around diversification of agriculture.

It is a matter of concern that over the years, the use of germplasm for breeding new varieties of different crops has gone down. This trend is global and that is why a global initiative on plant breeding has been initiated by the FAO through the support of the Bill and Melinda Gates Foundation (BMGF), which aims to reverse the trend. It is apparent that there is some complacency in our efforts for plant breeding. One of the reasons for this is the thinking that biotechnology can solve all problems. It is important to understand that biotechnology can only supplement, and not replace, plant breeding. In the 1980s and 1990s, the Indian wheat programme recorded an annual genetic gain of 1% p.a. This has stagnated since the release of wheat variety PB 343. The challenge is how to improve yield further. The same challenge is also now faced by the CIMMYT. In this context, advances through hybrid technology are encouraging. Scientists, both in the public and private sectors, gave hybrid technology to the world for cotton, pearl millet and castor, and also for pigeon pea, in collaboration with the International Crops Research Institute for the Semi-Arid Tropics (ICRISAT).

In the case of rice, China was the first to release hybrids. A yield gain of 1 t/ha could be achieved through this technology. China is developing super-hybrid rice, with a target yield level of 15 t/ha. This kind of effort is needed in India also where rice productivity is still below 3 t/ha. The private sector can play a major role, as the public sector has not been able to deliver expected output in hybrid seed production. We have 42 million ha under rice, but hybrid rice area is only 2.2 million ha. In the USA, single-cross hybrid maize technology provided higher productivity (7–8 t). The Bt hybrid maize can now yield up to 12 t/ha. Though new technologies are available, there is a need to make available seeds of these hybrids to farmers (WESS, 2013).

It was for this reason that a mission project on hybrids was initiated under the National Agricultural Technology Project (NATP), which resulted in the release of single-cross hybrid of maize for the first time. As a result, maize production doubled in one decade. However, the area under hybrid maize, particularly single-cross and quality protein maize (QPM) hybrids, is low, currently around 60%, which must be accelerated further to increase maize production. For this a strengthened PPP approach is needed.

PPP is essential for future growth in agriculture. We need to create an enabling environment and government should set out well-thought-out policies and a strategy for providing incentives. There is an obvious need for building mutual trust among different stakeholders. This is indeed a grey area for which we need to sit around the table and discuss successful models of PPP and achieve a better understanding of them. Currently, total acreage under GM crops is around 140 million ha. India, presently, has only Bt cotton. The Philippines has already released Bt corn as a food crop. In India, both Bt brinjal and Bt

corn have lately been permitted for field tests. In the future there may be resistance to transgenic technologies. In any case, these technologies are in the country's best interests. Even Europe is importing Bt cotton, soybean and corn for use as animal feed. In the author's opinion, there should be no concerns about the release of GM food crops in India provided appropriate testing procedures are followed. The partnership of Mahyco with Monsanto, coupled with an enabling environment created by ICAR and the Department of Biotechnology (DBT), both for testing and release, led to the release of Bt cotton in India. The government of India is moving in a positive direction with GM crops. In the last 15 years, 11 million ha are under Bt cotton; there is no better example of faster adoption anywhere in the world. As a result, cotton production almost doubled and productivity increased significantly. Currently, cotton exports alone earn $3.0 billion/year. Before Bt cotton technology, there were practically no tangible cotton exports.

There is another approach to enhancing food and nutritional security. In the north, both rice and sugarcane were not grown. Groundnut was not grown in Gujarat, which is at present the number one state for groundnut production. Potatoes were never grown in the Indo-Gangetic plains but now are grown in a large area. Maize in eastern India has begun to yield now more than 8 t/ha, which is indeed very impressive. Chickpea, a crop of north India, can be grown in Tamil Nadu because of the availability of short-duration varieties. Pigeonpea is being grown in the north and west due to the release of hybrids and short-duration varieties. A niche for soybean has been found in Madhya Pradesh, which is now the number-one oilseed crop in the country. It is indeed a matter of concern as to why India exports soybean, which could help to ensure nutritional security. However, until the country makes use of soybean as a food, it will have to export soybean meal, worth around US$3 billion p.a.

India needs to move forward and plan research into up-stream areas of strategic importance. It needs to make sure that its knowledge gets translated into products that can benefit the end-user. This is what is called 'translational research', for which it is necessary to work with farmers in a participatory mode, as was demonstrated through IPM in rice in Indonesia, which led to increased rice production and a decline in pesticide consumption by 50% in five years. We have to understand the problems of our farmers and integrate their traditional knowledge with scientific knowledge. We have to make sure that they are able to use their resources judiciously – alternate furrow irrigation in cotton can reduce water use by 30%. We need innovative technologies; a decade ago, no-one thought that in rice-wheat production systems one could use zero-till drill and achieve conservation agriculture; now, over 3.5 million ha in the Indo-Gangetic plains are under zero-till. This success could be extended to an area of 10 million ha of rice-wheat in India.

We should look for newer options such as precision farming, which is possible through efficient farm mechanization. Farmers are even adopting laser levelling to improve water-use efficiency. Here again, the role of the private sector is important.

Somehow, over the years, our extension system has become weak. The dissemination losses are higher due to less-competent people becoming involved in the extension services. In this context, the private sector can, again, play an important role. For example, the establishment of agri-clinics with the involvement of technology agents who can provide much-needed vocational training for much-needed custom hire services to farmers is an important aspect. The role of NGOs is also to be encouraged.

It is indeed heartening that the World Bank report of 2008 clearly brought out that there cannot be sustained and inclusive development unless high priority and funding is given to agriculture. Fortunately, agriculture has again taken centre-stage despite being neglected in the recent past. We need more capital investment in agriculture, as was the case soon after independence. A lot of good infrastructure – the highest dam, the longest canal, the best fertilizer factories in the cooperative sector, and markets/ *mandis* – has been created. This was possible since almost 18% of GDP was spent on capital investment. Unfortunately, over the last two decades, this support has declined to 9%. Now, the private sector extends support to build the much-needed infrastructure. For this to continue, the government needs to create an enabling environment to catalyse the private sector.

India is blamed for providing subsidy to farmers. It must be understood that agricultural subsidy in India is linked to productivity; and in the developed nations much higher support is provided for storage, marketing and export. The subsidy given is around 6.5%, and up to 10% is the acceptable limit according to the WTO. Hence, much-needed support for resource-poor farmers must continue in the overall interests of the nation. However, we must not be complacent and we need to continue scaling up efforts both up-stream and out-stream. We have to see that technologies reach the end-users quickly; we must build stronger partnerships between public and private institutions to ensure this happens; we need policy makers to provide an enabling environment and support to catalyse this process as a matter of national priority. All these measures would help to accelerate much-needed productivity in agriculture to achieve food and nutritional security on a long-term basis.

References

Nandkumar, T., Gulati, A., Sharma, P. and Ganguly, K. (2010) *Food and Nutrition Security Status in India: Opportunities for Investment Partnerships*. ADB Sustainable Development Working Paper Series. ADB, Mandaluyong City, Philippines. Available at: https://www.adb.org/publications/food-and-nutrition-security-status-india-opportunities-investment-partnerships (accessed 21 May 2018).

Sen, N. (2001) Meeting report, Indian Science Congress. *Current Science* 80(5), 607–609.

The Hindu (2001) India will be Hunger-Free in 20 Years: Vision Document. 4 January.

UNCSD (2012) Zero hunger challenge: transforming our food systems to transform our world. UN Conference on Sustainable Development. Available at: http://www.un.org/en/zerohunger/challenge.shtml (accessed 21 May 2018).

WESS (2013) *UN World Economic and Social Survey 2013. Sustainable Development Challenges*. Department of Economic and Social Affairs of the United Nations Secretariat (UN/DESA). United Nations. Available at: http://www.un.org/en/development/desa/publications/world-economic-and-social-survey-2013-sustainable-development-challenges.html (accessed 21 May 2018).

5

The White Revolution and Livestock Production

Animal husbandry, dairying and fisheries activities, along with agriculture, have been an integral part of human life since the start of civilization. These activities, as well as contributing to the food basket have helped maintain ecological balance. Due to conducive climate and topography, the animal husbandry, dairy and fisheries sectors have played a prominent socioeconomic role in India. Traditional, cultural and religious beliefs have also contributed to the continuance of these activities. In general, the livestock sector's role is significant in generating household income and gainful employment in the rural sector, particularly among landless, small and marginal farmers, and women, along with providing cheap and nutritious food for millions of people. Hence, livestock production and agriculture have been intrinsically linked, and both are crucial for overall food security.

The world's population is currently around 7.4 billion, and India has 1.34 billion, 18% of the total. The latest World Bank report indicates that India has the largest number of poor people in any single country. Around 363 million people (around 21%) still live below the poverty line. The prevalence of stunting among children under 5 is around 48%, which is very high according to UNICEF (2013). The International Food Policy Research Institute (IFPRI) reported that India ranks 100 out of 118 countries on the Global Hunger Index (GHI) (Paroda, 2017). Hence, an adequate, balanced and nutritious diet is critical to achieve household nutritional security, and

this is an important SDG. Both food and nutritional security are major challenges. The fast-growing population, expanding urbanization, shrinking agricultural land and the adverse impact of climate change are emerging challenges for future food and livelihood security. As per the 2011 census, 54.6% of the population is engaged in agriculture and allied activities and contributes 17.4% of GVA. The total foodgrain production during 2017/18 was 277.49 million t. More food production does not guarantee food security for all sections of the population unless economic access is ensured. The four 'pillars' of food security defined by the FAO are: food availability, food access, food stability and food utilization. These all must be addressed in any national effort to reach hunger reduction targets and to improve household food and nutritional security.

Animal Husbandry: An Important Integral Component of the Indian Economy

The livestock sector plays a critical role in the welfare of India's rural population; it contributes 9% of GDP and employs 8% of the labour force. This sector is emerging as an important growth lever for the Indian economy. As a component of the agricultural sector, its share of GDP has risen gradually, while the crop sector has declined. In recent years, livestock output has grown at a rate of about 5% p.a., higher than the growth in the agricultural sector. Fortunately, the livestock sector makes multifaceted contributions to the Indian economy. The contribution of the sector during 2013/14 to agricultural GDP/GVA was 26.1%. In some states, like Punjab, Haryana and Uttar Pradesh, it was higher, between 30% and 40%. The value of its output at current prices (2013–2014) was about 6.24 trillion, of which milk and milk products alone contributed more than 4 trillion. Livestock's contribution is well beyond direct food production, as it supports the livelihood of more than two thirds of the rural population. It provides a flow of essential food products, draught power, manure, employment, income and export earnings. Livestock wealth distribution is more egalitarian compared to land. Hence, from the equity and livelihood perspective, it is considered an important component in poverty alleviation programmes.

As already stated, the role of livestock for both the income and nutritional security of the people is paramount. An old adage states that 'land rich in livestock will never be poor and a land poor in livestock can never be rich'. Animals serve as a source of protein and essential nutrients. The importance of animal protein is well recognized as it contains essential amino acids, which are deficient in cereals. Meat, milk and eggs provide protein with a wide range of amino acids, which humans need, as well as bioavailable micro-nutrients such as iron, zinc, vitamin A, vitamin B12 and calcium. Many malnourished people are deficient in these. Thus, interventions aiming at enhanced livestock production have an impact on the pace of poverty and hunger reduction.

Driven by structural changes in agriculture and food consumption patterns, the utility of livestock is undergoing a steady transformation. The non-food functions of livestock are becoming weaker. The importance of livestock as a source of 'draught power' has declined considerably due to the mechanization of agricultural operations and reduction in farm size. Use of dung manure, often used as fuel in rural areas, is being replaced by chemical fertilizers. On the other hand, its importance as a source of quality food has increased. Sustained income and economic growth, a fast-growing urban population, a burgeoning middle class, changing lifestyles,

an increasing proportion of women in the workforce, improvements in transportation and storage practices, and the rise of supermarkets, especially in cities and towns, are fuelling increased consumption of animal food products.

Unlike other developed countries, the livestock production system in India is unique; it is the best example of 'production by masses rather than mass production'. Smallholders and landless labourers are the backbone of Indian livestock production. In recent years, a slow transformation is being witnessed from smallholder to semi-commercial or commercial modes of livestock production, especially in the Green Revolution states of northern India. This transformation requires adequate readiness in terms of technology, machinery, input-delivery mechanisms, value addition and marketing. Adequate availability of trained manpower in commercial livestock production is a prerequisite for such a transformation to be successful.

India – A Leader in the Livestock Sector

India's livestock sector is one of the largest in the world. The country owns 516.5 million animals, of which 199.1 million are bovines, 105.3 million are buffaloes, 71.6 million are sheep and 140.5 million are goats. India is second in the world for its number of goats, approximately 25% of its livestock. It is also the second-largest poultry market in the world, producing 82.93 billion eggs and 649 million t of poultry meat. When the populations of livestock and poultry are combined, they almost equal the human population. India is the leader not only in livestock production; thanks to the White Revolution it is the top milk producer in the world with 176.35 million t produced in 2017–18 (about 18.5% of the world's milk production). The per capita milk availability in 2016–17 was around 337 g, which was well above the Indian Council of Medical Research's (ICMR) recommended level. The growth rate in milk production has been between 6% and 9% in recent years. The dairy sector has also become an important secondary source of income for millions of rural households engaged in agriculture. The total output worth has been higher than the value of foodgrains. Animals provide nutrient-rich food products, draught power, dung as organic manure and domestic fuel, and hides and skins, and are a regular source of income for rural households. They are natural capital, which can be easily reproduced and act as a living 'bank', an insurance against income shocks of crop failure and natural calamities. Overall, India's livestock sector is growing fast and emerging as a major contributor in the global market (Birthal *et al.*, 2006; DAH and DF, 2016–17)

Constraints of Low Productivity

The animal production system in India is predominantly part of a mixed crop-livestock farming system, vital for the security and survival of a large number of poor people. In such a system, livestock generates income and provides employment, draught power and manure. This production system assumes special significance in the present context of sustained economic growth, rising income, increasing urbanization and changes in tastes and preferences, which have led to dietary changes that reflect the importance of milk, meat, eggs and fish. The estimated growth rates of production are: 4.6% for milk, 5.69% for meat and 4.56% for eggs. The trends in the production of milk, meat, wool, eggs and fish are given in Box 5.1.

Population-driven production enhancement is certainly not a viable option for the future; technology-driven productivity enhancement is the most pragmatic approach for sustainable livestock production. Despite holding the number-one position in milk production in the world, India's milk productivity remains one of the lowest compared to many leading countries of the world. The milk production/cow/year in developed countries like the USA, Denmark, Sweden, Finland and The Netherlands is above 7500 kg; in India the average milk production/cow/year is around 1200 kg; thus there is scope to improve productivity. Regarding individual animal milk productivity, the national average productivity of exotic, crossbred, indigenous and non-descript cows in 2015/16 was at 11.2, 7.3, 3.4 and 2.2 kg/day, respectively. During the same period, on average, an indigenous buffalo and non-descript buffalo produced 5.8 kg and 3.8 kg milk/day, respectively (Alejandrino *et al.*, 1999; Manoharan *et al.*, 2003; Mutibvu *et al.*, 2012).

The strategy for increasing milk, meat and egg production should focus on increasing individual animal productivity rather than animal population. This calls for continuance of well-proven technologies coupled with improvement in the productivity of the vast population of generally low-producing cattle and buffalo using emerging reproductive and molecular technologies, including multiple ovulation and embryo transfer (MOET), use of sexed semen and cloning for production, and faster multiplication of superior germplasm of elite animals. A well-planned genetic improvement plan is needed to improve dairy animal productivity and production. In the past, efforts to implement systematic breed improvement programmes in the country using progeny testing were not very effective, and genetic improvement with traditional breeding programmes was rather unsuccessful. Recent developments in the field of animal breeding and genetics, coupled with biotechnology, have opened up new opportunities to breed animals of high genetic merit with high efficiency. Adopting a genomic selection strategy would pave the way for systematic improvement of our

---

Box 5.1. Critical gaps in livestock productivity.

- Livestock production is mostly resource-driven rather than demand-driven.
- The mindset of people is to manage animals only under low/negligible input systems.
- A huge non-descript livestock population exists.
- There is a slow pace of genetic improvement in the native useful breeds.
- There is poor productivity of local breeds of animals.
- There is inadequate feed and fodder supply to optimize production.
- There is indiscriminate cross-breeding of animals.
- There is reduced availability of vaccines, cold chains and other health measures.
- Marketing of animal products is disorganized.
- Problems in spreading new technologies include weak forward and backward linkages due to a weak extension system.

sector (Bhatia and Arora, 2005; Birthal *et al.*, 2006; DAH and DF, 2016–17).

In this context, therefore, a correct assessment of veterinary human resource requirements is essential for meaningfully predicting and planning with regard to the development of veterinary staff and to make decisions on the future number, type, capacity and programmes of educational and training institutions. Developing courses and programmes in the areas of biotechnology and applied areas of disease diagnosis, surveillance, vaccine production, nutritional management etc. would help to ensure that the necessary manpower is available. We also need to reorientate our veterinary education system to increase the number of graduates who are of 'gold class' and who are job creators, not job seekers. The emphasis at universities should be on vocational training and creation of a cadre of para-vets. We need both formal and informal training to meet the growing needs of the sector (DAH and DF, 2016–17).

The inadequacy of the present extension service has a negative impact on the performance of livestock farms and the quality of produce, especially on clean-milk production and value-addition aspects. Hence, direct involvement of scientists and technology developers in technology transfer is critical for effective dissemination of knowledge. It could be said that while India has plenty in some aspects of livestock production, it lacks in other areas, and this requires concerted efforts in a holistic manner to bridge the gap.

The livestock sector does not receive the policy and financial attention it deserves. It receives only about 12% of the total public expenditure on agriculture and allied sectors, which is disproportionate to its contribution to agricultural GDP. The sector has been neglected by financial institutions. The share of livestock in the total agricultural credit has hardly ever exceeded 4% (short-term, medium-term and long-term). The institutional mechanisms to protect animals against risk are not strong enough. Currently, only 6% of animals (excluding poultry) are provided with insurance cover. Only about 5% of farm households in India can access information on livestock technology. This indicates apathy on the part of financial and information delivery systems (DAH and DF, 2016–17).

There are about 1.34 billion people in India and around 120 million cows and buffalo. There is a long tradition of milk production and consumption (based on small units), cooperative structures and a tradition of selling milk over the counter. Sixty-seven per cent of the population lives in the countryside while only 33% lives in cities. However, with the fast-growing trend of urbanization, the situation is changing. The dairy market is fragmented and the involvement of the government is limited. In the near future, the dairy chain itself will change. The pace of the change is a multifactorial issue, mostly depending on government rules and their implementation. Food safety and food security are very much at the top of the agenda in India and will remain so in the near future as feeding the huge population is critical for sustainable growth. India is number one for milk production in the world. In 2015/16, the growth rate of milk production was 6.28% with total production reaching 156 million t. Now, per person milk availability is 337 g, on average, while at the world level it is 229 g/person. It is worth mentioning that in comparison with the years 2011–2014, the growth in milk production during 2014–2017 was 16.9% (DAH and DF, 2016–17).

The Way Forward

The White Revolution indubitably made a great contribution to the Indian dairy industry but there is more to be done. A 'third way' of dairy development driven by demand and value-chain evolution, with a systems approach, may be the answer. Any such new initiative needs to focus on food-borne and zoonotic diseases originating from dairy animals. With emerging market opportunities, diagnosis, treatment and control of transboundary diseases need more investment and resources. Better farm hygiene and environmental health are critical for the sustainable growth and well-being of humans and animals. Animal welfare, feed and nutrition of dairy animals should be adequately addressed. Gender mainstreaming and healthcare for women of farming families need to be given special consideration.

The livestock sector, which recorded a growth of about 5.0% p.a. in the last decade, has great potential for future growth. Similarly, the fisheries sector has also recorded growth of more than 5%. There are major challenges for the animal husbandry sector, which, if addressed appropriately, would help in accelerating growth in livestock sector. To achieve this growth, the following road map is suggested:

- Urgent conservation of valuable indigenous breeds is needed in different parts of the country by establishing specific breed farms for much-required genetic improvement of local breeds.
- Reorientation of research priorities – in the past, research efforts were cattle-centred. This has yielded some results but at the cost of other promising livestock species like buffalo, sheep and goat. Henceforth, live-stock research needs to internalize certain crucial technical parameters. Priority may also be given to breeding buffalo, sheep and goats and to their scientific management. Buffalo hold great promise in the Indian context. Their feed-conversion efficiency is remarkable.
- There is a need to further expand the artificial insemination programme from the present level of about 25% to around 50% of the breedable bovine population so as to improve milk production through genetic improvement.
- Major expenditure (almost 30–40%) for dairy/livestock farming is on feed and fodder. Hence, increased production and productivity of forage crops through good-quality seed is critical. A rolling plan for seed production through effective coordination is warranted.
- Health, nutrition and vaccination support merits special consideration in the livestock sector. In the area of health, greater emphasis should be placed on smaller animals.
- A well-planned and operational livestock disease control programme involving PPP must be put in place. A comprehensive package of disease awareness, management and control measures is needed to properly educate farmers.
- Establishment of animal health clinics around villages would be necessary. Also, an animal insurance scheme, once implemented, would insulate livestock farmers against possible risks.
- Some immediate intervention is needed in the production and processing areas, such as: (i) easing import restrictions on feed like corn,

sexed semen, vaccines and feed additives; (ii) compulsory backward integration and modernization of slaughterhouses; (iii) establishment of good market infrastructure, including cold-storage facilities; (iv) effective disease surveillance mechanisms and establishment of disease-free zones for increased meat export; and (v) strengthening PPP.
- Access to markets is critical to speed up the commercialization of livestock products. Except for poultry and, to some extent, milk, markets for livestock and livestock products are still underdeveloped. Further, these are often dominated by informal market intermediaries, who exploit producers. Establishing farmers' self-help groups, cooperatives and producers' companies would help farmer–consumer links.
- State agricultural universities/veterinary universities must strengthen their capability to produce trained veterinarians and para-vets to provide effective health cover as well as to ensure effective transfer of technologies to end-users.

Looking Beyond

It is important to understand the multiple dimensions of the milk sector in India. The value of milk produced in the country reached Rs 4,958,410 million in 2014/15, thus making it one of the most important agricultural commodities. For the first time, the value of milk produced exceeded the value of total foodgrains (cereals plus pulses), which stood at Rs 4,868,460 million, according to the estimates of the Central Statistics Office. Moreover, milk production enhanced the economic status significantly and improved the lives of farmers in rural areas. The dairy sector provides employment to the landless and landowners alike and the income so generated helps check rural to urban migration. In addition, milk helps in addressing both malnutrition and poverty. However, the growth of the dairy sector in a setting of relatively weak infrastructure and governance has given rise to certain challenges. To meet these, the sector requires renewed attention and higher investment from government and from agricultural research as well as from the development community.

References

Alejandrino, A.L., Asaad, C.O., Malabayabas, B., De Vera, A.C., Herrera, M.S. *et al.* (1999) Constraints on dairy cattle productivity at the smallholder level in the Philippines. *Preventive Veterinary Medicine* 38(2–3), 167–178.

Bhatia, S. and Arora, R. (2005) Biodiversity and conservation of Indian sheep genetic resources – an overview. *Asian-Australasian Journal of Animal Science 2005* 18(10), 1387–1402.

Birthal, P.S. and Jha, A.K. (2005) Economic losses due to various constraints in dairy production in India. *Indian Journal of Animal Sciences* 75(12), 1470–1475.

Birthal, P.S., Taneja, V.K. and Thorpe, W. (eds) (2006) Smallholder Livestock Production in India: Opportunities and Challenges. Proceedings of an ICAR–ILRI international workshop held at the National Agricultural Science Complex, New Delhi, India, 31 January–1 February. NCAP (National Centre for Agricultural Economics and Policy Research)-ICAR (Indian Council of Agricultural Research), New Delhi and ILRI (International Livestock Research Institute), Nairobi.

Business Line (2016) Agriculture: from make in India to make in rural India through dairying. *Business India*, 17 January.

CGR (2012) *Animal Genetic Resources Conservation-Breeding-Regulatory Framework*. GRIP course, Centre for Genetic Resources, Uppsala, The Netherlands, 15 May.

DAH and DF (2016–17) *Annual Report 2016–17*. Department of Animal Husbandry, Dairy and Fisheries. Ministry of Agriculture and Farmers' Welfare, Government of India.

DAH and DF (2017–18) *Annual Report 2017–18*. Department of Animal Husbandry, Dairy and Fisheries. Ministry of Agriculture and Farmers' Welfare, Government of India.

Kurien, V. (2007) India's milk revolution: investing in rural producer organizations. In: Narayan, D. and Glinskaya, E. (eds) *Ending Poverty in South Asia: Ideas that Work*. World Bank, Washington D.C.

Manoharan, R., Selvakumar, K.N., Pandian, A. and Saravana, S. (2003) Constraints in milk production faced by the farmers in Pondicherry union territory. *Indian Journal of Animal Research* 37(1), 68–70.

Mathur, B.N. (2000) Current problems and challenges confronting the dairy industry in India. *Asian-Australasian Journal of Animal Science* 13, 447–452.

MoA and FW (2016) *Agricultural Statistics at a Glance 2015*. Directorate of Economics and Statistics, Department of Agriculture, Cooperation and Farmers' Welfare, Ministry of Agriculture and Farmers' Welfare, Government of India, New Delhi.

Mutibvu, T., Maburutse, B.E., Mbiriri, D.T. and Kashangura, M.T. (2012) Constraints and opportunities for increased livestock production in communal areas: a case study of Simbe, Zimbabwe. *Livestock Research for Rural Development* 24(9), 353–368. Available at: http://www.lrrd.org/lrrd24/9/muti24165.htm (accessed 22 May 2018).

NAAS (2016) Policy Paper No. 81: Climate Resilient Livestock Production. National Academy of Agricultural Sciences, New Delhi.

Paroda, R.S. (2017) Indian Agriculture for Achieving Sustainable Development Goals. Strategy paper. Trust for Advancement of Agricultural Sciences, Pusa Campus, New Delhi, 30 October.

Pendleton, A. and Narayanan, P. (2012) The white revolution: milk in India. Taking liberties: poor people, free trade and trade justice. *Christian Aid*.

Singh, R. (2014) Tips to overcome problems in cattle rearing. *The Hindu*, 10 December.

Singh, K.M. and Meena, M.S. (2012) Livestock value chains: prospects, challenges and policy implications. In: Bhatt, B.P., Sikka, A.K., Mukherjee, J., Islam, A. and Dey, A. (eds) *Status of Agricultural Development in Eastern India*. ICAR Research Complex for Eastern Region, Patna, Bihar, India, pp. 493–508.

Thornton, P.K. (2010) Livestock production: recent trends, future prospects. *Philosophical Transactions of the Royal Society of London. Series B, Biological Sciences* 365(1554), 2853–2867. DOI: 10.1098/rstb.2010.0134.

UNICEF (2013) Key facts and figures on nutrition. UNICEF report: Improving Child Nutrition: The Achievable Imperative for Global Progress. *Journal of Nutrition* 140(2), 348–354.

6

Aquaculture Development and the Blue Revolution

Role of Aquaculture

India is blessed with huge open-water resources – seas, rivers, lakes, reservoirs and wetlands. Aquaculture in freshwater ponds and tanks covering 2.43 million ha contributes a large share of the total fish production of 11.4 million t (approx 60%). The country has a coastline of 8118 km and nearly 2 million sq. km of exclusive economic zone (EEZ), and 500,000 sq. km of continental shelf (IASRI, 2016). From these marine resources, India has an estimated fisheries potential of 4.21 million t. It has an extensive river and canal system consisting of 14 major rivers, 44 medium rivers and numerous small rivers and streams. India experienced a 14-fold increase in fish production in the past six-and-a-half decades.

Indian fisheries and aquaculture is an important sector of food production, providing nutritional security to the food basket, contributing to agricultural exports and engaging about 14 million people in different activities. With diverse resources ranging from deep seas to lakes in the mountains, and more than 10% of the global biodiversity in terms of fish and shellfish species, the country has shown continuous and sustained increments in fish production since its independence. Constituting about 6.3% of global fish production, the sector contributes 0.9% of GDP and 5.43% of agricultural GDP. Paradigm shifts in terms of increasing contributions from the inland sector and from aquaculture have been significant over the years. With high growth rates, the different facets of marine fisheries, coastal aquaculture, inland fisheries, freshwater aquaculture and coldwater fisheries contribute greatly to the food, health, economy, exports, employment and tourism of the country.

The 429 Fish Farmers Development Agencies (FFDAs) and 39 Brackishwater Fish Farmers Development Agencies (BFDAs) are promoting freshwater and coastal aquaculture. With annual carp seed production of 45 billion and shrimp seed of about 14 billion, the country has been largely able to satisfy the increasing demand for seed for aquaculture of carp in freshwater and shrimps in land-based coastal aquaculture. With increasing emphasis on species diversification in the recent past, there has been greater focus on development of technologies for breeding and mass-scale seed production of several freshwater, brackishwater and marine finfish and shellfish species for aquaculture in ponds and cages (Ayyappan et al., 2011). Along with food-fish culture, ornamental fish culture and high-value fish farming are gaining importance.

With over 240,000 fishing craft operating around the coast, six major fishing harbours, 62 minor fishing harbours and 1511 landing centres, the needs of over 3.9 million fisherfolk are being met. Fish and fish products have emerged as the largest group in India's agricultural exports, with 1.05 million t and Rs 378.7 billion in value.

This accounts for around 10% of the total exports of the country and nearly 20% of agricultural exports. More than 50 different types of fish and shellfish products are exported to 75 countries around the world.

The aquaculture sector is recognized as the 'sunshine' sector of Indian agriculture. It helps in increasing food supply, generating adequate employment and raising nutritional levels. It has huge export potential and is a major source of foreign exchange earnings for the country. Freshwater aquaculture in India has evolved from confinement to the east-Indian states during the 1950s to the present vibrant industry that has spread over the entire country. Total fish production has experienced phenomenal increase, from 0.75 million t in 1950–51 to 10.79 million t in 2015–16. Dominance of marine-capture fishery, with 60% share of total fish production in 1990–91, has been reversed, with 7.21 million t (66.8%) now coming from inland fisheries (2015–16), of which more than 80% is from aquaculture.

The Blue Revolution

India is now predominant in aquaculture production, globally, occupying second position only after China. This quantity is almost fully consumed in the domestic market, except for shrimps and freshwater prawns, which are mainly exported. Specifically, freshwater aquaculture experienced a 15-fold growth in the past three-and-a-half decades, i.e. 0.37 million t in 1980 to about 5.7 million t at present. About 40% of the population does not eat fish, since they are vegetarian, and the remaining 60% consume fish. It has been estimated that the inland fishery resources have production potential of about 15 million t. Against this potential, the actual production was 6.58 million t in 2014–15, thus suggesting considerable scope for inland aquaculture.

The Blue Revolution, encompassing multidimensional activities, focuses mainly on increasing production from aquaculture and other fisheries resources, both inland and marine. The vast fishery resources (Table 6.1) offer immense opportunities to enhance fish production through aquaculture-system diversification, species diversification, proper management,

Table 6.1. Details of fishery resources in India. (From: DAH and DF, 2016–17)

Fisheries sector	
Global position	3rd in fisheries, 2nd in aquaculture
Contribution of fisheries to GDP	0.90%
Contribution to agricultural GDP	5.43%
Per capita fish availability	9 kg
Annual export earnings	Rs 378.7 billion (US$5.78 billion)
Employment in the sector	14.5 million
Marine	
Length of coast line	8118 km
Exclusive Economic Zone (EEZ)	2.02 million sq. km
Continental shelf	530,000 sq. km
Number of fish-landing centres	1537
Number of fishing villages	3432
Inland	
Total inland water bodies	7.3 million ha
Rivers and canals	195,000 km
Reservoirs	2.9 million ha
Tanks and ponds	2.4 million ha
Floodplain lakes/derelict waters	798,000 ha
Brackishwater	1.1 million ha

introduction of new and advanced technologies in both the marine and inland sectors, adoption of scientific practices and application of suitable fish-health management strategies. In marine fisheries and capture fisheries, growth has proved elusive for many reasons. One reason is that marine fishing activity remains confined to coastal waters, leaving most of the EEZ, measuring 2.02 million sq. km, underexplored. The narrow coastal belt of up to 50 m depth, is being over-exploited by traditional fishermen (Ayyappan et al., 2011), causing considerable depletion of fish stocks. Motorized fishing vessels are responsible for about 85% of the total marine catch, yet deep-sea fishing activity remains at low levels owing to the inadequacy of the fishing vessels.

The potential of inland aquaculture is still untapped in India, despite the fact that of the current 11.4 million t of total fish production, the inland fisheries account for two thirds of that total. A small number of farmers are currently practising inland aquaculture in states such as West Bengal, Andhra Pradesh, Odisha, Assam, Punjab, Haryana and Telangana, and a Blue Revolution is occurring. By increasing the coverage of the water area and the productivity of existing bodies of water by 50%, total production could be doubled. However, there are some critical gaps that need to be addressed to achieve a true Blue Revolution.

Given that India has a large number of water bodies, reservoirs, lakes, and ponds, inland aquaculture holds the key to the Blue Revolution in the country. Fish farming will have three benefits: (i) an increase in farmers' incomes; (ii) progress in the country's exports and GDP; and (iii) ensured nutritional and food security in the country. The country has demonstrated consistent 6–7% annual growth in aquaculture over the last three decades, which is unparallelled in most other agricultural sectors. In the last decade, where the average annual growth rate of fish and fish product exports in the world has been 7.5%, India has witnessed an average annual growth rate of 14.8%. Looking at the potential for the development of fisheries, Prime Minister Shri Narendra Modi has called for a Blue Revolution. The government has merged all existing schemes and started a new Rs 30 billion umbrella scheme: 'Blue Revolution: Integrated Development and Management of Fisheries', which includes inland fisheries, aquaculture and marine fisheries comprising deep sea, mariculture and all the activities of the National Fisheries Development Board (NFDB). The Department of Animal Husbandry, Dairying and Fisheries (DAHDF) has prepared a National Fisheries Action Plan 2020 for the next five years to increase fish production and productivity. In the plan, all of the fisheries resources – ponds and tanks, wetlands, brackishwater, cold water, lakes, reservoirs, rivers, canals – are included. All states have been requested to prepare a SAP for the next five years to achieve the objective of Blue Revolution. The aim of the scheme is to increase fish production and productivity by 8% annually and to reach 15 million t by 2020. During the last two years, under fishermen welfare, construction of 9603 fishermen's houses have been completed, 20,705 fishermen have been trained and around 5 million fishermen have been provided with annual insurance assistance.

Species Diversification

The bulk of inland aquaculture production in India comes from three fish species – rohu, catla and mrigal. A handful of farmers are experimenting with exotic species such as silver carp, grass carp and common carp (Ayyappan and Jena, 2003). Recently the sector has witnessed interest in commercial farming of the exotic pangas catfish (Sahu and Sahoo, 2011). Thus, diversification is needed for efficient growth. The number of cultured species is less than ten, against more than 100 in China. Species diversity is an area that has not yet been explored and which has tremendous potential to increase production. Efforts are also being made to diversify the species mix in freshwater aquaculture by introducing high-value catfish like magur, freshwater prawns and regional species.

A Growing Industry

Major areas for the industry have been optimization of production and productivity; augmenting exports; generating employment, improving the welfare and socioeconomic status of fishermen; capture and culture, including inland and sea; aquaculture; gears; navigation; oceanography; aquarium management; breeding; processing; export and import of seafood; special products and by-products; and research. There exist several investment opportunities in the sector, but there are several challenges and issues facing fisheries such as accurate data on assessment of fishery resources and their potential in terms of fish production; development of sustainable technologies for finfish and shellfish culture; yield optimization; harvest and post-harvest operations; landing and berthing facilities for fishing vessels; and welfare of fishermen. The strong and sustaining ecological resource base, rational and preemptive policy, public and private investment and good governance hold the key for the sustainable growth of the sector. Full potential can be achieved through infrastructure, investment, technology intensification, diversification and value addition. Various issues relating to fishing activities in India need to be addressed in a time-bound manner, with mutual understanding and cooperation between the public and private sectors.

Fish constitute slightly more than half of all vertebrate species, some 28,000 species. In India, the potential of fish culture is yet to be fully exploited. Fish are a rich source of protein and have high nutritive value. Extensive development of aquaculture needs to be given priority after the Green Revolution in order to feed an ever-growing population. Success of fish culture depends, apart from other factors, on selection of suitable species, sufficient water supply and quality of land.

Technologies Developed

The systems and technology used in aquaculture have developed rapidly in the last 50 years. Much of the technology used is relatively simple, often based on small modifications that improve the growth and survival rates of the target species, e.g. improving food, seeds, oxygen levels and protection from predators. Simple systems of small freshwater ponds, used for raising herbivorous and filter-feeding fish, account for about half of global aquaculture production. Advances in hydrodynamics applied to pond and tank design have enabled the development of closed systems that have the advantage of isolating the aquaculture systems from natural aquatic systems, thus minimizing the risk of disease or genetic impact on external systems. Developments in engineering, some adapted from offshore oil rig construction, increase the possibilities of a progressive offshore expansion of aquaculture using robust cages. Culture-based capture fisheries involving the release of young fish into the wild to improve harvest (an operation also referred to as restocking, stock enhancement or ranching) have been suggested for increasing production in large open-water systems, i.e. reservoirs (Sugunan and Sinha, 2000). Sea ranching has just begun, and its long-term viability is being assessed. Advances have also been made in capture-based aquaculture involving the growing/fattening of young fish (e.g. groupers, tuna) captured from the wild. Major progress has also been made in the development of aquafeed technology, combining a large number of ingredients into very small pellets.

The selection of the aquaculture system or approach in a particular development is determined by several factors including: development goals/objectives and target beneficiaries; acceptability/marketability of culture species; availability and level of technology; availability of production inputs; support facilities and services; investment requirements; and environmental considerations.

Freshwater Fish Farming

Freshwater fish farming in India has been synonymous with carp farming until recently. The carp group comprises three Indian major carps (catla, rohu and mrigal), exotic carps (common, silver and grass) and minor carps, which constituted 76.5% of total inland production in 2012. While the contribution of exotic carp was significant in earlier years, recent years have witnessed reduced popularity of these among fish farmers, shrinking to only 9.55% in 2012. However, the gap has been filled by increased production of indigenous major carp. Added to that, the last decade has witnessed several new entrants into the Indian aquaculture system, both indigenous and exotic species, which have boosted fish production as well as farm income. Aquaculture activity supports individuals on a full-time basis, industrial activity for the corporate sectors, as well as many entrepreneurs. Nationwide successful demonstrations of the developed aquaculture technologies have brought about a Blue Revolution in the country. Today, technologies have been standardized to produce fish using almost all types of water bodies, be they reservoirs, rivers, derelict waters, ponds, tanks, canals and the cold waters of the hill region. A brief account of the most significant technologies that have revolutionized the aquaculture sector are described below.

Breeding and Seed Production

Carp breeding and seed production

Riverine seed collection was the major source of seed during the first half of the 20th century. Since the 1920s, a number of natural and synthetic inducing agents such as the pituitary gland, human chorionic gonadotropin, pregnant mare serum, mammalian gonadotropin-releasing hormone, luteinizing hormone releasing hormone (LHRH), and LHRH-analogue were tried to breed carp (Gupta and Rath, 2011). However, the epoch-making achievement came in the form of the successful induced breeding of Indian major carp with the use of pituitary gland extract, in 1957, which led to the foundation of aquaculture development in the country (Chaudhuri and Alikunhi, 1957). Subsequently, ampouling of the pituitary extract, refinement of breeding protocol, evolution of an array of hatchery technologies and hatchery models, and better broodstock management techniques have helped increase efficiency of carp seed production. With assured seed supply and development of seed-rearing techniques, aquaculture activity increased and led to increased fish production.

Development of synthetic inducing agents

Use of pituitary gland extract as an inducing agent brought problems of variable efficacy besides the cumbersome protocol of having to inject brooders twice. This often led to higher handling stress, improper synchronization, poor breeding response and higher post-breeding mortality. Carp breeding was made easier with the development of synthetic analogues of fish gonadotropins. Ovaprim, a synthetic analogue of salmon gonadotropin (SGnRH) was the first of its kind and revolutionized the seed production activity (Ayyappan et al., 2016). This was followed by a series of synthetic chemicals like Ovatide, Ovopel and WOVA-FH, which have helped the hatcheries cater to the ever-increasing seed requirement, despite significant rises in seed demand over the years.

Development of different hatchery models

Hapa breeding, the common method used for carp breeding until 1980, was having problems of dependence on environment. Gradually, the quest for improving the efficacy of induced breeding led to the development of a number of hatchery models with better-controlled facilities and more reliable results (Dwivedi and Zaidi, 1983; Gupta and Rath, 2011). These include the glass jar hatcheries and plastic bin hatcheries

developed during the 1980s. The Chinese circular carp hatcheries developed during the early 1990s, with simulation of all the natural conditions required for spawning, proved to be the most efficient model. The fibreglass reinforced plastic carp hatchery, a smaller and portable version of the Chinese eco-hatchery, was launched by ICAR-Central Institute of Freshwater Aquaculture (CIFA) in 2006. Its ease of portability has made it suitable for small and marginal seed producers and has made it possible to take the induced breeding technology to remote areas.

Multiple spawning

In order to avoid the maintenance of large populations of broodfish by hatchery owners, ICAR-CIFA standardized the multiple spawning technique during the 1990s (Gupta et al., 1995). With this technique, the same broodfish of major carp could be bred as many as four times between March and September, thereby stretching the breeding season (June–August). It has been able to demonstrate two- to threefold higher spawn recovery over conventional single breeding during a season.

Cryopreservation of carp milt

The cryopreservation technique to preserve fish milt was standardized during the 1990s by ICAR-CIFA. This technique made it possible to preserve Indian major carp semen for 18 hours at 4°C prior to artificial insemination. Four fish-semen cryobanks, two in Andhra Pradesh and two in Odisha, were established in 2009/10 and are being used for stock upgrade of Indian major carp in hatcheries with more than 40% hatchlings recovery. Today, the country is almost self-sufficient in the production of carp spawn through a network of more than 2000 freshwater hatcheries, including about 350 fibreglass reinforced plastic hatcheries established in the country through both private and public participation.

Seed production of diversified species

The aim to diversify freshwater aquaculture has led to significant developments in the breeding and seed production of several other species. These include several minor carp, freshwater prawns, catfish, murrels, climbing perch, etc. Breeding and seed production technologies have been standardized for many indigenous minor carp, i.e. *Labeo calbasu, L. fimbriatus, L. gonius, L. bata, Cirrhinus cirrhosa, C. reba* and barbs, i.e. *Puntius sarana, P. pulchellus* (Ayyappan *et al.*, 2016). Ease in seed production and assured availability of seed has ensured their wider domestication all over the country. Intercropping of these minor carp in the mainstream major carp production system has demonstrated a 30% increase in biomass production besides ensuring availability of varied protein.

Next to carp, panga (*Pangasionodon hypophthalmus*) has become a popular species spreading over more than 15,000 ha in Andhra Pradesh and a considerable area in Bihar and Chhattisgarh. It is a fast-growing species relying on a feed-based system and suitable for cage culture. Much of the seed of this species used in the country at present is sourced from neighbouring Bangladesh, although several commercial hatcheries have been established in West Bengal, Odisha and Andhra Pradesh. New hatcheries are coming up in several other states. Seed production of magur has been an important activity in Assam, West Bengal and Odisha, catering to a large culture area in West Bengal and the north-eastern states. Several other hatcheries in the plain areas are catering to the grow-out activity. Establishment of a pabda hatchery is an upcoming venture due to the popularity of this species among farmers in West Bengal and other north-eastern states during the last decade. A few hatcheries have been established in West Bengal. The giant freshwater prawn (*Macrobrachium rosenbergii*) has been successfully bred in captivity (Rao and Tripathi, 1993) and is cultured under monoculture and polyculture with carp. More than 35 freshwater prawn hatcheries were operating in the country; however, the number has fallen in recent years due to the decrease in culture area of the freshwater prawn with increasing interest in the farming of the exotic Pacific white shrimp in these freshwater areas. The wild collection of prawn juveniles has also been catering to the need for prawn seed to a large extent, especially in the states of West Bengal and Odisha.

Standardization of breeding protocols for different ornamental fish, both indigenous and exotic, has contributed to boost the ornamental trade (Swain *et al.*, 2011). Several programmes relating to ornamental fish have been devised to motivate local people in the north-eastern states of Meghalaya, Arunachal Pradesh, Assam and Tripura, and the plains. Several ornamental fish villages have come into existence in West Bengal, Odisha, Tamil Nadu, Kerala and Karnataka, and have proved to be a potential means to empower rural women.

Development of indigenous systems of culturing pearls from common freshwater mussels, *Lamellidens marginalis, L. corrianus* and *Parreysia corrugate* is another important technology in the country (Janaki Ram and Tripathi, 1992). The easy surgical procedure, developed for production of both image and round pearl, has attracted the attention of many fish farmers, as witnessed by an increase in demand for training in this area in recent years.

Seed rearing

Most of the basic technologies of seed rearing and grow-out production in carp were developed in the 1980s at the Pond Culture Division of the Central Inland Fisheries Research Institute (CIFRI) at Cuttack. Subsequently, the techniques have undergone several modifications and refinements with respect to stocking density, use of critical inputs like feed and fertilizers, and efficient pond management protocol (Ayyappan and Jena, 2003). The present-day packages of nursery and rearing practices, with simpler pond maintenance, efficient input use and high seed survival, have helped increase survival, up to 50–60% in the nursery and 60–80% in the rearing phase, thereby making seed rearing a highly viable activity in farming. Instead of going for grow-out farming, several farmers nowadays are opting for seed rearing as a full-time activity.

Use of bigger concrete tanks for seed rearing of carp at high density has also proved effective. With better control of the environment in such systems, nursery rearing of carp spawn could be done at a density of $1000–2000/m^2$ (10–20 m/ha), which has shown fry survival as high as 50% compared to 30% in an earthen pond (Jena and Das, 2011). Furthermore, use of aeration and water exchange in such systems have been proven to enhance seed survival by up to 60%. Use of such concrete seed-rearing systems has made it possible to achieve more than six times higher fry yield/unit area in every crop. Farmers are able to harvest three to four crops during a season, which makes this system highly viable economically.

At present, more than 45 billion fry are produced, making the country self-sufficient in fry production. However, there is a dearth in the production of fingerling, the right size for grow-out stocking. In recent years, special emphasis has been given, through many schemes and promotional policies, to promoting fingerling production in order to achieve adequacy of supply of yearlings. Protocols have been developed for the production of stunted fingerlings of carp to ensure all-year-round seed availability.

Development of Grow-out Farming Technologies

Composite fish culture technique

Growth evaluation and standardization of different species combinations with respect to stocking densities, species ratio and use of critical inputs has led to the development of a composite fish culture of the three indigenous major carp: catla, rohu and mrigal. Introduction of exotic species into Indian waters was another landmark to boost aquaculture production. Common carp (*Cyprinus carpio*) were brought in for a second time in 1957, while Chinese grass carp (*Ctenopharyngodon idella*) and silver carp (*Hypopthalmichthys molitrix*) were brought in in 1959. The higher growth rates of these exotic carp in tropical climates increased pond productivity and Indian fish farmers readily accepted them as candidates for culture.

Large-scale demonstrations of composite fish culture were undertaken in the country through a series of projects and schemes: the National Demonstration Project (NDP), Operational Research Project (ORP) in 1974–75, Central Inland Fisheries Research Institute (CIFRI)-International Development Research Centre (IDRC) Project on Rural Aquaculture in 1975–79, and the All India Coordinated

Research Project (AICRP) on Composite Fish Culture and Fish Seed Prospecting between the 1970s and 1984 (Jhingran, 1991). Demonstration of carp production levels of 3–5 t/ha p.a. through the AICRP virtually laid the foundation for scientific carp farming all over the country (Ayyappan and Jena, 2001). The project had a great impact in West Bengal where some of the villages turned into total seed production centres, while others produced large quantities of market-size fish. In 1975, the Ministry of Agriculture of the Union Government implemented the Fish Farmers Development Agency (FFDA) scheme on a pilot scale. With the success of the scheme, the Union Government launched the Inland Fisheries Project during the 1980s with assistance from the World Bank to develop fish farming in large areas in five potential states, West Bengal, Odisha, Bihar, Uttar Pradesh and Madhya Pradesh, which was implemented through FFDAs. Since then, the network of 429 FFDAs established in almost all the potential districts in different states of the country shouldered the responsibility to popularize aquaculture in the country.

Polyculture of Indian major carp alone, or along with exotic carp at lower to moderate stocking density, has been realizing production of 4–10 t/ha p.a. The technology of intensive carp polyculture, involving stocking of larger fingerlings (25–50 g) at higher densities of 15,000–25,000/ha, provision of high-quality critical inputs such as seed, feed and fertilizers, along with additional provision of aeration, water-exchange facilities and fish-health management, has been in vogue, demonstrating higher production levels of 10–15 t/ha p.a. (Jena and Das, 2011). Market-driven forces have become the deciding factor in the modification of the present-day commercial culture system. Higher demand for rohu has forced fish farmers to adopt a bi-species culture of rohu and catla in the Kolleru Lake region of Andhra Pradesh, with rohu constituting 90% of the stock. Farmers in this region have also started stocking stunted juveniles/yearlings of 100–300 g in a multi-stocking, multi-harvesting, grow-out system, and have been realizing higher yields. The poor market for exotic silver carp, grass carp and common carp has led to the gradual reduction in the culture of these species in recent years despite their higher growth rate than their Indian counterparts.

The Kolleru Lake region of Andhra Pradesh has become the hub of organized aquaculture activity. At present, more than 1 million t of fish are being produced from the Krishna and West Godavari districts alone, which is supplied through organized cold-chain marketing to almost all parts of the country. Production from the culture sector has also increased significantly in recent years in states like Chhattisgarh, Jharkhand, Bihar, Odisha and Assam, apart from the significant contribution of West Bengal.

Cage and pen culture

The last decade has witnessed a surge in cage and pen culture activity with many states' fishery departments, with support from the National Fisheries Development Board (NFDB), promoting large-scale cage farming in reservoirs. States like Chhattisgarh and Jharkhand have been undertaking cage culture of exotic *P. hypophthalmus* in different reservoirs in a big way. Similar attempts at cage farming have also been undertaken by the fishery department of Odisha in the Rushikulya and Mahanadi river system and in the Hirakud reservoir.

Seed rearing in cages and pens in open waters for the promotion of culture-based capture fishery are another avenue of entrepreneurship that is becoming popular across the country. Stock enhancement through pen culture has been demonstrated by ICAR-Central Inland Fisheries Research Institute (CIFRI) in selected wetlands of Assam, mainly aimed at culture-based fisheries among the beel fishers. Such attempts, through pens erected in wetlands, have improved fish productivity considerably (Bhowmick and Das, 2011).

Culture-based capture fisheries

The reservoir resources in India cover more than 3 million ha distributed in varied climatic environments congenial to fish growth. Over the years, these natural waters have suffered from dwindling stocks of indigenous fish owing to anthropogenic activity such as construction of dams and barrages, environmental pollution and over-exploitation. The productivity of many of the small and medium-sized reservoirs has remained as low as 20–25 kg/ha, while the average fish production potential of these resources is estimated at 250 kg/ha for reservoirs and about 350 kg/ha for wetlands. As a move to improve the culture-based capture fisheries in the small and medium-sized reservoirs, many state governments have come up with reservoir management policies. With assistance from NFDB, seed-stocking programmes have been undertaken with seed produced from brooders indigenous to the same system. Awareness created among the local fishermen has helped improve the capture fishery resources of many of these water bodies, raising productivity to 150 kg/ha.

Wastewater-fed aquaculture

Technology has been standardized for utilizing treated sewage for fish production by the erstwhile sewage-fed fish culture wing of CIFRI and the present Regional Research Centre of ICAR-CIFA at Rahara. With this technology of using treated sewage, a large portion of the wastelands around Kolkata has been brought under fish production activity. Sewage water is proven to be an alternative to fertilizer for enhancing yield in paddy-cum-fish-farming systems and is being used in many areas of West Bengal.

Integrated farming systems

Several models of integrating fish culture with livestock have been developed over the years and have proved advantageous (Gopakumar *et al.*, 2000). Fish-cum-duck integrated farming using Indian major carp (IMC) as the fish component along with Khaki Campbell duck is a model accepted by rural farmers. Similarly, pig-cum-fish culture in north-eastern states and paddy-cum-fish culture in West Bengal and the north-eastern region have been accepted as popular farming systems.

Brackishwater Aquaculture

Although traditional fish farming in the *bheries* of West Bengal and *pokkali* fields of Kerala has a long history, scientific farming in the country was initiated only in the early 1990s (Rao and Ravichandran, 2001). The initial two decades of brackishwater farming were largely based on a single species of shrimp, the black tiger prawn (*Penaeus monodon*), owing to its high export market (Ponniah, 2011). Demonstration of the technology of semi-intensive farming with production levels of 4–6 t/ha in 4–5 months, coupled with institutional credit facilities and subsidies from the Marine Product Export Development Authority (MPEDA), helped in the development of commercial shrimp farming. Further, the entry of several big industrial houses to the sector has made for a vibrant and high-tech enterprise. Substantial investment was made towards the development of state-of-the-art hatcheries, feed plants, mega-size farms with modern equipment and gadgets, modern laboratories, processing plants and communication networks.

Shrimp farming

The rapid development of shrimp aquaculture in the coastal areas of the country also raised some environmental concerns, and the need for a regulatory mechanism to control the indiscriminate growth of aquaculture has become apparent, and the Coastal Aquaculture Authority (CAA) has started functioning. Unregulated growth also led to severe disease outbreaks, affecting farming since the mid-1990s. However, the disease problems could be managed to a great extent through the adoption of good management practices, which led to sustained increases in shrimp production in subsequent years.

The introduction of the Pacific white shrimp, *P. vannamei*, to the culture system in India has now seen total dominance of the species in shrimp farming, registering a coverage of almost 90%. The species has attracted farmers' attention due to its fast growth, the possibility of culture at higher stocking densities, the low incidence of native diseases, the availability of specific pathogen free (SPF) domesticated strains, the culture feasibility in a wide range of salinity of 0.5–45%, and the ready international market for the shrimps (Ayyappan and Jena, 2001). Production levels of 8–10 t/ha/crop of four months is a common practice. Of the 487,000 t of farmed shrimps produced in India during

2015/16, about 400,000 t are Pacific white shrimp. The area under culture has seen an increase of about 12%. Farmed shrimp accounts for 38% of the quantity and 64.5% of the value (Rs 244.26 billion) of total seafood exports worth Rs 378.7 billion (US$ 5.78 billion) in export revenue.

Species diversification

Brackishwater aquaculture over the years has been based on value rather than volume. However, in order to achieve sustainable growth of the sector, aspects of diversification are being emphasized now. At present, emphasis is given to culture of several other potential finfish and shellfish species – Indian white shrimp (*Fenneropenaeus* (*Penaeus*) *indicus*), seabass (*Lates calcarifer*), grey mullet (*Mugil cephalus*), milkfish (*Chanos chanos*), pearl-spot (*Etroplus suratensis*) and mud crabs (*Scylla serrata* and *S. tranquebarica*). The technologies for breeding and seed production of most of these species have already been standardized (Arasu *et al.*, 2009), and their successful farming has been demonstrated in recent years. However, it is necessary that mission-mode programmes are taken up for large-scale adoption of these technologies.

Future perspectives

In order to double the export earnings to the tune of US$10 billion in the next five years, it is necessary that additional areas are also brought under farming control, considering that vertical expansion may not be an ideal proposition due to environmental issues. For bringing the unutilized/underutilized areas under farming, aspects of environmental soundness, social acceptability, equity and resource conservation need to be addressed. Sustainable approaches like biosecured zero-water-exchange shrimp farming technology (BZEST), probiotic and biofloc-based farming technology, organic shrimp farming and pond-based crab fattening need to be considered to increase production and productivity.

The future action plan, therefore, necessitates promotion of innovative, ecofriendly and sustainable technologies; establishment of healthy brood banks and state-of-the-art hatchery facilities for quality seed production; species and system diversification for judicious utilization of available land/water resources; selective breeding of native Indian white shrimp; comprehensive health management; GIS-based mapping of potential brackishwater areas; and introduction of a comprehensive fisheries policy for coastal and inland saline areas.

Mariculture

Development of mariculture in India is in its infancy. The ICAR-CMFRI during the past five decades, however, has developed several technologies pertaining to mussels, oysters, pearl oysters and seaweeds, and, more recently, to the cage culture of marine fin fishes.

Culture of molluscs

The earliest mariculture attempts were made in 1958–59 with the standardization of culture techniques for green mussel (*Perna viridis*) and brown mussel (*P. indica*) (Kuriakose and Appukutan, 1996). Adoption of rack, long-line and raft culture methods, especially in the states of Kerala, Karnataka and Goa, has pushed cultured mussel production from almost zero in 1996 to about 15,000 t, annually, at present. Edible oyster (*Crassostrea madrasensis*) farming, which was initiated in the 1970s with the natural collection of spat, could not get impetus until the late 1990s. Subsequently, the popularization of the rack-and-ren method of culture in different estuaries and backwaters saw production increase to the current level of 17,000 t. The golden pearl oyster, *Pinctada fucata*, producing golden pearls, and the black lip pearl oyster, *P. margaritifera*, producing black pearls, are the species producing gem-quality pearls. The technology of pearl culture involving implantation of nuclear beads was developed in 1973 (Alagarswami, 1974). Onshore production of good-quality pearls larger than 6 mm diameter from *P. fucata* was also achieved. The technology for seed production of commercially important mussels, oysters, pearl oysters and clams was developed during the 1980s, which, however, is yet to be taken up on a commercial scale.

Culture of seaweeds

The culture of seaweeds in the country is mostly confined to cultivation of agarophyte, *Gracillaria edulis* (Gopakumar et al., 2007), due to its high regenerative capacity and commercial value. Recently, the culture of the carrageenan-yielding seaweed *Kappaphycus alvarezii* has become popular due to its fast growth and low susceptibility to grazing by fishes, and is being cultivated in selected coastal districts of Tamil Nadu, producing about 20,000 t (wet weight) annually.

Culture of marine ornamental fishes

Globally, a lucrative marine ornamental fish trade has emerged in recent years, which has become a low-volume, high-value industry. A long-term sustainable trade of marine ornamental fish could be developed only through hatchery-produced fish. Technologies of breeding, seed production and culture of over 20 marine ornamental species, which include the high-valued clowns and damsels, have been developed and extended for their commercial production (Gopakumar et al., 2011).

Open-sea cage culture

Sea cage farming in the country was initiated in 2007 with seabass (*Lates calcarifer*). The potential of the farming practice was demonstrated on both the west and east coasts of India. After achieving success with seabass, cage farming was extended to cobia (*Rachycentron canadum*), groupers (*Epinephelus coioides*), pompano (*Trachinotus blochii, T. mookalee*), red snapper (*Lutjanus argentimaculatus*) and lobster (*Panulirus homarus*). Successful open-sea cage culture of seabass, pompano and cobia is being demonstrated in 6 m-diameter × 5 m-deep circular HDPE/GI cages with production levels of 25–35 kg/m^3/eight months (Ayyappan et al., 2016). Currently, cage farming is practised in over 1000 cages by the farmers of Gujarat, Goa, Karnataka, Kerala, Tamil Nadu and Andhra Pradesh. CMFRI has also successfully developed and standardized the breeding and seed production protocols of important finfish species like cobia, silver pompano, Indian pompano, orange-spotted grouper and pink-ear bream. It is envisaged that selective intensification, scaling up of the culture systems and formulation of an appropriate policy framework would lead to a substantial increase in the production of marine fish through sea-based cage culture in the coming years, which can contribute substantially to the Blue Revolution in the country.

Genetic Improvement

Over the years, efforts have been made to exploit the genetic potential of freshwater fish and prawns, particularly through genome manipulations in the early 1980s and through selective breeding from the 1990s onwards (Reddy et al., 1999). The Jayanti rohu, produced through the selective breeding programme, has shown a 17% higher growth response/generation after eight generations. Being the major component of the species composition, this improved rohu has exhibited phenomenal increase in the vertical productivity of culture ponds. Apart from growth, disease resistance against Aeromonas hydrophila was also included in the rohu breeding programme from 2004 (Sahoo et al., 2011). At present, ten multiplier units are in operation in different states for the dissemination of the Jayanti rohu, which has significantly increased the fish production in the country. With convincing success in selective breeding, similar programmes have been taken up for growth improvement of catla and freshwater prawn. A new programme has been proposed for the initiation of the selective breeding of Indian white shrimps, Fenneropenaeus indicus.

Availability of Feed Ingredients and Formulated Feed

Use of feed concentrate for freshwater farming was almost negligible, except for the sinking pellets produced for freshwater prawns. However, in recent years, awareness is increasing among fish farmers. Availability of commercially formulated, balanced feed has supported the feed-based aquaculture system and contributed significantly to boost fish production. At present, nearly 80 feed mills are producing supplementary feed for the sector in both sinking and extruded floating pellet form, which is based on the nutritional requirement of the desired species. The use of floating pellets has not only ensured better control of the feed ration and feeding strategy, but also is helping to curb the pollution caused by excessive feeding. Such feeding practice has become popular in the commercial aquaculture of major carp, catfish like P. hypophthalmus, and freshwater prawn. A number of feeds have been developed for different life stages of shrimp species like P. monodon and P. vannamei, and important cultivable finfish like Lates calcarifer. Cost-effective feeds have also been developed for both freshwater and marine ornamental fish species.

Several feed mills have been established in the country, producing customized feed according to the needs of the sector. Nearly 44 million t of concentrate feed are manufactured in the country at present for the different animal husbandry sectors, of which about 20% is currently being used for aquaculture. Considering the stiff competition for feed ingredients that aquaculture is going to face in the future, efforts are needed to develop alternative strategies to produce farm-made feed from non-conventional local resources. Further, greater emphasis is needed on harnessing the natural productivity for feeding the cultured fish.

Disease Surveillance and Management

The increased occurrence of pathogens comes as a sequel to the intensification of aquaculture practice. The freshwater aquaculture sector has witnessed several incidences of diseases that have created havoc. The incidence of ulcerative disease syndrome is one such example. However, timely R&D intervention through development of CIFAX, a therapeutic for the same, has helped save the fish crop from this disease (Ayyappan et al., 2016). A study on argulosis has yielded results that are useful to control the parasite and save the huge crop loss that otherwise would occur due to argulosis. Several diagnostics and therapeutics have been developed against important pathogens. The National Surveillance Programme on Aquatic Animal Diseases (NSPAAD) has operated in the last five years in 17 states and two union territories covering 115 districts. The programme, financially supported by the National Fisheries Development Board (NFDB) and implemented with the involvement of 25 partner organizations, has been largely helping both the freshwater and brackishwater aquaculture sectors to keep a close watch on disease problems and is proving to be a timely warning of disease situations for the farmers and also the governments. Emergency response systems are in place to tackle transboundary disease issues. The establishment of a National Repository of Fish Cell Lines, with possession of over 50 cell lines at ICAR-NBFGR, is a step forward for the advanced study of viral diseases (Goswami et al., 2014).

The Way Forward

The consistent annual growth rate of 6–7% experienced in the past three decades has been instrumental not only in narrowing the demand–supply gap of fish for the domestic market but also in boosting exports to an impressive level. Aquaculture production further promises to grow in the future. The state of Andhra Pradesh has been the most important aquaculture state, both for freshwater aquaculture and coastal shrimp farming. Considering the growth trend of the recent past, it is envisaged that brackishwater aquaculture production (shrimp), would also continue to grow in Odisha, West Bengal and Gujarat. Clear business opportunities also exist in Andhra Pradesh with regard to the production of fish feed, post-harvest processing, quality testing laboratories, etc. In both brackishwater and freshwater aquaculture it is necessary to strengthen the complete value-chain approach, since there exists a need for better inputs and infrastructure for post-harvest processing. Awareness creation and promotion of better farm management practices are essential, and state governments need to realize the necessity. Especially with regard to domestic market access, which mainly contributes to the demand for freshwater aquaculture products, lack of infrastructure and processing facilities hamper market access. It is equally safe to enter the Indian aquaculture market with a focus

on shrimp production for export markets as it is the domestic market through freshwater aquaculture production. The latter will be, however, more beneficial for food security and for meeting the increasing demand of fish in the days to come. The area in the country needs to expand horizontally and at the same time the vertical increase in productivity needs to be addressed. However, due to reducing per capita land-holding and increased demand for water, additional allocation of large areas for aquaculture to address the deficit in fish production would need a strategic action plan for effective and efficient use of resources. Hence, there is an urgent need for increasing both research and development efforts to enhance the productivity of existing water bodies through vertical expansion and with increased water-use efficiency. Besides aquaculture production, equal emphasis will be required on sustaining capture fisheries, both marine and inland, and focusing on culture-based fisheries in reservoirs and other large open-water bodies in inland areas. Aquaculture is being considered a 'sunrise' sector for increasing production; the R&D perspectives must focus, primarily, on species and system diversification by bringing more potential finfish/shellfish species of promise into freshwater, brackishwater and open-sea farming; development of cost-effective feed for all live stages of presently farmed and potential species to be brought under farming; genetic improvement of cultivable species for growth and disease resistance; development of culture systems for increasing water productivity and water-use efficiency; and effective fish health management measures including preparedness for risk of transboundary and emerging diseases. Increasing focus on post-harvest processing and value addition would not only help in extending the market reach of fish and fisheries products, but also in raising the income level of farmers. In order to keep up the pace of farming enterprises, it is essential for the sector to build efficient human resources with adequate knowledge and skills and to get much-needed policy support from the government, as well as enhanced alleviation for institutional credit. With these in place, growth in the fishery sector will certainly accelerate.

References

Alagarswami, K. (1974) Development of cultured pearls in India. *Current Science* 43(7), 205–207.

Arasu, A.R.T., Kailasam, M. and Sundaray, J.K. (2009) *Asian Seabass, Fish Seed Production and Culture*. CIBA Special Publication 42, Central Institute of Brackishwater Aquaculture, Chennai, India.

Ayyappan, S. and Jena, J.K. (2001) Sustainable freshwater aquaculture in India. In: Pandian, T.J. (ed.) *Sustainable Indian Fisheries*. National Academy of Agricultural Sciences, New Delhi, pp. 88–133.

Ayyappan, S. and Jena, J.K. (2003) Grow-out production of carps in India. *Journal of Applied Aquaculture* 13(3/4), 251–282.

Ayyappan, S., Sugunan, V.V., Jena, J.K. and Gopalakrishnan, A. (2011) Indian fisheries. In: Ayyappan, S., Moza, U., Gopalakrishban, A., Meenakumari, B., Jena, J.K. and Pandey, A.K. (Tech. Coord.) *Handbook of Fisheries and Aquaculture*. Indian Council of Agricultural Research, New Delhi, pp. 1–31.

Ayyappan, S., Jena, J.K., Lakra, W.S, Srinivasa Gopal, T.K., Gopalakrishnan, A., Vass, K.K., Sahoo, P.K. and Chakrabarti, R. (2016) Fisheries sciences. In: Singh, R.B. (ed.) *100 years of Agricultural Sciences in India*. National Academy of Agricultural Sciences, New Delhi, pp. 258–304.

Bhowmick, U. and Das, A.K. (2011) Cage and pen culture. In: Ayyappan, S., Moza, U., Gopalakrishban, A., Meenakumari, B., Jena, J.K. and Pandey, A.K. (Tech. Coord.) *Handbook of Fisheries and Aquaculture*. Indian Council of Agricultural Research, New Delhi, pp. 469–499.

Chaudhuri, H. and Alikunhi, K.H. (1957) Observations on the spawning in Indian carps by hormone injection. *Current Science* 26(12), 381–382.

DAH and DF (2016–17) *Annual Report 2016–17*. Department of Animal Husbandry, Dairy and Fisheries. Ministry of Agriculture and Farmers' Welfare, Government of India.

Dwivedi, S.N. and Zaidi, G.S. (1983) Development of carp hatcheries in India. *Fishing Chimes*, 29–47.

Gopakumar, G., Nair, K.R.M. and Kripa, V. (2007) Mariculture research in India – status, constraints and prospects. In: Modayil, M.J. and Pillai, N.G.K. (eds) *Status and Perspectives in Marine Fisheries Research in India*. CMFRI Diamond Jubilee Publication, Central Marine Fisheries Research Institute, Kochi, India, pp. 316–361,

Gopakumar, G., Madhu, K., Madhu, R., Anil, M.K. and Ignatius, B. (2011) Marine ornamental fish culture – packages of practices. CMFRI Special Publication No. 101. Kochi, India,

Gopakumar, K., Ayyappan, S. and Jena, J.K. (2000) Present status of integrated fish farming in India and wastewater treatment through aquaculture. In: Kumar, M. (ed.) *National Workshop on Wastewater Treatment and Integrated Aquaculture Production*. SARDI, Adelaide, Australia, pp. 22–37.

Goswami, M., Nagpure, N.S. and Jena, J.K. (2014) Fish cell line repository: an enduring effort for conservation. *Current Science* 107(5), 738–739.

Gupta, S.D. and Rath, S.C. (2011) Carp breeding and seed production. In: Ayyappan, S., Moza, U., Gopalakrishban, A., Meenakumari, B., Jena, J.K. and Pandey, A.K. (Tech. Coord.) *Handbook of Fisheries and Aquaculture*. Indian Council of Agricultural Research, New Delhi, pp. 358–379.

Gupta, S.D., Rath, S.C., Dasgupta, S. and Tripathi, S.D. (1995) A first report on quadruple spawning of *Catla catla* (Ham.). *Veterinarski arhiv* 65(5), 143–148.

IASRI (2016) *Agricultural Research Data Book*. Indian Agricultural Statistics Research Institute, New Delhi.

Janaki Ram, K. and Tripathi, S.D. (1992) A Manual on Freshwater Pearl Culture. Series 1. Central Institute of Freshwater Aquaculture, Bhubaneswar, India.

Jena, J.K. and Das, P.C. (2011) Carp culture. In: Ayyappan, S., Moza, U., Gopalakrishban, A., Meenakumari, B., Jena, J.K. and Pandey, A.K. (Tech. Coord.) *Handbook of Fisheries and Aquaculture*. Indian Council of Agricultural Research, New Delhi, pp. 380–400.

Jhingran, V.G. (1991) *Fish and Fisheries of India*. Hindustan Publishing Corporation, New Delhi.

Kuriakose, P.S. and Appukutan, K.K. (1996) Technology of mussel culture. *Bulletin of the Central Marine Fisheries Research Institute* 48, 70–75.

Ponniah, A.G. (2011) Shrimp farming. In: Ayyappan, S., Moza, U., Gopalakrishban, A., Meenakumari, B., Jena, J.K. and Pandey, A.K. (Tech. Coord.) *Handbook of Fisheries and Aquaculture*. Indian Council of Agricultural Research, New Delhi, pp. 548–560.

Rao, G.R.M. and Ravichandran, P. (2001) Sustainable brackishwater aquaculture. In: Pandian, T.J. (ed.) *Sustainable Indian Fisheries*. National Academy of Agricultural Sciences, New Delhi, pp. 134–151.

Rao, K.J. and Tripathi, S.D. (1993) *A Manual on Giant Freshwater Prawn Hatchery*. Central Institute of Freshwater Aquaculture, Bhubaneswar, India.

Reddy, P.V.G.K., Gjerde, B., Mahapatra, K.D., Jana, R.K., Saha, J.N., Rye M. and Meher, P.K. (1999) *Manual on Selective Breeding Procedures for Asian Carps* (Indo-Norwegian collaboration). Central Institute of Freshwater Aquaculture, Bhubaneswar, India, pp. 36–40.

Sahoo, P.K., Rauta, P.R., Mohnaty, B.R., Mahapatra, K.D., Saha, J.N., Rye, M. and Eknath, A.E. (2011) Selection for improved resistance to *Aeromonas hydrophila* in Indian major carp *Labeo rohita*: survival and innate immune responses in first generation of resistant and susceptible lines. *Fish Shellfish Immunology* 31, 432–438.

Sahu, A.K. and Sahoo, S.K. (2011) Catfish breeding and culture. In: Ayyappan, S., Moza, U., Gopalakrishban, A., Meenakumari, B., Jena, J.K. and Pandey, A.K. (Tech. Coord.) *Handbook of Fisheries and Aquaculture*. Indian Council of Agricultural Research, New Delhi, pp. 401–412.

Sugunan, V.V. and Sinha, M. (2000) Guidelines for small reservoir fisheries management in India. *Bulletin of Central Inland Capture Fisheries Research Institute* 93, 31.

Swain, S.K., Madhu, K., Madu, R. and Gopakumar, G. (2011) Ornamental fish breeding and culture. In: Ayyappan, S., Moza, U., Gopalakrishban, A., Meenakumari, B., Jena, J.K. and Pandey, A.K. (Tech. Coord.) *Handbook of Fisheries and Aquaculture*. Indian Council of Agricultural Research, New Delhi, pp. 500–532.

7

Increasing Productivity Growth Rate in Agriculture

In order to obtain a sustained growth rate of 8%, India must accelerate its agricultural growth from the existing level of 2% to 4%. Hence, a mission-mode programme for faster agricultural growth needs to be introduced as a matter of priority. It will require a dynamic approach focused on planned, coordinated and monitored strategies. 'Business as usual' will not suffice (GoI, 2013a). For meeting the achievable targets, which are not so easy to achieve under existing challenges, the following ten strategic actions are proposed to be rigorously pursued.

Increased Capital Investment in Agriculture

Capital investment in agriculture needs to be increased from the current level of less than 10% to at least 15–20%. Investment in infrastructure in rural areas, especially in the eastern and north-eastern areas, such as in roads, markets, linking farmers to markets (LFM), watersheds, building of modern silos around big *mandis/towns*, building of godowns, cold chains for storage and value addition of perishable items, goods trains and air cargo services for quick and efficient transportation/export, would help to accelerate growth in agriculture. Public sector investment is the only option at this critical juncture as expected investment by the private sector is not forthcoming, and without creating minimum infrastructure, the potential of non-Green Revolution regions will not be harnessed to the desired extent.

Supply of Growth-oriented Inputs at Farmers' Doorsteps

Growth-oriented agriculture requires concerted effort in the following areas:

- Supply of quality seeds must be ensured through enhancing minimum replacement rates of hybrids (up to 100%), cereals (20%) and oilseed and pulses (10%).
- India's present use of NPK fertilizers (107 kg/ha) is less than half of China's (245 kg/ha). Hence, annual mineral fertilizer consumption rates need to be enhanced by at least 5% with greater emphasis on overcoming the existing imbalance of the NPK ratio. Fertilizer use should be determined on the basis of soil analysis and decision-support systems. Against the national average, Rajasthan, Chhattisgarh, Odisha, Assam and most of the north-eastern states are using less than half (50 kg/ha) the fertilizer dose; these states need to overcome this major productivity constraint. The aim should also be to overcome micro-nutrient deficiencies of sulphur, zinc, iron etc. Also, the target of enhancing current use of biofertilizers from less than 5% will have to be doubled soon.

Greater emphasis needs to be given to increased nutrient-use efficiency through good agronomic practices (GoI, 2013b).

- Supply of biocontrol agents and biopesticides for enhanced use in vegetables, pulses, rice, maize, sorghum, sugarcane and cotton etc. needs special emphasis.
- Use of efficient farm machinery and equipment for timely operations has to be promoted through large-scale fabrication, by providing subsidy and credit with easy access (e.g. zero-till drill, raised-bed planter, sugarcane planter, rice transplanter, happy turbo seeder, wheel hand hoe etc.). Precision farming is the key to sustainable agriculture both in irrigated and rainfed farming systems.

A well-targeted, state-wise/crop-wise action programme for input availability should be devised and implemented as a priority.

Improving Productivity

Early in the Green Revolution (1969–1975), high-yielding, semi-dwarf varieties of wheat and rice were introduced to farmers, along with increased use of fertilizers, agricultural chemicals, machinery and irrigation. Subsidies were provided for seed, water and fertilizer. There were tremendous productivity gains and, by 1980, the country produced enough rice and wheat to meet basic needs. The economic benefits to farmers, however, were strongly skewed towards a few northern states where the Green Revolution had started. To create greater equity in access to resources, from 1975 through the 1980s, the Indian government disseminated crop production inputs and technologies to farmers in more states. An unintended consequence of this was less discriminating use of inputs, including water, pesticides and fertilizers, leading to land degradation and reduced groundwater in some areas of the country. These experiences drove home the message that simply increasing inputs or expanding irrigation and crop land is not the answer to food security concerns; rather, farmers need to increase productivity without degrading the resource base. This requires choosing the right seeds and the right types and amounts of inputs for local conditions and applying them in the right way to ensure that productivity growth is truly sustainable. Measuring total factor productivity (TFP) is

one of the ways in which the government determines whether productivity changes are due to application of improved technologies or just increased use of water and other inputs, which helps guide policy decisions. In addition to technology-driven efficiencies, TFP growth can be driven by shifting to crops with a higher economic benefit. Both of these factors have played a role in India's agricultural productivity growth (Chand and Parappurathu, 2012). Chand *et al.* (2011) assessed the contribution of different productivity-enhancing factors to TFP growth for a variety of Indian crops. They found that public investment in agricultural research constituted a significant source of TFP growth in 11 out of 15 crops. Public investment in extension and technology transfer contributed positively to TFP enhancement in only two crops, which likely reflects suboptimal investment. They suggest that improvements in both the investment levels and quality of extension services are needed. In addition, to achieve the 4% growth p.a. in agricultural GDP, which is the government's NITI Ayog target, greater emphasis should be placed on the development of livestock, horticulture and fisheries (Rada, 2013). When the Indian finance minister, Arun Jaitley, presented the government's 2014/15 budget in July 2014, he said: 'The government is committed to sustaining 4% growth in agricultural GDP and for this we will bring a technology-driven second Green Revolution with focus on higher productivity, including a "Protein Revolution" and expanded "Blue Revolution" of inland fisheries' (Jaitley, 2014).

India ranks number one, globally, in milk, tea and pulse production, and second in wheat, rice, groundnut, rapeseed, mustard and vegetable and fruit production. In sugarcane, also, it is second only to Brazil. India has the largest cattle and buffalo population, is second-largest in goat population and third-largest in sheep population. Ironically, it is behind in crop and animal productivity, which has to be increased through efficient management of natural resources and by adopting precision farming and good management practices. Some concrete actions are necessary in the following areas:

Increasing wheat production

Increased wheat production is synonymous with assured food security. India, at present, has become number two, after surpassing the USA (number four) and having almost the same area under wheat. Russia is third (www.statista.com). India can become number one (over China) in the near future, maybe in the next five to ten years, if an aggressive and well-planned strategy for increasing wheat production using area expansion and enhanced productivity approaches, with greater emphasis on the eastern and central regions, is implemented. Adoption of conservation agriculture, precision farming, balanced use of NPK fertilizers and improved varieties (including higher seed replacement) would accelerate production growth, which is almost stagnating in most states. Strategically, increased durum wheat production in Madhya Pradesh would accelerate prospects for exports too.

Stabilizing rice areas and production

India has the maximum area under rice (around 43.2/2.6 million ha) with productivity of 2 t/ha and production level plateaued at around 110 million t. In China, from around 30 million ha, paddy (not rice) production is 180 million t. China's productivity level is almost twice (6 t/ha) that of India's. Major yield gains have come from hybrid technology in China, which covers around 55% of the area with 1 t/ha advantage over high-yielding rice varieties. India has to evolve a new strategy by which the area, especially that which is under rainfed rice (having low productivity), can be reduced with a simultaneous increase in yield using hybrid rice, IPM and conservation agriculture, including direct-seed technology. The main focus should be on the rainfed, lowland rice-producing states in the eastern region, including eastern Uttar Pradesh, Bihar, Odisha, West Bengal and Assam. Hybrid rice technology has demonstrated good potential in this region and hence more area could be covered. Also, conservation agriculture (zero-till drill), rice transplanter, super-granulated urea application, IPM (as in Indonesia) would help stabilize rice production as well as reduce the area to be used for crop diversification, especially for horticulture, including vegetable production.

Increasing rice productivity in eastern India

The seven eastern states account for 56% of India's crop land used for rice production, but they produce only 48% of the nation's crop. In contrast, Punjab accounts for 6% of land under rice cultivation and produces 11% of the nation's rice crop. Improved rice productivity would have a significant impact on household food security and income generation, and on the region's economic growth. In the eastern region, developing and using the best seeds for different types of soil and climatic conditions is a priority for increasing productivity. India's ICAR-Indian Institute of Rice Research coordinates multi-site evaluations of promising experimental hybrids from both public and private researchers at 25–30 locations representing different agroclimatic zones of the country. For example, water conditions vary throughout the country and hybrid seeds have been identified that are best used in either water submergence, flood, drought or salinity-prone areas. In eastern India, hybrid seeds developed through this process have achieved, on average, 30% higher yields in farmers' fields than existing inbred rice varieties. Pusa RH-10, a hybrid basmati rice developed by the ICAR-IARI shows even higher increases under rainfed upland conditions, and the aromatic flavour and texture is very popular. Since the private sector has been more effective than the government in hybrid rice seed production, a public–private partnership between IARI and the non-profit Indian Foundation Seed and Services Association was formed, which resulted in the faster spread of the Pusa RH-10 hybrid. Evaluations indicate that seed production was highly lucrative for farmers. A particular benefit was the generation of additional employment of 65 person-days/ha, and most of the producers are women. Bringing Green Revolution in Eastern India (BGREI) and similar science-based initiatives, and PPP, creates numerous benefits and has a positive, cascading impact on the economy (GHI, 2012; 2014)

Enhancing maize production

Yield potential of maize can be enhanced by promoting single-cross hybrids. Since their use began in 2001, maize production and productivity levels have doubled. Maize has a good future potential in India. Area expansion by almost 1 million ha in eastern Uttar Pradesh, Bihar, Jharkhand and West Bengal in the last decade is a positive indication. Maize production can be doubled, provided the hybrid maize area is increased. Besides local demand for food and fodder, there is ample scope for its export as animal feed for pig and poultry production to southeast Asia as well as China. All north-eastern states have good potential for maize production. In fact, maize is envisaged to be a major cereal with great potential in the global market in the future. Amongst cereals, maize is currently achieving the highest productivity, which is higher than 4%.

Sugarcane for biofuel

After Brazil, India is the second-largest sugarcane-growing country. As against 5 million ha in Brazil, the area under sugarcane in India has fluctuated between 4 and 4.5 million ha since 2000. While Brazil has exploited sugarcane for biofuel production for automobiles (almost 50%), India still has to explore this option, beside its internal demand for sugar. India has the best R&D infrastructure and is known globally for nobalization of canes, resulting in short duration, drought- and disease-tolerant varieties; enabling its spread to central, northern and western India. Unfortunately, productivity in major sugarcane states in northern India (Uttar Pradesh, Haryana, and Punjab) is lower than the national average, despite good varieties and production technologies. Productivity is also low in the second-largest sugarcane-growing state, Maharashtra. This scenario has to be changed through a mission-mode approach, for which technological options exist. In view of spiralling prices of petrol, it is high time we have a policy reorientation towards using sugarcane for biofuel. India can become the number-one country in sugarcane and also biofuel production, as sufficient scope exists both for its horizontal and vertical expansion. Hence there is an urgent need to move quickly in that direction, with clear policy and support from the private sector.

Increasing pulse production

India is the major producer and consumer of pulses. Unfortunately, productivity is less than 1 t/ha. A major effort is therefore required to adopt a comprehensive, well-planned mission-mode approach to enhance pulse production. In this context, the recommendations of an expert committee report submitted to the Technology Mission on Oilseeds and Pulses (TMOP), Department of Agriculture and Cooperation (DoAC), in June 2000, have now been implemented under the National Food Security Mission (NFSM). Improved short-duration, disease-resistant varieties are being popularized through large-scale field demonstrations to overcome an existing 25–30% yield gap (around 4 million t). Short-duration varieties need to be promoted in new areas, namely chickpea in the south, urd-bean (blackgram) in rice fallows in the coastal regions of Andhra Pradesh, Odisha and West Bengal, pigeonpea in the north-west (Haryana, Gujarat, Maharashtra and Rajasthan), and mung-bean (greengram) in the north (western Uttar Pradesh, Haryana and Punjab). Use of sulphur in deficient regions, higher replacement rates for improved variety of seeds, micro-irrigation, IPM approaches and one life-saving irrigation have enhanced pulse production to almost 4 million t, which is the level of current deficiency. Hybrid pigeon pea technology with greater yield benefits in Gujarat, Rajasthan and Haryana is another new option.

Increasing oilseed production

Soybean

In the last 50 years, from nowhere, soybean has emerged as the number-one oilseed crop ahead of groundnut. Besides production of oil, it has an export value of above US$ 3 billion p.a., mainly from soya meal. This has been possible due to excellent coordination between the R&D and processing/marketing sectors. Lately, a sense of complacency has emerged. India is currently the number-five producer, after the USA, Brazil, China and Argentina. Though its area is almost the same as in China and Argentina, productivity is almost half (1.3 t/ha). Compared to Madhya Pradesh, yield levels are higher in Rajasthan, Maharashtra and Andhra Pradesh. A little push and coordination at national level would help accelerate soybean production. North-eastern states are also an important niche area for soybean but require suitable varieties and proper policy support. Another important policy-related issue is the use of GM soybean. Major soybean countries, such as the USA, Brazil and Argentina, have greatly benefitted from genetic gains from GM technology. Why India should remain behind is incomprehensible. Hence, policy direction towards the use of soybean as a food crop, the promotion of GM soybean, and its spread in north-western India to diversify rice-wheat cropping systems, is critical in order to harness the potential that soybean offers.

Groundnut

Improved varieties, higher rates of seed replacement, using sulphur and plastic mulching, and IPM can result in significant yield improvements in Andhra Pradesh, Karnataka, Maharashtra and Madhya Pradesh. Gujarat, Tamil Nadu and Rajasthan are lately ahead in productivity. With the current area (6 million ha), a production target of 10 million t can be achieved with well-targeted efforts. Odisha and Bihar offer new area options with higher productivity, which must be explored. In groundnut, control of weeds, one supplemental irrigation, and sowing in permanent raised beds with plastic mulching would make all the difference. These efforts would ensure that groundnut becomes again the number-one oilseed crop.

Rapeseed and mustard

Expansion of area in the eastern states (West Bengal, Assam, Bihar) and the north-eastern states would help to achieve higher production. Hybrid technology could be exploited in the northern and western states. Support for one irrigation, preferably using sprinklers, higher doses of fertilizers and IPM would give better yields, if Rajasthan, Haryana, Uttar Pradesh, Madhya Pradesh and West Bengal were catalysed to adopt these good agronomic practices.

Sunflower

So far, the potential of sunflower has not been fully exploited in India. Improved early maturing hybrids of sunflower in the northern states of Haryana, Punjab and western Uttar Pradesh can help accelerate production. New niche areas with higher productivity potential for this promising crop could be Bihar, West Bengal, Assam and Odisha. A major constraint is non-availability of hybrid seeds, for which both the public and

private sectors have to be catalysed by setting targets for seed supply of selected promising hybrids.

Hybrid castor and safflower

These crops have great export potential. They are mainly grown in the western states of Gujarat, Maharashtra and Rajasthan (mainly castor). Promoting the use of improved hybrids and, where possible, use of one supplemental irrigation would make a significant difference. Fortunately, good hybrids are also available for large-scale coverage.

Making Grey Areas Green

To achieve the 'Evergreen Revolution', special emphasis needs to be laid on rainfed agriculture, which still covers around 55% of areas. This is necessary for sustainability and for the improved livelihood and income of resource-poor farmers who have no means of risk management unless practices of diversified agriculture are adopted, such as a silvipastoral approach through crop–livestock integration, agri-horticulture (Maharashtra) and agroforestry (mainly growing of trees around bunds (poplar) or in fields (*khejri*)). Crop and livestock insurance and linking farmers to markets (LFM) need to be major strategic policy interventions by government.

Rainfed agriculture demands a paradigm shift towards integrated natural resource management (INRM) through conservation agriculture practices (Brazil, USA and Canada) and productivity increases through hybrid technology (most rainfed crops such as maize, sorghum, pearl millet, sunflower, castor and cotton), in which India is the world leader and has reported significant gains in the past. With new options of hybrids in crops such as pigeon pea, safflower and rapeseed-mustard, this option would further widen but would require a well-planned, time-bound action for seed availability on farmers' doorsteps. Even subsidized hybrid seeds would be in the national interest, as was the case when the Green Revolution was achieved, for which subsidy of HYV seeds of wheat and rice was given.

Timely operations, weed management, IPM, sprinkler or drip irrigation systems and use of biofertilizers would also need a coordinated area-wise/ecoregion-wise approach. Community management of all watersheds developed during the last three decades would require to be revisited and technical backstopping by research institutions and SAUs in each region would be needed to make them effective in overcoming risk factors. Crop/livestock insurance and availability of easy credit at low interest rates would require new policy decisions; and to avoid the distress sale of produce, an LFM approach with greater cooperative effort on post-harvest processing and value addition would be the option. In rainfed agriculture, as technological gains are relatively less important, it is critical that technology dissemination losses are minimized and best-possible extension services are provided through a new self-employed cadre of technology agents, and through promoting 'agri-clinics'. A new thrust is needed on vocational training of young agricultural graduates, linking their services to farmers on a custom-hire basis through bankable projects. Timely and efficient technology dissemination for resilient agriculture is the main key to achieving the Evergreen Revolution. Present efforts of the government through an Authority for Rainfed Area Development along the lines of the NDDB, with greater autonomy and authority, linked with adequate resources, will go a long way towards converting grey areas to green. The resilience of irrigated agriculture has already been achieved, mainly through *rabi* production, which has now surpassed *kharif* production. Taking advantage of good precipitation in many states, India must attain resilience in *kharif* production also. For example, much of the rainfall is lost on farms in Odisha, Bihar and West Bengal for want of field bunds – as practised in Rajasthan and Gujarat. A simple practice of having just one-foot-high bunds along fields would ensure almost 80% of on-farm water harvesting in eastern states, thus stopping soil and water erosion and flooding in eastern India.

Emphasis on a New Area Approach

Past experiences have demonstrated that a new area approach can lead to faster progress mainly due to complete adoption of technological packages. Examples are rice in north India, groundnut in Gujarat, soybean in Madhya Pradesh and maize in Bihar. Such an approach is still relevant and can yield greater dividends, provided it is adopted on a scientific basis, such as GIS-based land-use planning for crop diversification. Examples are: hybrid rice in eastern India, soybean in eastern and north-eastern regions and sunflower in the north. Scientific land-use planning has invariably been a weak link in the past; this gap needs to be bridged, and India has all the scientific capability and institutional support it needs.

Thrust on Horticulture

Maharashtra can be cited as the best example of the promotion of horticulture in India. Right policy decisions, technical guidance and funding support for initial establishments are required. Availability of disease-free good planting material for fruit crops and hybrid seeds for vegetable crops, bred by both public and private institutions, opportunities for linking farmers to markets, and processing and value addition are all critical for growth in the horticulture sector. India has all the potential to be the number-one fruit and vegetable producer in the world. For this we have to learn from the experiences of Brazil and China and put in place an enabling policy environment to support farmers engaged in horticulture.

India has the advantage of different weather conditions (temperate, sub-temperate and tropical) for growing a variety of fruits. The spectacular achievements of increased production of potatoes, bananas, apples, oranges, mangoes and grapes is a reflection of this fact. India also has comparative advantage of geographic locations, good technologies, cheap labour and a strong private sector. Fortunately, it has a robust internal market with increasing demand for fruits and vegetables by a fast-expanding middle-class with higher income and buying power. In the context of globalization, it can accelerate its exports through proper grading, processing, packaging etc., as has been achieved by Brazil. It needs to be linked with foreign markets, for which its embassies should have agricultural attachés. The National Horticulture Board (NHB) and the National Agricultural Cooperative Marketing Federation (NAFED) would have to be given a new mandate in the present context.

Growth rates in agriculture through vegetables and fruit would be much faster. In potato, India can capitalize by exporting true potato seed (TPS) to other countries, the trade in which is currently dominated by European countries. It could accelerate potato production for export, as best-possible technologies and short-duration varieties are available in India, and the country has a comparatively higher growth rate for potato production than many other countries. India could also move to number two or three in potato production from its current number-five position. This would require proper planning and coordination from production to consumption (including export). Similarly, India is one of the largest producers of oranges, but most varieties are for table use only. There is considerable scope for diversification of varieties suitable for juice and processing. Export potential also exists for vegetables, vegetable seeds, flowers (especially roses and orchids) and pulp of tomato, mango, banana, potato chips, tapioca and sugarbeet. In brief, it needs to embark upon a special horticulture production programme along the lines of the Special Foodgrains Production Programme initiated and monitored by the Planning Commission (now NITI Aayog).

Promoting Inland Aquaculture

Growth in marine fish production is globally on a decline, with slow growth in India, whereas a remarkable growth rate (6–7%) has been achieved in inland aquaculture, which contributes almost 66.8% of total fish production, around 11.4 million t. This has resulted in a major export of shrimp and fish abroad (around US$3 billion annually), a remarkable achievement. Gujarat, Haryana, Madhya Pradesh and Rajasthan are showing good progress along with Andhra Pradesh, the leading state. A special thrust on R&D is required, including support for brackish-water aquacultural production and supply of quality seed, rural-based fish processing, packing and cold-storage facilities, and transportation, as well as export promotion/internal market-orientated efforts. Other states can learn from the progressive fish farmers of Andhra Pradesh in linking with markets in West Bengal, Odisha, Bihar and Delhi.

Emphasis on Livestock Development

India has the largest cattle and buffalo population, the second-largest population of goats, and the third-largest population of sheep in the world. The total livestock population is around 512 million (DAH and DF, 2012). India is number one in milk production (176.35 million t) (NDDB, 2016; DAH and DF, 2017–18), yet it has not been able to compete globally in the export of milk products, meat and even live animals, as has been achieved by Australia, New Zealand and The Netherlands. India has 729 million poultry and production of eggs is 88.0 billion. Supply of fodder and feed; the use of silvipastoral systems in rainfed areas (especially in Rajasthan, Madhya Pradesh, Maharashtra, and Andhra Pradesh); the establishment of artificial insemination centres; the use of seed semen livestock clinics; the supply of good-quality vaccines; insurance of livestock (at least of valuable, productive ones); establishment of modern abattoirs; and processing, packaging, storage and marketing (including export) facilities would all lead to much faster growth in the livestock sector for which India has a comparative advantage; for example, it can be a major producer of mozarella cheese, being the largest buffalo milk producer; yet its share in the global market is negligible.

Another important activity is crop–livestock integration, especially in arid regions. In some areas, such as Rajasthan and Maharashtra, a major thrust is needed on silvipastoral practices using agroforestry and use of rangeland pastures and legumes, besides growing drought-tolerant shrubs and trees such as *khejri* (*Prosopis* sp.) and *babool* (*Acacia* sp.). In the Rajasthan Canal area, especially Bikaner and Jaisalmer divisions, a review of arable cropping is needed, to be replaced by pastures and livestock with a view to long-term sustainability and profitability. An appropriate mid-course correction with policy reform and support for livestock development-related programmes, through the Livestock Mission – including insurance, credit, healthcare and marketing – would be desirable. Also, there is a need to protect and improve local breeds, which are unique and valuable assets.

Improved On-farm Efficiency and Precision Farming

The most critical factor for faster growth in the future is input-use efficiency. It would demand timely operations, i.e. precision farming. Integrated natural resource management (INRM) would call for rational use of water, seeds, fertilizers and pesticides. Precision farming through greater emphasis on mechanization (both in irrigated and rainfed farming systems) would be the major recourse in future. Large-scale manufacturing of equipment and machinery/tools (such as zero-till drill, happy turbo seeders, planters, seed and fertilizer drills, sprinklers, mechanical harvesters, combines and small tools for weeding, harvesting, threshing, cleaning etc.) and their easy availability would help to accelerate future growth in agriculture. The Mission on Small Farm Mechanization, with skilled human resources (especially youth) and incentives to entrepreneurs for scaling innovations, would go a long way in ensuring conservation agriculture for sustainable intensification (CASI).

The Need for Critical Policy Interventions

For the agricultural sector, India has been fortunate in the past to receive policy support at the highest level. India's first prime minister, Pandit Jawahar Lal Nehru, had said 'Everything else can wait but not agriculture.' All his successors have accorded a high priority to agriculture including the present PM, Shri Narendra Modi. An enabling policy environment is critical for future growth and development. It is therefore necessary that we continue having appropriate policy interventions in future to reach a growth rate target of 4% in the agricultural sector. As such, the following areas need specific attention to accelerate growth in India:

- enhanced capital investment – to be raised to 15–20%;
- the creation of an enabling environment to link farmers to markets (LFM) – an initiative like e-NAM;
- credit availability for farmers at low interest rates (around 4%);
- the announcement of an attractive minimum support price (MSP) well in advance for essential and strategic crops/commodities;
- a new policy, in view of globalization, on agricultural exports to capitalize on comparative advantages – linked with well-organized market intelligence;
- major incentives for greater use of growth-linked agricultural inputs (seeds, fertilizers (both mineral and biofertilizers), pesticides (both chemical and biopesticides), small farm machinery and equipment, etc.);
- continued support for buffer stocking of essential commodities at a threshold level (around 15–20 million t for cereals), with creation of ultra-modern silos and cold-storage facilities;
- insurance of crops and livestock with premium rates that are affordable by resource-poor farmers;
- incentive-oriented and simplified laws and procedures for the establishment of small-scale cooperatives for processing and value addition of farm produce in rural areas; and
- accelerated pace of consolidation of land holdings in states lagging behind and future land-use planning for diversified agriculture on a scientific and ecoregional basis.

References

Chand, R. and Parappurathu, S. (2012) Temporal and spatial variations in agricultural growth and its determinants. *Review of Rural Affairs, Economic & Political Weekly Supplement* 26 (27), 55.

Chand, R., Praduman, K. and Kumar, S. (2011) Policy Paper 25: Total Factor Productivity and Contribution of Research Investment to Agricultural Growth in India. National Centre for Agricultural Economics and Policy Research, New Delhi.

DAH and DF (2012) *19th Livestock Census – 2012 All India Report*. Ministry of Agriculture, Department of Animal Husbandry, Dairying and Fisheries, Krishi Bhawan, New Delhi.

DAH and DF (2017–18) Annual Report 2017–18. Department of Animal Husbandry, Dairy and Fisheries. Ministry of Agriculture and Farmers' Welfare, Government of India.

GHI (2012) *2012 GAP Report: Measuring Global Agricultural Productivity*. Global Harvest Initiative, Washington, DC.

GHI (2014) *2014 GAP Report – Measuring Agricultural Productivity Growth in India*. Global Harvest Initiative, Washington, DC.

GOI (Government of India) (2013a) *State of Indian Agriculture 2012–2013*. Ministry of Agriculture, New Delhi.

GOI (Government of India) (2013b) *Twelfth Five Year Plan (2012–2017)*. *Economic Sectors, Volume II*. Planning Commission, New Delhi.

Jaitley, A. (2014) *Finance Minister Presentation of 2014–2015 Budget to Parliament*. Minister of Finance. Available at: https://www.indiabudget.gov.in/budget2014-2015/glance.asp (accessed 24 May 2018).

NDDB (2016) *Annual Report 2015–16*. National Dairy Development Board, Anand (Gujarat).

Rada, N.E. (2013) Agricultural Growth in India: Examining the Post-Green Revolution Transition. Selected paper for presentation at the Agricultural & Applied Economics Association's 2013 AAEA & CAES Joint Annual Meeting, Washington, DC, 4–6 August.

8

Reorienting Agricultural Research for Development for Sustainable Agriculture

———————————

The sharp increases in food prices that have occurred in global and national markets in recent years, and the resulting increase in the number of hungry and malnourished people, has sharpened the awareness of policy makers and of the general public to the fragility of the food system. This awareness must be translated into political will and effective action to render the system better-prepared to respond to long-term demand for growth, to be more resilient against various risks that confront agriculture, and to ensure that the ever-growing population will be able to produce and/or have access to adequate food today and in the future. There is a need to address new challenges that transcend the traditional decision-making remit of producers, consumers and policy makers.

Agriculture has remained an integral part of the socioeconomic fabric of rural India since time immemorial, and occupies centre-stage in the Indian economy as it sustains the livelihood of over 70% of rural households and provides employment for around 50% of the population. Despite wide variations in growth performance during seven decades, since independence, primarily due to the subsistence nature of farming and its dependence on the monsoon, the country still witnessed several innovations in agricultural research and extension leading to a quantum leap in productivity and production. One of the most evident scenarios – the Green Revolution – was brought on by a science-led synergistic extension approach capitalizing on the use of genetic resources, irrigation, fertilizer, appropriate policies and farmers' hard work. This innovation fired up growth in the agriculture sector and led to an unprecedented transformation in the development of the country. Increased agricultural productivity, rapid industrial growth and expansion of a non-formal rural economy resulted in higher per capita GDP, ensuring the food security of the nation. The continued pace of research and innovation is becoming a major challenge, especially with a growing population, which has reached 1.3 billion, resulting in a decline of productivity, deterioration of natural resources, impact on climate change and, above all, fatigue of the existing research and extension system, largely in the public sector.

Challenges Ahead

Food demand versus small farm holdings

With an expected extra 2–3 billion people to feed over the next 40 years, targeted efforts will be needed to achieve 70% more food being available to keep pace with demand. At the same time, climate change is already affecting agriculture in many developing countries, including India, and the effects will become increasingly challenging in future. Both R&D and innovations by farmers enabled India to harvest a record 265.04 million t in 2014, which declined to around 251.57 million t during two consecutive droughts. By 2030, India would need to produce 70% more foodgrains, and this in the context of multiple challenges such as depleting water resources, diversion of human capital from agriculture, shrinking farm size, soil degradation, indiscriminate and imbalanced use of chemical inputs and the overarching effects of climate change. A low investment in agricultural research for development further complicates the problem. Therefore, ensuring availability of and economic access to food, in both quantity and quality (nutrition), for the poorest of the poor in the country, remains a daunting challenge. To this end, the GCARD Road Map, developed through the interaction of diverse stakeholders from around the world at the Global Conference on Agricultural Research for Development held at Montpellier, France, in 2010, highlights the urgent changes required in agricultural research for development (AR4D) globally, to address the needs of resource-poor smallholder farmers and consumers. The road map contains a six-point plan for transforming agricultural R&D: (i) a collective focus on key priorities determined and shaped by science and society; (ii) a true and effective partnership between research and those it serves; (iii) increased investment to meet the huge challenges ahead and to ensure the required development returns from AR4D; (iv) greater capacity to generate, share and make use of agricultural knowledge for development change among all actors; (v) effective linkages that embed research in the wider development context and actions enabling developmental change; and (vi) better demonstration and awareness of the development impact on and returns from agricultural innovation. It envisages a major paradigm shift in the farming system with greater emphasis on 'innovations for greater impact on smallholder farmers', requiring partnerships among stakeholders and their capacity building, arresting natural resource degradation, and mitigating climate change (World Bank, 2008; GCARD, 2010).

Ever-increasing population growth is interlinked with fast-declining and degrading land, water, biodiversity, environment and other natural resources; which are three to five times more stressed in India compared to the rest of the world due to population, and economic and political pressures. The country has already reached the limit of land available for agriculture and hence limited scope exists for horizontal expansion. Inefficient use and mismanagement of production resources, especially land, water, energy and over-use of agrochemicals, have vastly reduced fertility and damaged soil health. Soils have lately become both hungry and thirsty. To a greater extent, lack of political will and populist policies of providing free or subsidized inputs like seeds, fertilizers, water and energy, have exacerbated the problem further.

Harnessing Opportunities

Innovations in Natural Resource Management

One of the main causes of slow growth in agriculture is relatively poor dissemination of emerging technologies relevant to smallholder farmers. We have to reach them, and for that suitable strategies are to be worked out. Also, innovations are needed to meet the major challenge of resource scarcity by reducing the cost of inputs on the one hand, and improving the livelihood of resource-poor smallholder farmers and bringing in transformation in their socioeconomic status on the other. To liberate the nation from hunger and poverty, while sustaining existing natural resources, policy makers would need to have a renewed focus and commitment to rejuvenate agricultural innovation systems and effective extension processes through additional funding and strict monitoring of AR4D. Without this, the task of achieving inclusive growth will be elusive.

Strengthening Collaboration and Partnerships

The Green Revolution was an outcome of partnership between the National Agricultural Research System (NARS), international centres like CIMMYT, IRRI and the extension system including enthusiastic farmers. Regional and global networks and partnerships for knowledge sharing and enhanced capacity development of all stakeholders is a must for outscaling innovations. It has been increasingly realized that under the changing scenario of production to consumption, the linear approach in technology development and deployment will not serve the purpose of addressing SDGs. Therefore, for inclusive growth in agriculture through large-scale uptake of new technologies, a major paradigm shift in approach, from R&D to AR4D, needs to be there, involving greater participation of all stakeholders. The past experience from regional organizations/programmes like the Asia-Pacific Association of Agricultural Research Institutions (APAARI), the South Asian Association for Regional Cooperation (SAARC), the Association of Southeast Asian Nations (ASEAN), the Rice Wheat Consortium (RWC) and the Cereal Systems Initiative for South Asia (CSISA) reveals that regional partnerships are extremely important to catalyse adoption of new technologies, mainly through sharing of knowledge and success stories around good agricultural practices (Weatherhogg *et al.*, 2001; Paroda *et al.*, 2007; Beniwal *et al.*, 2010; Paroda, 2017).

Linking Research with Extension

In the present context, the agricultural sector has to be more science-oriented and technology-driven. The research orientation has to be sensitive to local needs and be meeting the aspirations of both farmers and consumers. There needs to be a closer working relationship between research and extension organizations. The scientists involved in basic, strategic, applied and adaptive research, together with subject matter specialists, extension workers and farmers, are to be an integral component of the knowledge dissemination and agricultural advisory system. The interface between research and technology transfer is indeed very critical for converting outputs into outcomes. In fact, strong linkage is required between 'lab and land' and between village and institutions. This can be achieved by a paradigm shift from top-down to bottom-up, for technology generation, refinement and adoption.

Furthermore, the research agenda of the institutions should be better organized for technology development and dissemination. For making an impact on agricultural research, there must be strong links between researchers, extension agencies, farmers and other stakeholders. In all the institutions, technology transfer programmes

need to be an integral part of technology development to empower farmers with proper knowledge. Hence, farmer participatory research has to be given a strong focus. Also, a major role has to be played vigorously by the state's extension machinery (Beniwal *et al.*, 2010; Paroda, 2017).

Empowering Women for Inclusive Growth

It is well-recognized that women's empowerment is important both for agricultural growth and household nutritional security. Globally, about 43% of women are engaged in agriculture. In India, 60% of farming operations are performed by women. Therefore, agriculture can be a primary driver for women's empowerment. Innovations would improve their work efficiency and would also ensure their overall household development and nutritional security. However, women in agriculture are invariably deprived of the right agricultural knowledge and access to technology and credit in order to overcome on-farm drudgery and market-related constraints. Often they are even deprived of their rights to land and resources. Unless their rights are addressed, adverse effects on their performance will continue to be felt. *The State of Food and Agriculture* report of 2010–11 by the Food and Agriculture Organization of the United Nations (FAO, 2010–11) indicated that reducing the gender gap between male and female farmers would raise yields on farms by almost 20–30%. As a consequence, it is expected that this would lead to a reduction of 12–17% in undernourished people globally, which, in turn, would translate into reducing the number of hungry people by 100–150 million. Hence, the generation of technologies relevant to women farmers, and their adoption, should become a key part of the agenda for future agricultural growth (World Bank, 2008; Kabeer, 2012).

Retaining Youth in Agriculture

An ageing population of farmers and declining interest among rural youth to take up agriculture as a profession are challenges to the future of agriculture not only in India but also in other parts of the world. A large section of youth invariably prefers to migrate to cities to seek employment, especially government jobs. Hence, a major challenge today is how to retain rural youth in agriculture, which certainly cannot be ignored. The skewed interest of rural youth in agriculture is directly related to the existing poor physical amenities, socioeconomic conditions and lack of enabling environment. Serious issues of poor soil health, non-availability of water for irrigation, climate uncertainties, low-paid employment, inadequate credit facilities, low profit margins and lack of insurance against crop failure discourage youth from getting involved in agriculture. Social factors like perceived low esteem, including by parents, force youth out or prevent them from entering agriculture. Concerted effort is vital to stimulate the interest of youth. Proper incentives for rural youth involvement in agricultural education, research and extension, and linking them to expanding markets, would have a positive effect on youth pursuing agriculture as a profession. Also, training should be given to them at district/taluka level (World Bank, 2008; Paroda, 2017).

Earlier, seed, pesticide, fertilizer and farm machinery were the only potential sectors to employ agriculture graduates/rural youth. New opportunities are emerging in IT-linked agri-extension, seed technology, biotechnology, food processing, cold storage, packaging, supply chain management, insurance and farm credit, and other skill development programmes, enabling youth to take up 'start-up' programmes and equip them to be in the mainstream. In addition, large-scale production of tissue culture-based, high-quality planting materials; field-level pest identification and disease diagnosis; use of portable biochemical and molecular diagnosis kits; post-harvest processing at farm level; artificial insemination facilities and know-how; and vermicomposting are some relatively simple and low-cost technologies in which rural youth can be trained to improve farm productivity and employment potential. In this context, we now need to give more emphasis to vocational training of youth (male and female) for relevant skill acquisition and greater confidence-building, to serve as 'technology agents' as well as efficient knowledge/service providers on a custom-hire basis. It is high time that efforts are made at all levels to engage youth in activities around 'plough to plate' and to make farming a lucrative profession (Paroda, 2017). Knowledge-based agriculture around secondary and speciality agriculture can enhance opportunities for additional income. Thus, the social status of youth in society can improve a great deal.

Future Road Map: the Need for a Paradigm Shift

The success of the Green Revolution was mainly due to a 'holy alliance' among researchers, extension specialists and farmers. The technology dissemination approach adopted was top-down and centred around individual farmers. Faster adoption of technology was also on account of miracle seeds of wheat and rice, promoted largely by a public extension system that has become relatively weak. On the contrary, new innovations around natural resource management require a 'bottom-up' approach, involving farmers' participation, while ensuring confidence building among farming communities in taking risks and making agriculture more scientific and resilient. In the process, sharing of knowledge about good agricultural practices, without dissemination loss, and incentives for critical inputs becomes crucial for achieving success in the future. Partnership among key stakeholders is essential to promote growth in agriculture. In the process, complacency that has crept into the public extension/advisory services needs to be overridden. Also, a paradigm shift is needed from the present NARI system to that of the NARES. This will require active involvement of stakeholders – farmers, NGOs, the private sector, scientists and policy makers. Another paradigm shift has to be in the extension approach towards translational research to ensure outscaling innovations for greater impact on higher productivity and income. The extension approach has to be intensified around farming communities rather than individual farmers. Natural resource management-related innovations require more lead time to assess their impact on farmers' fields, unlike the impact of high-yielding varieties on crop productivity. This throws up a new institutional challenge for reforms to the existing extension system, which is mostly dependent on public organizations. The role of the private sector, especially through involvement of youth and gender in agriculture, becomes most relevant in the present situation. Empowering youth through vocational training and building a cadre of technology agents to provide technical support as well as custom-hire services to smallholder farmers would go a long way in linking research with extension and accelerating agricultural growth. In other words, we need to link 'land with lab', village with institute and scientists with society to ensure faster adoption of production-enhancing and resource-saving technologies that would benefit both the producers and consumers. In the process, agriculture technology agents would become job creators and not job seekers and would provide, on farmers' doorsteps, best technologies and quality inputs. Another strategy could be to create agri-clinics, where technology agents could join hands to ensure a 'single-window' system of advisory services to farmers. A good farmer is knowledge-hungry and concentrates on gathering and using scientific information in order to improve his agricultural practice and enhance production, rather than merely being dependent on government subsidies (CGIAR, 2009; Paroda, 2017).

The Way Forward

The changing and strengthening of agricultural research for development requires participatory coordination and awareness by research institutions and organizations of ongoing R&D activities and creating an enabling environment for continuous improvement of scientific and professional material and technical resources for development and promotion of demand-driven innovations in agriculture and associated areas. In the future, public institutions will continue to dominate agricultural R&D; however, the private sector could demonstrate its ability to successfully increase agricultural productivity. Therefore, the government should create policies that encourage private sector enterprises to invest in R&D and elaborate successful cooperation between the public and private sectors, which would encourage both agricultural innovation transfer and agribusiness advice. This innovative system promises to integrate agricultural producers and farmers into a market-orientated agricultural economy. The time has come for Indian agriculture to liberate the country from the twin scourge of hunger and poverty and of malnutrition of children and women. The nation needs to continue to feed the ever-growing population with adequate food whilst ensuring good nutrition. Accelerated science and innovation-led agricultural growth must be inclusive and must address the needs and aspirations of resource-poor smallholder farmers in the country. Under the growing challenges of resource degradation, escalating input crisis and costs, plus the overarching effects of climate change, major future gains in foodgrain production largely depend on a paradigm shift from integrated germplasm improvement to integrated natural resource management. The future AR4D efforts by NARS need to be reoriented towards the involving farmers' participation. We need to find innovative ways to effectively disseminate knowledge and lay greater emphasis on outscaling innovations to impact the livelihood of smallholder farmers. An effective approach could be to revitalize government extension machinery, which is currently at the lowest level of functionality. Henceforth, a 'farmer first' approach has to be the goal of the NARES to bridge the income divide between farmers and non-farmers, and should benefit producers and consumers. To ensure this, developing countries like India must enhance their investments (triple) in AR4D to address emerging challenges and ensure food, nutritional and environmental security for all.

References

Beniwal, S., Maru, A., Khalikulov, Z. and Ahmadov, H. (2010) Central Asia and the South Caucasus: Challenges, Opportunities, Priority Needs and Actions Required for Improving Agricultural Research for Development. Central Asia and the Caucasus Association of Agricultural Research Institutions (CACAARI), Tashkent, Uzbekistan.

CGIAR (2009) Towards a Strategy and Results Framework for the CGIAR. Draft report by the Strategy Team, 21 October. CGIAR, Montpellier, France.

FAO (2010–11) The State of Food and Agriculture: Women in Agriculture. Food and Agriculture Organization of the United Nations, Rome. Available at: http://www.fao.org/publications/sofa/2010-11/en/ (accessed 24 May 2018).

GCARD (2010) Improving the Livelihoods of the Resource-poor Smallholder Farmers and Producers in Developing Countries: An Urgent Appeal for Action by GCARD. The Global Conference on Agricultural Research for Development (GCARD) 2010, Montpellier, France, 28–31 March.

Kabeer, N. (2012) *Women's Economic Empowerment and Inclusive Growth: Labour Markets and Enterprise Development*. Working paper 2012/1. School of Oriental and African Studies, University of London.

Paroda, R.S. (2017) *Reorienting Agricultural Research for Development Agenda for Sustainable Livelihood Security of Smallholder Farmers*. Strategy paper. Trust for Advancement of Agricultural Sciences, Pusa Campus, New Delhi.

Paroda, R.S., Beniwal, S., Gupta, R., Khalikulov, Z. and Mirzabaev, A. (eds) (2007) Final Report of the Expert Consultation in Regional Research Needs Assessment in Central Asia and the Caucasus, 7–9 March. GFAR, CACAARI, ICARDA-CAC, Tashkent, Uzbekistan.

Weatherhogg, J., Dixon, J. and de Alwis, T. (2001) Global farming systems study: challenges and priorities to 2030. Regional analysis – South Asia. Consultation document. FAO, Rome.

World Bank (2008) World Development Report: Agriculture for Development. The World Bank, Washington, DC.

9

Strategies for Scaling Innovations for Impact on Smallholder Farmers

_____

The Need for Innovation

Accelerating agricultural growth is an important goal for most of the nations in achieving the Sustainable Development Goals, especially to remove poverty, achieve zero hunger and ensure environmental security. Those developing nations that have reorientated their agricultural research for development agenda towards scaling of innovations have made much faster progress. The greater the emphasis on agricultural research for innovation, the higher has been the growth of agricultural GDP (Fan, 2013).

In fact, the Green Revolution was, in itself, an innovation-led initiative around the use of high-yielding dwarf wheat and rice varieties that responded favourably to higher inputs, leading to a quantum jump in productivity. The criteria for success had been: (i) political support; (ii) good institutions and human resources; (iii) availability of critical inputs (seeds, water, fertilizer etc.); (iv) enlightened extension workers and hardworking farmers; and (v) partnership among the stakeholders.

The emerging second-generation challenges of the Green Revolution are: factor productivity decline, depleting natural resources, increasing cost of inputs, higher incidence of diseases and pests, less profit for farmers, and the adverse impact of climate change. Obviously, increasing the income of 80% of farmers who are small and marginal, having holdings of less than 2 ha, would require technologies by which they can save cost on inputs and have more income through higher productivity and by linking to markets. Thus, scaling of innovations like hybrid technology, conservation agriculture, micro-irrigation, integrated nutrient management (INM), IPM, adoption of GM food crops and protected cultivation become high priority. For this to happen, enabling policy, strong PPP and innovative extension systems to transfer the right knowledge, especially around secondary and speciality agriculture, will be needed. Moreover, innovation without incentives and rewards and a congenial policy environment, including an IPR regime, would not be possible, and innovative institutional as well as policy-related initiatives would be needed to make a difference, as was experienced during the Green Revolution era.

For any innovation to be scaled out, it is critical to assess its economic feasibility and potential for large-scale adoption as well as its impact. Moreover, many innovations are farmer-led, and need to be assessed, validated, refined and outscaled in order to harness their expected benefits to the farming community at large (TAAS, 2013). For this, farmers' participatory approach is needed with the active involvement of scientists and provision of incentives, especially in the form of bankable projects, with availability of credit at low interest rates of less than 4% (Saxena, 2017).

Past Initiatives

The ICAR, having the mandate for research, extension and education, had been engaged in providing national public goods that help accelerate agricultural growth by disseminating appropriate technologies to farmers. In the process, KVKs, now numbering 700, have been instrumental in providing front-line extension for scaling new technologies that have helped farmers increase production as well as income.

Besides front-line demonstrations, a large number of farmers in each district have been provided with access to new seeds, planting materials, good agronomic practices and training for skills development. These institutional systems have helped considerably in achieving faster growth in different sectors of agriculture.

The provision of a revolving fund for enhancing the availability of seeds of improved varieties/hybrids, faster multiplication of planting material, and fabrication of tools and implements etc. have all helped in accelerating the growth of Indian agriculture. To ensure the effective involvement of the private seed sector, the ICAR provided the breeder seeds of parental lines of hybrids of crop varieties for faster multiplication and distribution of seeds. This led to faster growth of the seed sector in India. Various mechanisms of incentives and rewards were also put in place in the late 1990s based on the Johl Committee Report (1995). Somehow, these incentives have not reached the real performers due to bureaucratic hurdles and resistance to change. To encourage private sector involvement in R&D, provisions of IPR and the Protection of Plant Variety & Farmers Right Act (PPV&FRA) were enacted at the beginning of the new millennium.

Incentives and rewards were also an integral component of the prestigious World Bank project named the National Agricultural Technology Project (NATP), negotiated in 1998, followed by greater thrust of PPP for scaling innovations under the second phase of the project, named the National Agriculture Innovation Project (NAIP). It eventually helped in outscaling many useful innovations for the benefit of end-users – farmers, producers and consumers (ICAR, 2006; 2014).

Some of the recent initiatives for scaling innovations by the Ministry of Agriculture and Farmers' Welfare are: Attracting Rural Youth in Agriculture (ARYA), Mera Gaon Mera Gaurav, National Skill Qualification Framework, skill training, Value Addition and Technology Incubation Centres in Agriculture (VATICA), knowledge systems and homestead agricultural management in tribal areas, nutri-sensitive agricultural resources and innovations (NARI), climate-smart villages, and web and mobile advisory services. The potential role of farmer-producer organizations (FPOs) in innovation upscaling is also important. There is a need for figuring out complementarity between the public and private research organizations for scaling out agricultural innovations for smallholder farmers and overcoming major barriers in such collaborations. Unlike the public sector, the private sector concentrates on fewer technologies and invests heavily in those. The key constraint with the private sector in R&D is longer duration (7–10 years for varietal development) and continued investment during that period. The major preconditions for scaling out innovations are: they should be needs-based and relevant; the originator should have proof of concept, be compliant with regulations and able to show the cost of such compliance; and there must be clear incentive and a sense of urgency (Bhooshan, 2017).

The Department of Biotechnology (DBT) has also taken up several initiatives for scaling innovations through the Biotechnology Industry Research Assistance Council (BIRAC), a platform to nurture industry–academia connectivity. Other initiatives include biotechnology parks, bioincubators and science clusters. Even though there are different schemes for agricultural biotechnology such as the Biotech-KISAN scheme, performance is not on a par with other sectors like health. There are also schemes to encourage scientists/faculties to move to entrepreneurship. The key challenges for the entrepreneurs are lack of financing and market access. DBT had started several initiatives such as Students' Innovations for Advancement of Research Explorations (SITARE), eYUVA (creating entrepreneurial culture in universities) and BioNEST (nurturing entrepreneurship by establishing bioincubation centres) for supporting entrepreneurs. There is a need to accelerate the entrepreneurship fund and a possibility of social immersion programmes for incubates to assess market needs (Swarup, 2017).

Similarly, various initiatives by the National Innovation Foundation (NIF) for promoting grassroots entrepreneurship include the Micro Venture Innovation Fund, the Grassroot Technological Innovation Acquisition Fund and the establishment of the NIF Incubation and Entrepreneurship Council. Innovations are also encouraged by organizing exhibitions and through awards and scholarships. Participatory research and decentralized fabrication and services are essential for improving technologies for outscaling in India (Zehr, 2017).

Innovations need to be considered in totality, and innovations policy institutions are essential in developing a strategy for their outscaling. The ICAR needs to strengthen existing policies, institutions and incentives for upscaling and outscaling of innovations. The existing policies and mechanisms need thorough review in the present context. There is an obvious need to have competent human resources with marketing expertise so as to commercialize the technologies and take them to small farmers. The BIRAC model of DBT is a platform for innovations; a similar model needs to be developed in ICAR farm producer organizations (FPOs) and this could be a good option for promoting agricultural innovation and commercialization of technologies. While planning for upscaling and outscaling, adequate care needs to be taken to avoid planning fallacy (underestimation of time and resources required). Planners need to think whether an innovation can really be engineered and applied to the present-day requirements of smallholder farmers (Gulati *et al.*, 2006; Swarup, 2017; Pal *et al.*, 2017).

Considering that agriculture is important in eliminating poverty and hunger, and addressing concerns of climate change and pursuing SDGs, the importance of relevant innovations for inclusive agricultural and economic growth cannot be overemphasized (Fan, 2013). In order to create an environment for innovation, emphasis on capacity development is essential for India to progress and compete, especially in the era of globalization. Hence, generating new innovations to meet emerging challenges, involving both public and private sectors, is the need of the hour. Effective impact of innovation is possible only through the implementation of relevant policies and with the required IPR regime in place. In this context, the National Intellectual Property Rights Policy and the implementation of the Protection of Plant Varieties and Farmers' Rights Authority (PPV&FRA) by the government of India would certainly accelerate the pace of innovation (Saxena, 2017).

Relevant Experiences Abroad

Since industrialization, agricultural innovation in the developing countries has predominantly been fostered by the public sector. But with the commercialization of agriculture, the private sector, including multinational companies with a base in developed countries, has been a major player. In the USA, public sector universities became R&D laboratories for private companies after the enactment of the Bayh-Dole Act of 1980 (Bayh-Dole Act, 1980), which allowed universities and other non-profit institutions to have ownership rights of discoveries that were the result of federally funded research. This facilitated the transfer of technologies to the private sector through the establishment of science parks and incubators. Europe also followed a similar institutional framework to facilitate new innovations and their faster dissemination. Lately, greater emphasis on innovation in China, mainly publicly funded, has transformed its economy through greater participation of the private sector and foreign companies for outscaling innovations.

In the pre-WTO era, public sector institutions in developing countries played a major role in generation of national public goods through agricultural research. In the post-WTO era, with economic liberalization, private sector investment in agricultural innovations has helped towards their faster delivery and adoption. Some of the developing countries moved faster in promoting the culture of cooperation in R&D. MLSCF (Malaysia), Fundacion (Chile), CENTEV (Brazil), Technoserve (Mozambique), Timbali (South Africa) and IAA-IPB (Indonesia) are such public sector-funded research institutions that have worked in partnership with the private sector to upscale and outscale innovations. Also, some multi-stakeholder platforms were developed for scaling innovations (e.g. MasAgro, Mexico) (Fan, 2013).

An Enabling Environment

Earlier agri-innovations had a relatively simple process/cycle of development and dissemination through the public extension system, to the benefit of end-users, mainly farmers. However, more recently, new and more efficient players have entered the process. The emphasis on the commercialization of technologies and resource generation has necessitated the involvement of new actors, mainly private sector companies, in the commercialization of research products. These new initiatives are mainly being guided by the profit motive and are finding favour because of efficient and faster delivery mechanisms, though sometimes more costly for smallholder farmers. On the other hand, some rural innovations by enterprising farmers are also recognized as potential options for solving location-specific problems, but they need validation, further refinement and outscaling to benefit larger farming communities. Mainstreaming of such innovations is, therefore, a challenge that needs to be recognized and resolved by appropriate incentive-and-reward mechanisms and by institutions/PPPs. Thus, innovations have moved away from the conventional innovation systems (linear transfer of technology) to those of agricultural innovation systems (multi-stakeholder platforms) and also to farmers' innovation systems – grassroots innovations (Saxena, 2017).

Both policy and institutional support are key enabling factors for faster dissemination of technologies. In this regard, the recent National Intellectual Property Rights Policy advocates promotion of a holistic and conducive environment for catalysing intellectual property for socioeconomic development and protecting public interests. Other national initiatives such as Make in India, Skill India, Start-Up India, Smart Cities and Digital India are also supportive of agricultural innovations. The flagship programmes of the government like Start-Up India aim to build a strong ecosystem for nurturing innovations for faster adoption. Under this, the Atal Innovation Mission (AIM) is an action plan envisaged with a major focus on the promotion of entrepreneurship and innovation in the agricultural sector and other sectors. These initiatives and enabling policies may appear to be more relevant to industry, but the resulting innovations will have a cascading and demonstrable impact on AR4D (Saxena, 2017; Pal *et al.*, 2017).

The Need for Outscaling of Innovations

There are several technologies that need to be outscaled. In the dairy sector, such technologies include animal identification, precision animal feeding, advanced reproductive technologies, disease diagnosis innovations, technologies for detection of adulterants in milk and milk products and small-scale farm machines (such as mobile milkers). There are now four generations of technology for improving reproductive health and these must be scaled out. Artificial insemination and semen sexing can make a major impact on milk productivity. The Kerala and Kolar models of community milking, and technology for value-added dairy foods, are now standardized and need immediate interventions for their outscaling. In order to better understand technology and its spread, the mindset of 'managing livestock under zero or low input' should be changed to 'commercial enterprise' (Mani, 2017).

Scaling out innovations in the case of agro-processing and value addition also needs to be given due attention. Exploitation of value-added products from agrobiomass like lignin and algae, food products of bioprocessing and chemical processing, and composite fruit coating can generate immense benefits for farmers and rural entrepreneurs. Most of these processes are restricted to labs and require scaling up. There is a need for government support in upscaling innovations in this sector through R&D and the establishment of incubation centres. Such support needs to be proactive and facilitate integration with industry (Bhooshan, 2017).

There is a need for upscaling and outscaling for small-farm mechanization technologies in India. Greater attention needs to be paid to the involvement of industry in commercialization. Contract research on a urea ammonium nitrate application system, funded by the Department of Fertilizers and National Fertilizers Ltd, is a good example of success. Unique facilities such as design innovation centres, a collaborative initiative between the Indian Institute of Technology (IIT), Kanpur, and the IARI, are a promising model for incubation, design improvements and start-up facilitation. There is also a need for more public funding for research into agricultural mechanization, the establishment of national centres in different zones for mechanization, scaling up innovations through PPP, linking grass roots innovations to institutional innovations, and the establishment of design innovation centres at different institutions (Jat, 2017).

Innovations that are in progress for producing high-quality, high-value agricultural produce include: plastic mulching coupled with fertigation; walk-in poly-tunnels for vegetables; insect-proof net houses; shade-net structures; vegetable farming under rain shelters; naturally ventilated poly-houses; climate-controlled, hi-tech greenhouses for disease-free nursery raising; and hi-tech soil-less production. Protected vegetable cultivation has been very successful in Ladakh and other places. However, there are key constraints that include: high initial cost; poor-quality material; high cost of input; lack of guidance, knowledge and marketing; a nematode problem; and lack of refrigerated vehicles. In order to outscale such technologies, there is a need for further R&D to develop crop varieties/hybrids suitable for protected cultivation; skilled human resources development; establishment of a Bureau of Indian Standards (BIS) standard for polyhouse material and its testing facilities; cluster approach; and streamlining of subsidy. Low-cost polyhouse, mulching and fertigation have proved to be more popular because of their low cost (Singh, 2017).

There are various options for outscaling innovations in natural resource management. Innovation is an amalgamation of technology, local adaptation, social inclusivity and access by end-users. It is important to understand the big challenges associated with 'half innovations' and the successful conversion of half innovations into full innovations based on local needs. Major requirements for outscaling NRM innovations include long-term investment, portfolio of policies and practices, patience, capacity, innovation-led business models and robust *ex ante* analysis of return on investment. Scientific social responsibility/science-corporate social responsibility needs to be given due importance. Since these NRM-based innovations generate a lot of social and environmental good, there is an urgent need for greater public investment in their promotion and use (Jat, 2017).

There are various technologies outscaled by public institutions like IARI, and the strategies adopted for outscaling innovations include technology commercialization through PPP, access to knowledge and information through PUSA KRISHI-App, partnerships for enhancing service provision, and linking farmers with markets through Farmer Producer Organizations (Beej India Ltd). Issues, concerns and challenges include a disconnect between production and marketing, licensing issues with industries, lack of exclusive funding support for agro-start-ups, insufficient delegation of powers to cutting-edge institutions, lack of strong actions against IP violation, lack of trained professionals, and technology-readiness. The way forward could be demand-driven R&D with more industry/research/academia interaction, technology transfer and integration with incubation for start-up, virtual marketing, and use of mobile/internet technologies.

Greater emphasis is required on providing incentives to researchers to whom ICAR has taken steps to grant incentives for patenting, innovation and partnership, and to establish an IP management structure involving institutes, zonal technology management units, and national platforms – Agrinnovate India Ltd – to interface with the private sector. Emphasis is also needed on the role of vision, skills, incentives, resources and action plans for innovation. There is a need to establish a central cell/platform to screen innovations at ICAR level. There is also a need to create an innovation fund to promote and commercialize new technology. Research efforts need to be exercised in a systems approach rather than based on disciplines/commodities.

There are some major innovations that currently need to be outscaled as a matter of priority. These are: (i) hybrid rice – the current area coverage in the last two decades is hardly 2 million ha, whereas scope exists for at least 10 million ha in the next decade; (ii) single-cross maize hybrids – the area covered is presently less than 60%, whereas scope exists for almost 90% of maize area; (iii) conservation agriculture – under the rice-wheat cropping system, the current area is about 3.5 million ha, whereas scope exists for almost 8 million ha in the Indo-Gengetic plains alone. Conservation agriculture innovation also has vast scope under rainfed farming covering around 55% of the total 144 million ha of cultivable land in the country; (iv) protected cultivation – the current area under protected cultivation is 40,000 ha, compared to 2 million ha in China; (v) micro-irrigation – out of a total irrigated area of 64.7 million ha, the area so far covered under micro-irrigation is only 7.7 million ha, which can certainly be doubled in the next decade. Hence, it is evident that to harness the benefits of these innovations, concerted efforts are urgently needed. There are many more useful innovations that need to be outscaled and require a critical evaluation as to how this can be done quickly for maximum impact on the livelihood of smallholder farmers (Jat, 2017; Singh, 2017).

Policies for Innovations

For successful scaling of innovations, there is an urgent need to put in place policies such as: (i) facilitation of farmers' collectives like FPOs with a proper legal framework, establishment of a cadre of agri-business professionals at village level, credit to farmers across the value chain, machine-rental services etc.; (ii) research policies aimed at promotion of agro-ecology-based research, research for trade policy, agro-processing, value-chain development, sustainable livelihood, and new funding models for encouraging research by state government; (iii) pricing policies around fixing a minimum support price (MSP), inclusion of efficiency, compensation for risk and ecosystem services; and (iv) policies for investment in agriculture rather than subsidies and for promoting private investment. There is also an urgent need to attract the private sector into development of wholesale markets, warehouses and cold storages, agro-processing infrastructure, canal irrigation and agricultural extension. NARS has undergone various policy reforms in research, intellectual property rights and technology transfer (Saxena, 2017; Pal et al., 2017).

There is an urgent need for human capital for the development of innovations and the invent-innovate-invest continuum, and of concepts of skill, speed and scale in the innovation system. The country needs to place greater emphasis on human capital development, particularly for building entrepreneurship for which availability of adequate funds is essential. There are concerns over the abysmal state of credit and information access by farmers in India. Such concerns relate to livestock sector insurance (presently granted only for a year and for only high-yielding animals) and issues around taxation of dairy, fishery and poultry enterprises. There is an urgent need to develop a value-chain approach, both in research and in policy.

Even though technological innovations are abundant, institutional failures lead to lower adoption. The problem of lack of appropriate policies, institutions and technologies was also present at the cusp of the Green Revolution during the 1960s. Whereas government has an important role to play even now, the innovation system has become multisectoral, involving other actors. Therefore there is an urgent need for institutional and policy reforms that are more appropriate in today's context. Institution or policy failures need to be revisited and more suitable policies and institutions developed. The lack of internal capacity for negotiating complex trade and other international treaties needs to be addressed. The USA has much stronger private sector activity in venture capital, whereas European countries have a number of public sector business models for scaling up and scaling out innovations in agriculture. Who bears the risk with innovations and how risk and reward are shared between the public and the private sector are issues that need a clear business model. Government should also shift from being mainly directive to being more facilitative in the promotion of innovation. This would require a cost-effective regulation system for investment and the commercialization of economical, efficient and productive innovations (Gulati et al., 2006; Fan, 2013).

Major policies for scaling up innovations should centre around incentive-(non-subsidy)-and-reward systems, enabling a policy environment for faster adoption of innovations, increased resource allocation for agricultural research and development (at least 1% of agricultural GDP), scaling innovations through PPP and policy and institutional reforms for large-scale adoption of efficient resource-saving technologies such as conservation agriculture (CA).

Different state-led institutional and policy reforms for outscaling innovations in micro-irrigation and water management are worth critical assessment. The states of Rajasthan, Andhra Pradesh, Maharashtra and Gujarat have about 45% of their area under micro-irrigation, whereas potential for this technology is estimated at 8.6 million ha over the country. The Andhra Pradesh model of micro-irrigation, the Karnataka PPP model and the Rajasthan model of allowing only micro-irrigation and a ban on floor irrigation in the canal area are some of the successful examples replicated in other states.

Intensifying Agri-innovations

The evolution of NARES has primarily been based on social commitment with a motive to provide national public goods and to serve a large number of resource-poor small farmers. Thus innovations had been an integral part of the Indian AR4D system from the beginning. Somehow, they have not been subjected to evaluation of their socioeconomic impact, as has been done in the developed world. Taking a cue from the technologically advanced economies, India realized that encouraging innovation is a key factor in generating agri-business opportunities and increasing farmers' incomes. It is in this context that ICAR responded and prepared guidelines for agri-IPRs management and commercialization in 2006 (Johl Committee Report, 1995; ICAR, 2006) and also initiated a National Agriculture Innovation Project (NAIP) with funding from the World Bank. Also, guidelines for incentives and rewards for outscaling innovations and resource generation were put in place. Somehow, the pace of promoting innovations and allowing the right incentives for researchers has remained slow (Saxena, 2017).

A countrywide network of institute technology management units (ITMUs) has been created for the management of agri-innovations and agri-intellectual properties in all ICAR institutes, duly supported by the zonal technology management and business planning and development units at five selected ICAR institutes. This new initiative helped to kick-start innovation awareness of the importance of commercialization. The ICAR *Rules and Guidelines for Professional Service Functions* were accordingly published for smooth implementation of an Indian agri-IPR network in 2014 (ICAR, 2014). Eventually, many agri-technologies and services from NARES were successfully commercialized. The Business Planning and Development Unit

(BPD), under the NAIP, and the agri-business incubators (ABI) established under the National Agriculture Innovation Fund (NAIF) were an experiment aimed at the commercialization of new innovations. Accordingly, the entire process of innovation generation and commercialization, involving PPP, led to an intangible treasure of experience, which needs to be intensified for better management of agri-innovations in the future (Bhooshan, 2017).

The strength of an innovation is generally considered in terms of its commercial and societal value. As such, a large number of agricultural innovations identified and commercialized during the previous decade need to be outscaled. India needs to innovate further in order to address the emerging challenges in agriculture. Problems such as nutritional security, climate change and declining profitability are some of the issues that need attention. The role of the private sector, though well-realized and appreciated for input development and delivery, is still to be appreciated and expanded to non-conventional areas, and the role and importance of PPP has to be promoted further. There should be adequate provision for dissemination of innovations for management of natural resources and sustainable farm practices, which are significant for agriculture yet may not attract the attention of the private sector (Gulati *et al.*, 2006; Zehr, 2017).

In a short time, India has successfully commercialized some new technologies. In addition, ICAR has built the needed capability of handling innovation and IPR-related issues in the future. However, there are many innovations and technologies that remain under-utilized. Important among these are animal health, protected cultivation, micro-irrigation, watershed development, hybrid rice, GM seeds, bioagents, farm machinery and post-harvest technology. There is an urgent need to revisit technology dissemination and commercialization mechanisms and associated policies in the context of scaling current and future technologies. In-depth analysis of commercialization mechanisms at the system and organization levels needs to be carried out for effective upscaling and outscaling of agricultural innovations. There is a need to look at the incentive-and-reward system for innovators so that a clear road map is put in place in an enabling environment, with appropriate policies and necessary incentives (Saxena, 2017).

The Way Forward

The following are important suggested action points for developing strategies for scaling innovations for impact on smallholder farmers:

- Innovations have played and will continue to play a significant role in agricultural transformation globally. The innovation process involves multiple stakeholders and the right policy environment is needed to scale out innovations for impact in the broader national agricultural perspective.
- Agricultural research must move from a commodity-centric to a systems approach, and all stakeholders (farmers, private sector, NGOs) must be part of the research and innovation continuum. Hence, institutional/innovation platforms are needed to encourage much-needed scientist–farmer, and public–private partnerships.
- In order to achieve an innovation-driven agrarian economy, the innovation capacity of R&D systems, civil society organizations (CSOs) and farmers should be developed. For this purpose, intensity of public investment will have to be enhanced. Greater attention is needed on capacity development of those responsible for scaling innovations for successful commercialization.
- There is an urgent need to strengthen the existing technology-transfer system within NARS (front-line extension, Agri-Business Incubator, Agrinnovate India Ltd) and establish technology parks as well as transfer systems for commercialization both in ICAR and SAUs. Placement of adequate manpower is required, as well as financial resources and freedom to operate. Convergence of technology and diversification of extension and other service systems are also critical for outscaling innovations.
- Available innovations, including those that are farmer-led, must be assessed for needed validation, refinement and prioritization based on their commercial potential. This should also entail identification of suitable partners for the ventures. Financing, risk management and incentives for outscaling innovation are necessary to encourage potential entrepreneurs.
- An innovation platform would help to accelerate the scaling out of innovations and, therefore, an 'agri-innovation board' is required to be established in the Ministry of Agriculture and Farmers' Welfare. This board must be headed by an eminent agricultural scientist and its members drawn from different ministries, including finance, commerce and industry.
- The board should have a minimum of Rs 10 billion for financing the activities of scaling out agricultural innovations. This could be taken from existing funding support for innovation (Start-up India, Atal Innovation Scheme) or from a separate funding mechanism such as the National Innovation Fund (NIF) initiated by the Council of Scientific and Industrial Research (CSIR).
- Concerned ICAR institutes and SAUs must ensure that they provide skill-based certificate training for entrepreneurship and much-needed backstop services critical for successful scaling of innovations. The personnel, once trained, can work as para-innovators or technical service providers. To link with industry, ICAR would need to develop effective partnerships with organizations such as the Federation of Indian Chambers of Commerce and Industry (FICCI), the Associated Chambers of Commerce and Industry of India (ASSOCHAM) and the Confederation of Indian Industry (CII).
- Farmer producer organizations, self-help groups, cooperatives, producer companies etc. could be involved in outscaling innovations. These organizations should have easy access to technology, financial services including credit, and 'hand-holding' from public organizations for promoting demand-driven innovations in the broader national interest.
- Participation of the private sector in R&D and upscaling and outscaling of innovations needs an enabling policy environment, and access to public technology and other resources. In order to facilitate this, the government should move from a 'directive' to a 'facilitative' role. This may require revisiting existing regulations in order to provide a 'predictable and enabling' regulatory framework. Incentives and rewards for innovators need to be put in place to sustain interest in outscaling and much-needed technical backstopping.

References

Bayh–Dole Act (1980) The Bayh–Dole Act or Patent and Trademark Law Amendments Act (Pub. L. 96-517, 12 December). US Government.

Bhooshan, N. (2017) Outscaling of Public Sector Technologies. Paper presented by Dr Neeru Bhooshan, ICAR-Indian Agricultural Research Institute, New Delhi, during a dialogue on 'Incentives and strategies for scaling out innovations for smallholder farmers'. Organized by the Trust for Advancement of Agricultural Sciences, Pusa Campus, New Delhi, 30–31 October.

Fan, S. (2013) Ensuring Food and Nutrition Security in Asia: the Role of Agricultural Innovation. Strategy paper. Trust for Advancement of Agricultural Sciences, Pusa Campus, New Delhi, 11 January.

Gulati, A., Joshi, P.K. and Landes, M. (2006) Contract Farming in India: An Introduction (e-book). International Food Policy Research Institute (IFPRI), National Center for Agricultural Economics and Policy Research (NCAP) and US Agency for International Development (USAID), New Delhi and Washington, DC.

ICAR (2006) ICAR Guidelines for Intellectual Property Management and Technology Transfer/Commercialization. Indian Council of Agricultural Research, New Delhi.

ICAR (2014) ICAR Rules and Guidelines for Professional Service Functions (Training, Consultancy, Contract Research and Contract Service). Indian Council of Agricultural Research, New Delhi.

Jat, M.L. (2017) Outscaling Natural Resource Management Innovations. Paper presented by Dr M.L. Jat, International Maize and Wheat Improvement Center (CIMMYT), NASC Complex, Pusa, New Delhi, during a dialogue on 'Incentives and strategies for scaling out innovations for smallholder farmers'. Organized by the Trust for Advancement of Agricultural Sciences, Pusa Campus, New Delhi, 30–31 October.

Johl Committee Report (1995) S.S. Johl Committee Report of the Committee on Partnership, Resource Generation, Training, Consultancy, Contract Research/Contract Service and Incentive and Reward Systems. Indian Council of Agricultural Research, New Delhi.

Mani, I. (2017) Scaling Innovations for Small Farm Mechanization. Paper presented by Dr Indra Mani, ICAR-Indian Agricultural Research Institute, New Delhi, during a dialogue on 'Incentives and strategies for scaling out innovations for smallholder farmers'. Organized by the Trust for Advancement of Agricultural Sciences, Pusa Campus, New Delhi, 30–31 October.

Pal, S., Subhash, S.P. and Arathy, A. (2017) Upscaling agricultural innovations. Paper presented by Drs Suresh Pal, S.P. Subash and Ashok Arathy, ICAR-National Institute of Agricultural Economics and Policy Research, New Delhi, during a dialogue on 'Incentives and strategies for scaling out innovations for smallholder farmers'. Organized by the Trust for Advancement of Agricultural Sciences, Pusa Campus, New Delhi, 30–31 October.

Saxena, S. (2017) Incentives for patenting, innovation and partnership. Paper presented by Dr Sanjeev Saxena, Indian Council of Agricultural Research, New Delhi, during a dialogue on 'Incentives and strategies for scaling out innovations for smallholder farmers'. Organized by the Trust for Advancement of Agricultural Sciences, Pusa Campus, New Delhi, 30–31 October.

Singh, B. (2017) Outscaling protected cultivation: constraints and options. Paper presented by Dr Brahma Singh, former Director, Life Sciences, DRDO, New Delhi, during a dialogue on 'Incentives and strategies for scaling out innovations for smallholder farmers'. Organized by the Trust for Advancement of Agricultural Sciences, Pusa Campus, New Delhi, 30–31 October.

Swarup, R. (2017) Initiatives for scaling innovations in agricultural biotechnology. Paper presented by Dr Renu Swarup, Department of Biotechnology, Government of India, New Delhi, during a dialogue on 'Incentives and strategies for scaling out innovations for smallholder farmers'. Organized by the Trust for Advancement of Agricultural Sciences, Pusa Campus, New Delhi, 30–31 October.

TAAS (2013) Seventh Dr M.S. Swaminathan Award for Leadership in Agriculture. Trust for Advancement of Agricultural Sciences, Pusa Campus, New Delhi, 24 June.

Zehr, U. (2017) Strategies for promoting proprietary technologies. Paper presented by Dr Usha Zehr, Mahyco Foundation, Jalna/Mumbai, during a dialogue on 'Incentives and strategies for scaling out innovations for smallholder farmers'. Organized by the Trust for Advancement of Agricultural Sciences, Pusa Campus, New Delhi, 30–31 October.

10

Enhancing Productivity of Foodgrains

Globally, India is the third-largest producer of cereals, with only China and the USA ahead of it. India's population is likely to reach 1.5 billion by 2030 and therefore the challenge facing the country is to produce more and more from diminishing per capita arable land and irrigation water resources and increasing abiotic and biotic stresses. India produced 277.49 million t of foodgrains in 2017–18 to meet the needs of a current population of 1.34 billion. The current situation in India is that cereal production has to be doubled by 2050 in order to meet the needs of an expected population of 1.8 billion, in addition to meeting the needs of livestock and poultry. Since land is a shrinking resource for agriculture, the pathway for achieving these goals can only be higher productivity per unit of arable land and irrigation water. Factor productivity will have to be doubled, if the cost of production is to be reasonable and the prices of farm products are to be globally competitive. The average farm size is going down and nearly 80% of farm families belong to the marginal and small-farmer categories. Enhancing small-farm productivity, increasing small-farm income through crop-livestock-aquaculture integrated production systems and multiple livelihood opportunities through agro-processing and biomass utilization, are essential to meet food production targets and for reducing hunger, poverty, nutritional insecurity and rural unemployment. Some 55–60% of the Indian population continues to depend on agriculture and allied activities for their livelihood. Hence, growth of this sector is an essential prerequisite for overall economic growth.

Consumption and Demand Pattern

Sustained economic growth, increasing population and changing lifestyles are causing significant changes to the Indian food basket, away from staple foodgrains and towards high-value horticultural and animal products. While per capita consumption of foodgrains has declined, their total consumption has increased due to an ever-increasing population, as India adds almost one Australia (around 15 million) to its population every year. Also, changes in dietary pattern, towards animal products, have led to an increased demand for foodgrains as feed. Nonetheless, foodgrains, particularly rice and wheat, continue to be the main pillars of India's food security. On the supply side, stimulated by public investment in irrigation and rural infrastructure and the rapid spread of high-yielding varieties of rice and wheat, together with improved crop production practices, India has achieved an impressive growth in foodgrain production. Per capita annual production of foodgrains increased from 183 kg during the early 1970s to 207 kg by the mid-1990s, even though the country's population increased more than 50% during that period. After the mid-1990s, per capita foodgrain production started declining due to diversification of the food basket on account of availability of more vegetables, fruit, milk, meat, fish etc., associated with the affluence of the society due to increases in the incomes of the middle class. This changing scenario of consumption and production will have a significant influence on the demand and supply prospects in India. The trend in per capita consumption of foodgrains for rural and urban consumers in the period 1983–2012 and the projected demand during 2016–2022 are given in Table 10.1.

Per capita consumption of foodgrains (as direct demand) in 2004–05 by region and income group, separated for rural and urban populations, was used as a baseline consumption for projecting the future per capita consumption. Cereal consumption has shown a decline in both rural and urban areas. The decline is sharper in rural than in the urban areas. Per capita consumption of coarse cereals has shown much steeper decline than that of rice and wheat. Per capita consumption of foodgrains had declined from 139.9 kg in 1993–94 to 130.3 kg in 2004–05 to 115.4 kg in 2011–12 and is expected to be around 111.6 kg in 2021–22. The expected direct household demand for foodgrains would increase to 202 million t by 2021–22, comprising 97.4 million t of rice; 73.5 million t of wheat; 15.1 million t of coarse grains and 16.2 million t of pulses (Table 10.2) (Kumar *et al.*, 2009; DoA and FW, 2016–17).

Besides direct demand, there is also an important component of indirect demand, which includes seed, feed, industrial uses and wastage. Conventionally, indirect demand is assumed to be 12.5% of the total foodgrain production. The shares of seed, feed, wastage and other food uses have been computed as 9.5% of the total production of rice, 13.5% of wheat, 4.1% of coarse cereals and 10.8% of pulses. In addition, seed, feed, industrial use and wastage allowances have been projected as 36.9 million t in 2011–12; 39 million t in 2016–17 and 41.1 million t in 2021–22, which constitutes about 16% of the total foodgrain production of the country (Kumar *et al.*, 2009; Firdos Ahmad and Shaukat, 2012; DoA and FW, 2015–16, 2016–17).

Current Production Trends

For enhancing foodgrain production, two issues need urgent attention: (i) assisting farmers in disaster situations to restore crop production systems; and (ii) useful germplasm sources should be identified in the context of climate change. Between 1950 and 2007, production of foodgrains (comprising production of rice, wheat, coarse cereals, pulses and sugarcane) in the country increased at an average annual rate of 2.5% compared to the growth of the population, which averaged 2.1% during this period. Warding off 'doomsday' predictions of hunger and famine, India was in a situation, following the Green Revolution in the late 1960s, where it hardly

Table 10.1. Trend in per capita consumption of foodgrains in India. (From: Kumar *et al.*, 2009)

Category of population	Estimated (kg/year)				Projected (kg/year)	
	1983	1993–94	2004–05	2011–12	2016–17	2021–22
Rural	191.1	172.5	156.0	143.9	145.1	144.6
Urban	149.9	139.9	130.3	115.4	114.2	111.6
Rural + urban	181.6	163.8	148.8	134.8	134.6	133.6

Table 10.2. Total demand for foodgrains, 2004–2022 (million t). (From Kumar *et al.*, 2009)

Commodity	2004–05	2011–12	2016–17	2021–22
Rice	79.5	87.4	92.0	97.4
Wheat	57.7	67.2	71.9	73.5
Coarse cereals	13.4	14.2	14.5	15.1
Total cereals	**150.7**	**168.7**	**178.2**	**185.8**
Pulses	9.8	12.5	14.3	16.2
Total foodgrains	**160.5**	**181.2**	**192.6**	**202.0**

had to resort to foodgrain imports between 1976 and 2006. With good monsoon rains and various policy initiatives by the government, India's overall production of foodgrains has increased by 2.3 million t to an all-time high of 277.49 million t in 2017–18, ahead of its previous estimate of 275.68 million t. Record production has been achieved in wheat, rice, pulses and coarse cereals. The earlier record for foodgrain production was 265.04 million t, achieved in the crop year 2013–14. However, the output had dropped due to drought in two consecutive years to 251.57 million t in the crop year 2015–16. According to the data, rice production is projected to reach an all-time high of 110.15 million t in the crop year 2016–17, compared to 104.41 million t in 2015–16. The previous record for rice production was 106.65 million t in 2013–14. Wheat production (98.38 million t in the year 2016–17 compared to 92.29 million t in the previous year) had been at an all-time high. The earlier record was 95.85 million t achieved in 2013–14. A record production of coarse cereals was 44.19 million t in 2016–17, whereas the previous high was 43.39 million t in 2013–14. The production of pulses was also at an all-time high of 26 million t in 2016–17 compared with 19.25 million t achieved in 2013–14 (DoA and FW, 2015–16, 2016–17).

Among developing countries, China and India, together, accounted for over 30% of world cereal output in the early 1990s, contributing significantly to the global decline. At the same time, the output of their export crops rose ten times faster than foodgrains owing to the diversion of land and resources to export crops. The developed countries, which together accounted for about 40% of world cereal output, saw only an 18.6% rise in cereal output over the same period, or an annual growth rate of 1.3%, ahead of their own population growth but insufficient to meet their own rising domestic needs and to provide an adequate surplus for meeting the increasing deficit of the developing world.

Increase in agricultural production during the last decade had mainly been due to growth in productivity. However, in the case of some crops such as maize, gram, soybean and cotton, the growth in acreage was also substantial. Growth in acreage under cotton, oilseeds and pulses came at the expense of coarse cereals, particularly sorghum and pearl millet. Overall acreage

under coarse cereals declined from 28.94 million ha in 2004–05 to 24.71 million ha in 2014–15, indicating a drop of about 15%. Three important ways to boost foodgrain productivity are given in Box 10.1.

The phenomenal increase in foodgrains, from 196.81 million t in 2000–01 to an all-time high of 277.49 million t in 2017–18 has led to a surplus situation compared to domestic requirements, which has led to an increase in overall exports. The crop-wise trends in production and yield are described below.

Wheat production of 97.11 million t in 2017–18 is also a record that is higher than the previous record production of 95.85 million t achieved during 2013–14. Production of wheat during 2016–17 is also higher, by 4.03 million t, than the average wheat production, and higher by 4.36 million t compared to the 92.29 million t achieved during 2015–16. The area under wheat cultivation increased from 27.99 million ha in 2006–07 to 30.47 million ha in 2013–14, whereas the production increased from 75.81 million t in 2006–07 to an all-time record high of 96.64 million t in 2016–17, implying a significant improvement in productivity. More than a three-fold rise in wheat production was made possible by increasing the area under assured irrigation facilities, better seed treatment and adoption of newer varieties, effective rust management and timely sowing of crops to escape terminal heat stress. However, the unseasonal rains and hailstorms during February/March 2015 adversely affected the production of rabi crops. As a result, wheat production in 2014–15 was 88.94 million t as compared to 95.85 million t in 2013–14. The productivity of wheat, which was at 2708 kg/ha in 2006–07 increased to 2872 kg/ha in 2014–15. This rise in productivity resulted from a developmental focus on increasing the seed replacement rate (SRR) along with varietal replacement with high-yielding ones and seeds resistant to different biotic (especially rusts) and abiotic stresses, including multi-stress-tolerant cultivars. The improvement in productivity is equally ascribed to technological interventions, such as use of improved varieties, maintaining an optimum sowing time, line sowing, proper fertilizer application, timely irrigation etc. to contain rusting, cultivating rust-resistant varieties like DPW-621-50, PBW-550, DBW-17, HD-1105 etc. in the North-Western Plains Zone (NWPZ) and

Bihar, Gujarat, Assam, Uttarakhand, Himachal Pradesh, Jammu & Kashmir, eastern Uttar Pradesh, Maharashtra and Madhya Pradesh – it can easily add around 15 million t to wheat production. Similarly, attaining productivity equivalent to Haryana in the states of Punjab, western Uttar Pradesh and Rajasthan, where productivity is above the national average, will further add 15 million t. Thus, an additional 30 million t can be achieved mainly through better agronomic management (Hooda Committee Report, 2010).

To achieve this additional production of 30 million t, we need to emphasize timely sowing of recommended wheat varieties suitable for different agroclimatic conditions, ensuring timely availability of quality input and adoption of good agricultural practices. Use of resource conservation technologies – laser land levelling, surface seeding, zero-till sowing and furrow-irrigated raised-bed planting systems – need to be popularized. A seed replacement rate of 20% has to be targeted, especially in eastern India, and around 30–35% in other states. Also there is a need to promote durum wheat cultivation in Madhya Pradesh and Maharashtra. Furthermore, developing varieties tolerant to terminal heat and low input, requiring short-duration, and high-yielding varieties/hybrids, needs attention (Hooda Committee Report, 2010).

Maize

Among cereals, maize is a high-potential crop. In India, its production has increased from 15.1 million t in 2006–07 to 19.29 million t in 2008–09, mainly with the development and use of a number of single-cross hybrids. The productivity level of 2 t/ha has been attained, which amply demonstrates the future potential of maize in India. The expansion of area in eastern Uttar Pradesh, Bihar, Jharkhand and West Bengal by almost 1 million ha during the last decade is a positive indication as well. All north-eastern states have good potential for maize production (Hooda Committee Report, 2010; Singh, 2017). About 75% of *kharif* area is rainfed, while *rabi* (winter season) maize is predominantly grown in favourable ecologies. With increase in irrigation facilities, some areas of *kharif* (rainy season) maize are also cultivated with irrigation, due to its comparative advantage in productivity compared with other kharif crops. Extreme weather events due to climate change produce uneven rainfall, drought, flooding, temperature, high wind etc., which affect adversely maize productivity. Heat stress at flowering and grain-filling in spring maize cause substantial yield loss. Biotic stresses such as post-flowering stalk rot (PFSR), leaf blight, banded leaf and sheath blight (BLSB), downy mildew, ear rot and borers also affect maize productivity adversely (Singh, 2017).

Taking the average of the last 15–20 years, maize is experiencing a 5.39% growth rate. Maize, being a high-potential crop, can help immensely in meeting the additional food demand of 6 million t p.a. In states like Rajasthan, Uttar Pradesh, Madhya Pradesh, Gujarat, Jammu & Kashmir, Maharashtra, Himachal Pradesh and Jharkhand, where productivity levels are below the national average, the productivity level can be enhanced up to national average. It will add 3.2 million t to total production. Similarly, if the productivity of Karnataka, Tamil Nadu, Punjab and West Bengal, which is higher than the national average level, is enhanced equivalent to the productivity of Andhra Pradesh (4.6 t/ha), then additional production of 2.37 million t can be achieved, and thus total additional production of 5.57 million t of maize can be achieved (Hooda Committee Report, 2010). As quality seed availability of single-cross hybrids is low (around 40%), there is a need for strong PPP. Emphasis should be given to developing improved cultivars tolerant to biotic and abiotic stresses with resilience to climate change. Development of high-yielding single-cross hybrids, known as quality protein maize (QPM), offer good opportunity and the cheapest source of quality protein to address food and nutritional security. The thrust on the introduction of high-yielding single-cross maize hybrids is essentially required in eastern states of India and also in the areas where rice-wheat or rice-rice cropping systems are showing strain. Special attention is to be given to increase the area under winter maize in eastern Uttar Pradesh, Assam, Bihar and Jharkhand. Farm mechanization needs to be promoted to reduce the cost of production. Since maize has multiple uses – poultry feed, processed food, nutritional animal feed, besides its use for starch – collaboration with industries should be promoted (Hooda Committee Report, 2010). Innovative extension mechanisms are required to outscale technologies to counter low adoption of improved technologies to bridge productivity gaps (Singh, 2017).

In the next five to six years, the maize area may increase from 9.3 million ha to a maximum of 12 million ha, especially in the peninsula of India, owing to favourable ecologies in the region. The major increase in the kharif area of about 1.8 million ha and 1.2 million ha in rabi may come from Karnataka, Andhra Pradesh, Odisha and Tamil Nadu. To meet increased demand, the major focus should be to increase productivity rather than area, with an annual growth rate of 7–8%. The target to achieve an average productivity level of 5–6 t/ha is possible by enhancing the area under irrigation, crop diversification and improved agronomical practices during kharif (Singh, 2017).

Coarse cereals

Pearl millet and sorghum are important drought- and heat-tolerant crops. Pearl millet has high scope of production in the areas/states like south-west Haryana, Rajasthan and Gujarat besides a number of states in the eastern and southern parts of India. Sorghum is grown both in kharif and rabi seasons in Maharashtra for dual purpose (grain and fodder) whereas in Rajasthan and Haryana it is extensively grown for green fodder purpose with single-cut and multi-cut forage sorghum varieties. Constraints observed in coarse cereals include: (i) bird damage; (ii) inadequate availability of quality seed of hybrids; (iii) downy mildew problems, and poor adoption of recommended package of practices; and (iv) in the case of sorghum, lower profit and non-competitiveness with cotton, sunflower and pulses. The national average in pearl millet is 1.011 t/ha. If its productivity in Rajasthan, Maharashtra, Karnataka, Andhra Pradesh, Telangana and Jammu & Kashmir are brought to the level of the national average, an additional 2 million t of pearl millet production can be achieved. In the case of sorghum, the national average is 1 t/ha. An additional production of around 1.2 million t can be targeted if states like Maharashtra, Rajasthan, Tamil Nadu, Uttar Pradesh, Odisha and Haryana achieve yields equal to the national average, and Karnataka, Madhya Pradesh, Telangana and Gujarat attain

productivity equivalent to that of Andhra Pradesh (1.42 t/ha).

There is a need to develop high-yielding sorghum hybrids both for kharif and rabi; multi-cut sorghum-Sudan grass hybrids and dual-purpose, stay-green sorghum varieties; and downy mildew-resistant, drought-/heat-tolerant and short-duration hybrids in pearl millet. Application of appropriate agronomic practices; large-scale adoption of high-yielding early maturing varieties; strengthening of PPP for hybrid research and quality seed production and ensuring one supplemental/life-/crop-saving irrigation; and promotion of rabi sorghum with efficient agronomic practices need careful attention (Hooda Committee Report, 2010).

Pulses

India is the major producer and consumer of pulses. Unfortunately, productivity is still less than 1 t/ha. Attention is required to the adoption of a comprehensive and well-planned mission approach to accelerate pulse production with increase in average productivity of 0.7 t/ha to 1 t/ha and increasing area under pulses through inter- or intra-crop pulse production. There is lack of HYVs resistant to various diseases and pests and also tolerant to abiotic stresses; poor dissemination of improved technologies and adoption of a recommended package of practices; and non-availability of quality seed material. An additional production of 4.24 million t of pulses can be achieved if the productivity level of Maharashtra, Odisha, Rajasthan, Karnataka, Chhattisgarh and Tamil Nadu was enhanced equivalent to the national level, and productivity in states like Andhra Pradesh, Telangana, Uttar Pradesh, Bihar, Jharkhand, Madhya Pradesh and West Bengal was enhanced to the level of Haryana (1.032 t/ha). To achieve additional production of 4.24 million t, there is a need to promote hybrids particularly of pigeonpea for the north and western regions; improved varieties of kabuli gram in the north; popularization of improved short-duration, disease-resistant varieties of various pulses through large-scale field demonstrations to overcome existing yield gaps (25–30%); short-duration chickpea in the south; urdbean in rice fallows in the coastal region of Andhra Pradesh, Odisha and West Bengal; pigeon pea in the north-west region (Haryana, Gujarat and Rajasthan); and mungbean in the northern part (western Uttar Pradesh, Haryana and Punjab). Also, at least 50% of areas of the rice-wheat system need to be covered with new short-duration disease-resistant mungbean varieties in between (catch crop) the two crops. There was a need to develop drought- and disease-resistant varieties with special emphasis on kabuli gram and lentil. IPM must be promoted for a variety of pulse crops (Hooda Committee Report, 2010).

Pulses production has registered a remarkable increase from 14.2 million t in 2006/07 to a record level of 23.95 million t in 2017–18. The increase in the total production of pulses has been on account of improvements in the production levels of urdbean and gram. The production of pulses during 2014/15 was 17.82 million t. Productivity of pulses increased from 612 kg/ha in 2006/07 to 744 kg/ha in 2014/15. A major increase in the productivity of pulses was noticed in the states of Himachal Pradesh, Gujarat, Punjab, Madhya Pradesh, Chhattisgarh, Bihar and Jharkhand. The government gives priority to increasing the production of pulses (DoA and FW, 2017). Around 50% of the budget under the National Food Security Mission is allocated to pulses. In order to increase the production of pulses in the eastern states, pulses have been included under the BGREI scheme from 2015 to 2016 as part of demonstrations under the cropping system-based approach to targeting rice-fallow areas. Besides, new avenues are also being explored in collaboration with the ICAR, SAUs and other international organizations, i.e. ICRISAT, ICARDA, for addressing various researchable issues and demonstrating improved pulse-production technologies. Emphasis is also being placed on area expansion through promoting pulse cultivation in rice-fallow areas, intercropping of pulses with commercial crops, oilseeds, cereals etc. and productivity enhancement through demonstrations, INM, IPM and popularization and promotion of high-yielding varieties or hybrids. From 2016 to 2017, new initiatives like distribution of seed mini-kits, subsidy of production of quality seed, creation of seed hubs, strengthening the breeder seed production programme, strengthening/establishing production units of biofertilizers and biocontrol agents, and cluster frontline demonstrations through Krish Vigyan Kendras (KVKs) are being undertaken under NFSM to increase productivity and production of pulses in the country (DoA and FW 2016–17).

The major strategies and thrust areas suggested include: projection of realistic production targets, productivity enhancement, enhanced replacement rate for good-quality seeds of new high-yielding varieties, including hybrids, timely planting, proper weed control, application of sulphur and provision of one or two life-saving irrigations, using micro-irrigation methods, targeted seed production, critical inputs and their supply, sulphur application to raise productivity, use of micronutrients for balanced fertility management, area expansion approach, post-harvest technology, aggressive transfer of technology, price policy and marketing support, and policy related to operation of NFSM. Further, farmer-to-farmer seed exchange, educating the farmers about production of quality seed, provision of appropriate facilities like seed storage, handling charges and crop insurance etc. are required, as well as area expansion of short-duration varieties of chickpea in non-traditional areas of Andhra Pradesh, Bihar and Karnataka; inter-mixed cropping of blackgram, greengram, pigeonpea and chickpea in central and peninsular regions; introduction of short-duration pigeonpea with groundnut in Gujarat, Maharashtra, Andhra Pradesh, Telangana and Tamil Nadu; and lentils and peas in large areas in rice fallows in Uttar Pradesh, Bihar, West Bengal and Odisha. This strategy, adopted under NSFM during the last two years, with large-scale on-farm demonstrations, has yielded rich dividends. In the last two years, an increase of 4.5 million t of pulses was achieved due to a well-planned strategy, effective implementation and better monitoring. In fact, India has become self-sufficient in pulses for the first time, meeting its annual shortage of around 4 million t successfully (Paroda Committee Report, 2000).

Conclusion

Since independence, achieving self-sufficiency in foodgrains is one of the most significant achievements in Indian history. Despite a four-fold increase in population, from 33 million to 134 million, India has produced five times more foodgrains, from almost 50 million t to the present 277.49 million t. Considering demographic projections, India would need to produce around 350 million t by 2030, i.e. around 80 million t more – almost 6–7 million t p.a. additional foodgrains. This is indeed a major challenge but certainly achievable provided concerted efforts are made to expand both horizontally (area expansion) and vertically (productivity enhancement). Scope for both exists but would require a well-planned strategy, crop-wise, ecoregion-wise and demand-driven-wise, supported by enabling policies, increased investments in R&D and infrastructure, and proper access to both knowledge and technological inputs. Once these are in place, achieving the additional 80–100 million t of foodgrains, essential for national food and nutritional security, will be possible. Such a strategy would achieve the SDGs by 2030 – no poverty and zero hunger.

References

Agarwal, P.K., Hebbar, K.B., Venugopalan, M.V., Rani, S., Bala, A., Biswal, A. and Wani, S.P. (2008) Quantification of yield gaps in rain-fed rice, wheat, cotton and mustard in India. *Global Theme on Agroecosystems Report No. 43*. ICRISAT, Patancheru, Andhra Pradesh, India.

Chand, R., Acharya, S.S., Chengappa, P.G. and Deshpande, R.S. (2009) *Policies and Institutions, State of Indian Agriculture*. National Academy of Agricultural Sciences, New Delhi.

Chaudhary, S. (2015) *Three Ways to Boost Indian Agriculture*. World Economic Forum. Available at: https://www.weforum.org/agenda/2015/11/3-ways-to-boost-indian-agriculture/ (accessed 23 August 2018).

DoA and FW (2016–17) *Annual Report 2016–17*. Department of Agriculture, Cooperation and Farmers' Welfare, Ministry of Agriculture and Farmers' Welfare, Government of India, New Delhi.

DoA and FW (2017) *Commodity Profile for Pulses*. Department of Agriculture, Cooperation and Farmers' Welfare, Ministry of Agriculture and Farmers' Welfare, Government of India, New Delhi.

Firdos Ahmad, M.D. and Shaukat, H. (2012) The performance of India's food grains production: a pre- and post-reform assessment. *International Journal of Scientific and Research* 2(2), 1–15.

Hooda Committee Report (2010) Report of Working Group on Agriculture Production under the chairmanship of Dhupinder Singh Hooda to recommend strategies and an action plan for increasing agricultural production and productivity, including long-term policies to ensure sustained agricultural growth. Department of Agriculture and Cooperation, Ministry of Agriculture, Krishi Bhawan, New Delhi.

Kumar, P., Joshi, P. and Birthal, P.S. (2009) Demand projections for foodgrains in India. *Agricultural Economics Research Review* 22, 237–243.

MoA and FW (2009) *Agricultural Statistics at a Glance, 2009*. Directorate of Economics & Statistics, Department of Agriculture, Cooperation and Farmers' Welfare, Ministry of Agriculture and Farmers' Welfare, Government of India, New Delhi.

NITI Aayog (2015) *Raising Agricultural Productivity and Making Farming Remunerative for Farmers*. NITI Aayog, Government of India, New Delhi.

Paroda Committee Report (2000) *Expert Committee Report on Pulses*. Technology Mission on Oilseeds and Pulses, Department of Agriculture and Cooperation, Ministry of Agriculture, New Delhi.

Singh, N.N. (2017) Retrospect and Prospect of Doubling Maize Production and Farmers' Income. Strategy paper. Trust for Advancement of Agricultural Sciences, Pusa Campus, New Delhi.

Singh, R.K.P. and Kumar, P.R. (1998) Growth and instability in production of principal foodgrain crops: a case of backward economy. *Bangladesh Journal of Agricultural Economics* 21(1, 2), 1–20.

TAAS (2017) *Scaling Conservation Agriculture for Sustainable Intensification in South Asia – A Regional Policy Dialogue: Proceedings and Recommendations*. Trust for Advancement of Agricultural Sciences, IARI Campus, New Delhi.

World Bank (2008) *World Development Report: Agriculture for Development*. The World Bank, Washington, DC.

11

Horticulture for Food and Nutrition Security

Preamble

India supports more than 17% of the global population with only 2.4% land cover. The agricultural sector is an important contributor to the Indian economy (17.6% of GDP), besides providing nearly 54% of the country's employment. Despite several challenges, namely tumultuous weather, seasonal cyclones, occasional drought, demographic pressure, industrialization, urbanization, unprecedented use of insecticides and pesticides, and compulsion for the migration of people from rural to urban areas, especially for employment, the country witnessed record food grain production of 277.49 million t during 2017–18.

Food and nutritional security are the key SDGs. There has been appreciable progress on the food front including horticulture. Foodgrain production increased five-fold, horticulture nine-fold, milk six-fold and fish nine-fold in 2015–16, compared with production in 1950–51. However, economic access to nutritious food continues to be a cause of concern. Currently, more than 350 million people continue to suffer from malnutrition, which is a cause of various types of diseases and premature deaths of children and women. Therefore, the country can only be food-secure if the citizens have economic access to nutritious food to meet their physical needs. In this context, horticultural crops (fruits, vegetables, potatoes, tuber crops, mushrooms, plantation crops, spices etc.)

have emerged as the best options, not only to provide required nutrients but also to enhance access to food through enhanced farm profitability. The current trend shows that dietary habits are changing with increasing income, from cereal-based diets to cereal plus vegetables/fruit-based diets. Resultantly, there is a growing demand for fruit and vegetables. It has been recognized that the horticulture sector is the best option for agricultural diversification to ensure food, nutrition and healthcare. Horticulture provides a wider choice for farmers and also complements the food sector, i.e. with potato, tuber crops, banana and vegetables. A new paradigm shift in farming in the recent past has been towards horticulture-based farming systems to ensure greening, environmental services and to provide nutritious food while enhancing farm profitability (GoI, 2011).

Horticultural Development

At present, horticulture is considered to be a sunrise sector of the Indian economy. It is growing rapidly and offers good options for agricultural diversification. Horticulture is not merely a means of diversification but has become an integral part of food, nutritional security and poverty alleviation. This sector alone provides livelihood for 30–40% of India's population. As per the estimates of the Ministry of Agriculture and Farmers' Welfare released in January 2018, the

production of horticulture crops has already hit a record 305.4 million t in 2017–18, about 1.6% higher than the previous year's production (299.85 million t) and 8% higher than the previous five years' average (http://www.livemint.com). Even in 2016–17, with impressive production of 299.85 million t, horticulture contributed around 30% of agricultural GDP. Over the past few years, horticulture has made remarkable progress in terms of expansion of both area and production under different crops, increase in productivity, crop diversification, technological interventions for production, and post-harvest and forward linkages through value addition and marketing.

Among the eight different groups of the horticulture sector, fruits and vegetables form the single largest sub-sector accounting for 78.4% of the area and over 92% of the total production. Horticultural production has shown a quantum jump in the recent past. Estimated area and production during 2016–17 showed an increase of 17.4% and 32.3%, respectively, over 2011–12. This is suggestive of both area- and productivity-led growth. During the last decade, the horticulture sector as a whole registered a compound growth of over 6%. The increase had been more significant in fruits and vegetable crops. The present status of Indian horticulture is given in Box 11.1.

Horticulture for Nutrition and Health

Horticulture has emerged as an important sector in agriculture with coverage of nearly 24.9 million ha, having an annual production of 305.4 million t, surpassing even foodgrain production. The year 2017–18 marked the sixth straight

---

Box 11.1. Present status of Indian horticulture.

- Production of horticultural crops like vegetables and fruits has touched a record 305.4 million t in 2017–18, about 1.6% higher than the previous year and 8% higher than the previous five years' average, the agriculture ministry said in its report in January 2018. It was 299.85 million t in 2016–17. It has surpassed foodgrain production (277.49 million t) from a much smaller area (25.11 million ha). The record production during 2017–18 will mark the sixth straight year of horticulture production outstripping that of foodgrains, suggesting a structural change in Indian agriculture where farmers are increasingly growing perishable commercial crops due to a growing market and a quicker cashflow, as these crops require less time from sowing to market.
- Production increased from 167 million t in 2004–05 to 305.4 million t in 2017–18, showing an increase of 81%.
- Productivity of horticultural crops is much higher compared to productivity of foodgrains (11.94 t/ha against 2 t/ha). Productivity of horticultural crops increased by 31.9% between 2004–05 and 2016–17.
- Within horticulture, production of vegetables had been 181 million t in 2017–18, about 1% higher than the year before, while that of fruits was almost 95 million t, 2% higher than the previous year.
- Data showed that during the year, the area under different perishable crops stood at 24.9 million ha, about 0.3% higher than the year before. Between 2015–16 and 2017–18, productivity of horticultural crops rose from 11.7 t/ha to an estimated 12.3 t/ha.
- Disaggregated data on estimates of production of specific crops showed onions at 21.4 million t, about 4.5% lower than the year before, and potatoes at 49.3 million t, marginally higher than the 48.6 million t in 2016–17. Production of tomatoes is estimated to rise 7.7% in 2017–18 to 22.3 million t.
- Area increased from 18.7 million ha in 2005–06 to 25.11 million ha (about 16% of arable land) in 2016–17.
- Horticulture contributes 38% to the Gross Net Value (GNV) of agriculture.
- There has been a 45% increase in per capita fruit and vegetable availability.
- Exports increased by 315% over ten years.
- Price volatility continues to be a major risk in horticulture, with the price of onions, tomatoes and potatoes plunging below growing costs several times during 2015–16. While farmers in Madhya Pradesh were forced to sell onions for Rs 2/kg in June 2015–16, in northern India farmers from Punjab, Haryana and Uttar Pradesh were forced to dump their potato crop for want of buyers around the same time.

year that India's horticulture production outstripped foodgrain output, indicating a structural change in Indian agriculture. The fact that horticultural crops are now grown in over 10% of India's gross cropped area is indeed a success story. It also signals the success of small and marginal farmers in growing more fruits and vegetables, driven by higher demand. Changes in area and production clearly indicate that production gain is due both to area and productivity increase. Fig. 11.1 depicts the trend of horticultural crops in India, whereas Fig. 11.2 indicates annual production during the 12th Plan period and future projections (million t). Tables 11.1 and 11.2 indicate actual and projected production of horticultural crops for the 12th Plan period and the value of horticultural crops, respectively, while Table 11.3 provides wider options in terms of calories. Fruits and vegetables are also a rich source of vitamins, minerals, proteins and carbohydrates, which are essential in human nutrition. Hence, these are referred to as protective foods and assume great importance for the nutritional security of people. Thus cultivation of horticultural crops plays a vital role in national prosperity and is directly linked to health and well-being. The emphasis on horticulture is a recognition of the need for attaining nutritional security and for sustainable income. Healthier diets will improve the learning capacity of children

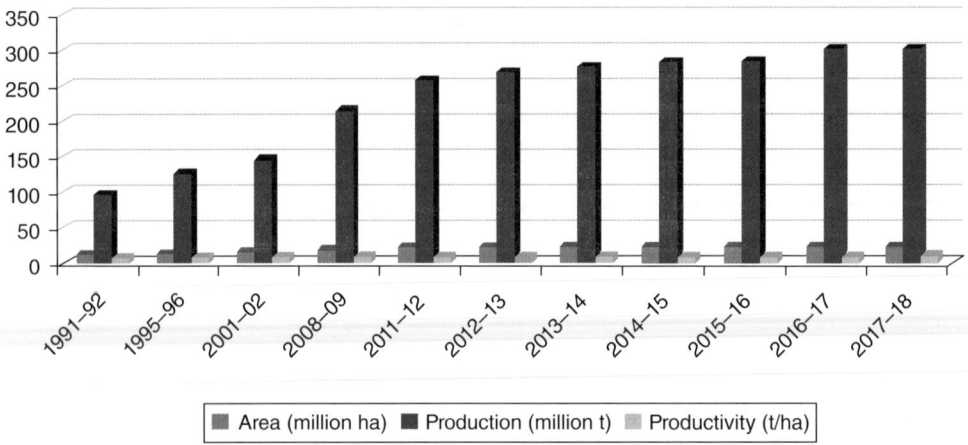

Fig. 11.1. Area, production and productivity of horticultural crops in India. (From: Ministry of Agriculture and Farmers' Welfare, Government of India)

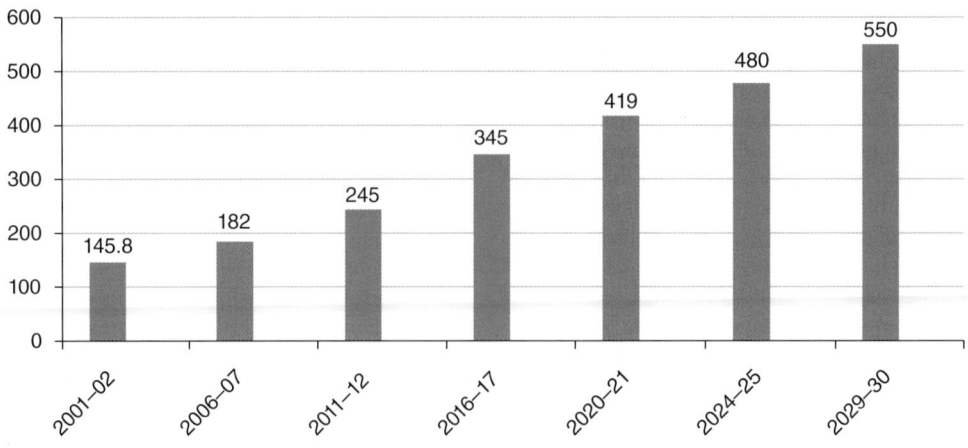

Fig. 11.2. Annual production during the 12th Plan period and future projections (million t). (From GoI, 2011)

Table 11.1. Actual and projected production of horticultural crops for the 12th Plan period.

Crop group	Base period production (2009–10)	Target production at the end of the 12th Plan (2016–17)	Compound annual growth rate (%)
	Production (million t)		
Fruit and nuts	71.40	104.00	6.5
Vegetables	123.80	199.19	8.7
Spices	4.01	5.14	4.0
Coconut	10.81	15.35	6.0
Plantation crops	1.54	2.18	6.0
Tuber crops	12.0	15.36	3.0
Flowers (cut and loose)	1.46	2.99	15.0
Miscellaneous crops	0.60	1.02	10.0
Total horticulture	**225.62**	**345.23**	**6.7**

Coconut conversion = 1450 nuts/t; cut flowers conversion = 15,000 = 1 t; estimated production = 295.35 million t (2016–17)

Table 11.2. Value of horticultural crops (at constant prices).

Crop group	Value of 2009–10 production (Rs million)	Value of 2016–17 production (Rs million)	Growth rate (%)*
Fruits	335,580	546,995.4	9.0
Vegetables	348,740	556,240.3	8.5
Spices	47,117	68,555.9	6.5
Coconut	30,484	47,128.5	7.8
Plantation crops	7,238	11,189.9	7.8
Tuber crops	451,200	719,664.0	8.5
Flowers	343,100	686,543.1	14.3
Miscellaneous	98,700	133,245.0	5.0
Total horticulture	**1,662,159**	**2,769,562.0**	**9.5**

*2004–05 = base period (GoI, 2011)

Table 11.3. Number of calories available in fruit and vegetables (per 100 g). (FAO, 2013)

Fruits	Calories (per 100 g)	Vegetables	Calories (per 100 g)
Apple/plum	56	Broccoli	25
Avocado pear	190	Brinjal	24
Banana	95	Cabbage	45
Chickoo	94	Carrot	48
Cherry	70	Cauliflower	30
Date	281	Fenugreek (*methi*)	49
Grape (black)	45	French beans	26
Guava	66	Lettuce	21
Pomegranate	77	Mushroom	18
Lychee	61	Onion	50
Mango	70	Pea	93
Orange	53	Potato	97
Strawberry	77	Spinach	26
Peach/pear	50	Tomato	21
Pineapple	46	Watermelon	26

and the working capacity of adults, leading to higher incomes and a reduction in poverty (http://www.agricoop.nic.in; MoA and FW, 1991–92 to 2016–17; GoI, 2011).

Horticultural crops provide ample opportunities for healthcare. According to the Food and Nutrition Board of the National Research Council, men and women between 23 and 50 require about 2800 and 2200 calories/day, respectively, to maintain good health. Pregnant women and lactating mothers will need an additional 300–500 calories/day (Peter, 2015). Thus, fruits and vegetables provide better options for meeting the energy requirements of humans (Table 11.3). It is pertinent to mention that fruits (aonla, bael, jamun, papaya), vegetables (carrot, cauliflower, onion, garlic, leafy vegetables), spices (ginger, turmeric, black pepper, fenugreek, ajowain) and ornamental plants (Ashoka, Ficus, Catharanthus) protect us against various kinds of diseases. Spices like turmeric, chilli and cumin have been recognized to protect against cancer. Noni (*Morinda citrifolia*) is recognized as the best for healthcare as it provides protection against various diseases including HIV. Virgin coconut oil protects from HIV and coconut water provides all the nutrients a child needs.

India is the second-largest producer of fruits after China, with an estimated production of 92.84 million t of fruits from an area of 6.48 million ha (as per NHB Review Committee meeting, 16 May 2017). A large variety of fruits are grown in India, of which mango, banana, citrus, guava, grape, pineapple and apple are the major ones. Apart from these, fruits like papaya, sapota, annona, phalsa, jackfruit, ber, pomegranate, in tropical and sub-tropical groups, and peach, pear, almond, walnut, apricot and strawberry, in the temperate group, are grown in a sizeable area. To some extent, they provide energy-rich food. Banana, jackfruit, annona, sapota and fig contain carbohydrates in the range 19–24% and are good sources of energy compared with potato, colocasia, baby corn, yam and green pea (15.9–24.6% carbohydrates and 79–125 kcal energy). Closely following this group of fruits as good sources of energy are mango, lychee, grape, ber, pomegranate, phalsa and jamun. Fresh avocado is the only high-energy fruit yielding 161–215 kcal per 100 g due to its high fat content (15–26%). But fruits and vegetables are indispensable as a source of vitamins

and minerals, which help in building resistance against diseases. Fruits and vegetables provide 90% of the required vitamin C and 60% of vitamin A. Mango and papaya are rich in pro-vitamin A and guava in vitamin C. Banana is a good source of carbohydrate (GoI, 2011).

Fruits yield larger quantities of food/ha compared to cereals. For example, the maximum paddy yield is 3 t/ha, whereas it is 22 t/ha in the case of banana, 45 t/ha in the case of pineapple and 40 t/ha in the case of grape. Much less area is required to obtain the calorific requirement per adult per year (1,100,000 kcal) from growing bananas (0.03 ha) or mangoes (0.16 ha) than from growing wheat (0.44 ha). Fruits are a rich source of organic acids that stimulate appetite and help digestion. Many fruits and vegetables possess laxative properties as they possess dietary fibre and pectin, stimulating intestinal activity. Due to poverty, micronutrient deficiency (vitamin A and iron deficiency anaemia) is posing a threat to large masses in Asia and the Pacific regions, which could be minimized through a horticulture intervention and awareness drive.

Towards Achieving Nutritional Security

Vegetables and fruits play a prominent role in prevention of several chronic diseases such as heart disease, cancer, cataracts, osteoporosis, diabetes etc. In order to have a protective effect, it is necessary to consume 400–600 g of fruits and vegetables every day (Peter, 2015). But the consumption level of fruits is low and widely variable from region to region in India. An increase in the intake of fruits along with vegetables will meet the required daily allowance of many nutrients. Although India is the second-largest producer of fruits and vegetables in the world, next to China, per capita consumption is only around 46 g and 130 g, respectively, against a minimum requirement of about 92 g and 300 g as recommended by the Indian Council of Medical Research and the National Institute of Nutrition, Hyderabad. With the present population level, the annual requirement of horticulture produce will be 419 million t by 2020/21 as against the present level of production of 299.85 million t (2016/17).

Challenges Ahead

The growing population in India is a big challenge for meeting food needs worldwide. According to predictions from the Food and Agriculture Organization of the United Nations (FAO), agricultural productivity in the world will sustain the growing population in 2030 but millions of people in developing countries will starve or remain hungry. By 2025, 83% of the expected global population of 8.5 billion will be in the developing world (FAO, 2011). The question before us is: Can we meet food needs and provide nutrition, healthcare, fuel and fibre to a growing population? The answer is: It is difficult, but not impossible. Past experiences build confidence. In the post-independence period, India made steady progress in agriculture when extra land and water was made available, and a few genes performed wonders in ushering in the Green Revolution. But the challenges before us now are much greater than before. In the prevailing circumstances of shrinking farming land, depleting water resources and changing climate, the situation has become complex. Optimistically, through the input of science and technology, challenges ahead could be converted into opportunities for sustainable production. Horticulture has proved to be the best means of diversification for higher land productivity, achieving a gross return/ha, but there is a need to make sustainable efforts in enhancing the production of fruits, vegetables, tubers and plantation crops to meet the growing demands of an ever-increasing population with nutritionally rich horticulture produce. Currently, climate change is posing a threat to horticultural crops due to erratic rainfall, greater demand for water and enhanced biotic and abiotic stresses. The challenges could be addressed through identification of genes tolerant to high temperature, flooding and drought; development of nutrient-efficient cultivars; and a production system for efficient use of nutrients and water. This would need highly prioritized research to address the impact of climate change. Concerted and integrated efforts with effectiveness and efficiency will be essential to meet the ever-increasing demand.

Technological Advancements

At the present time, several technological innovations have been advanced in the complete value chain involving technology for orchard establishment, availability of true-to-type planting material, plant architecture engineering and management, mulching, fruit thinning, INM, water management, IPM and disease management, post-harvest technology, processing and marketing. The positive changes in the horticulture sector have occurred because it has received importance from all stakeholders, the public sector, private sector and farmers during the last two to three decades. This is primarily the result of the realization that diversification to horticultural crops is now the major option to improve food and livelihood security. Under NARS, the R&D on horticulture has been strengthened using multicrop and multidisciplinary approaches of (i) genetic improvement; (ii) efficient crop management; and (iii) post-harvest management.

Genetic Improvement

In an endeavour to attain food and nutritional security, germplasm enhancement and its utilization is extremely important in providing strong backing for breeding programmes. Concerted efforts are being made by NARS for documentation, characterization, conservation and utilization of plant genetic resources in horticultural crops, which enabled the conservation of 72,000 accessions of cultivated, wild and related taxa. Now it has become necessary to identify accessions possessing high nutritional value and bioactive compounds that play a great role in nutritional security, and this shall be helpful in breeding varieties with special attributes. Efforts have been made to develop HYV and hybrids of different horticultural crops for different regions. More than 1800 improved high-yielding, high-quality varieties, coupled with disease- and pest-resistant varieties and hybrids, have been released by various institutes/universities for cultivation in diverse agroclimatic conditions of the country. Regular-bearing mango hybrids, export-quality grapes, multiple disease-resistant vegetable hybrids, high-value spices and tuber crops for industrial use have been developed. Improved varieties have revolutionized the horticultural sector. High-yielding Gauri Sankar and Sree Bhadra sweet potatoes have focused on minimizing malnutrition and improving nutritional security. Similarly, breeding to develop grape cultivars suitable for vine making, black pepper cultivars rich in aroma compound caryophyllene, and development of processing tomatoes etc. are some of the research programmes being carried out in various horticultural institutes. Varieties are being bred for processing qualities: Kufri Chipsona in potatoes for chips and the heat-tolerant variety Kufri Surya; high total soluble sugar (TSS) white onions, W448, in the National Research Centre for Onion and Garlic (NRCOG); and papaya varieties for table and papain production are some of the successful research attempts being carried out at various ICAR institutes.

Hybrid technology has revolutionized the production of vegetable crops and demand for hybrid seeds is continually increasing. Hybrids of tomato, chilli, cucumber and muskmelon are being produced at several locations in different states in the country. The All India Coordinated Vegetable Improvement Project (AICVIP) has so far recommended cultivation of over 45 hybrids. Besides, many hybrids of vegetable crops, developed and marketed by the private sector are also available to the farmers. At present, the area under high-yielding F_1 hybrids in important vegetable crops ranges from 17.8% to 31.5%, particularly in tomato (31.5%), cabbage (31.39%) and brinjal (17.8%), and areas under capsicum and chilli are also under expansion. High production, earliness, superior quality, uniform produce and resistance to biotic and abiotic stresses are the main advantages of F_1 hybrids. Keeping in view the dynamic needs, the research efforts in various institutes have focused on development of hybrids with multiple disease resistance, early maturity and utilizing the male sterility system. Cytoplasmic male sterile (CMS) lines have successfully been utilized to produce potential experimental crosses of onion and commercial hybrids of chilli. The parental lines of a number of hybrids developed have been sold on a non-exclusive basis to the seed companies with the aim to promote these hybrids among farmers.

Biotechnological tools have provided ample scope for breeders to improve diverse traits, including yield, disease resistance, abiotic stress tolerance and quality, more precisely and in reduced time. Use of meristem culture and micro-grafting is successful in citrus for elimination of viruses. Anthers of the capsicum variety Arka Gaurav and tomato hybrid Avinash 2 responded to culture with an embryogenic-like response without an intervening callus phase. Androgenesis has been successfully used for brinjal, pepper, cabbage, cauliflower, potato, asparagus and carrot, whereas gynogenesis has been successful in onion. Embryo rescue has been successfully employed in the production of hybrids of *Musa acuminata* × *Musa bulbisiana*, *Carica papaya* × *Carica cauliflora* and interspecific crosses in pineapple and seedless × seedless grape varieties. Use of molecular markers for crop profiling, fingerprinting, molecular taxonomy, identification of duplicates, hybrids, estimation of genetic fidelity and tagging of genes for marker-aided selections are gaining importance. Efforts are under way to fingerprint mango, banana, cashew nut, kiwifruit, walnut, grape, citrus etc. by different research centres. DNA sequence has been isolated for root-knot nematode resistance (Mi) gene in tomato and is being used to facilitate breeding this valued trait into new varieties and even other species. QTL mapping is in progress in many crops such as brinjal, tomato and capsicum, while association mapping (linkage disequilibrium) is used in the case of perennials such as black pepper, cardamom and coconut. Gene pyramiding for useful genes in one background variety of commerce is the mainstay of biotechnological research and is in progress in solanaceous vegetables. To tackle issues of managing disease resistance, resistance to insect pests, nutritional quality improvement and to extend shelf life of fruits and vegetables, efforts are being made to develop transgenics. A large number of transgenics with the *Cry-1* AB gene have been produced with resistance to the most damaging insects, usually lepidopterans. Nutritionally improved transgenic potatoes have also been obtained by transferring the amaranth seed albumin gene (AmA1) from *Amaranthus hypochondriacus* into potato. RNAi technology has succeeded in developing potato which does not sweeten at lower temperatures, and the RB gene transferred in two potato cultivars has given appreciable protection against late blight disease (GoI, 2011).

Efficient Crop Management

Future commercial fruit growing will depend on successful use of rootstocks for better scion compatibility, canopy architecture, fruit quality, nutrient absorption, water-use efficiency, biotic and abiotic stress tolerance, and adaptation under the influence of climate change. Suitable rootstocks and scions become essential to achieve targeted production. Citrus rootstock, Rangpur lime, can adapt to water stress and calcareous soils, and can resist *Phytophthora*. The popular rootstocks for grapes are Dogridge B and 110 R, which can sustain abiotic stresses like drought and soil salinity and provide vigour in the vine. In sapota, khirni (*Maninkara hexandra*) has proved drought-tolerant and productive in marginal soil. There have been technological changes in seed production, techniques for production of hybrid seeds using cytoplasmic male sterile lines (CMS), technologies for vegetative methods of propagation and *in vitro* propagation technologies, a success story in banana, potato, citrus and many other crops. Disease-free planting materials are essential for resource conservation, wherein it eliminates the infected plant material and reduces the cost of crop production. Polymerase Chain Reaction (PCR)-based diagnostic protocol has been developed for rapid detection of viruses and phytophthora in citrus, banana, potato, coconut and tuber crops. High-density planting technology has been standardized for many crops and also adopted for growing banana, pineapple, citrus, papaya, mango, cashew and a few other fruit crops. Technologies for high-density planting, canopy management and rejuvenation of old and senile orchards have been developed and successfully demonstrated for many fruit crops. Also meadow orcharding in guava is being adopted for higher productivity.

Among various inputs, fertilizers alone account for 20–30% of the total cost of production. Soil nutrient-based fertilizer application is useful in vegetable crops, but fruit trees rarely respond to nutrient needs based on soil tests; thus leaf nutrient standards have been developed for many fruit crops to enhance the efficiency of fertilizer, but the focus is now required on the use of biofertilizers, Vesicular-Arbuscular Mycorrhiza (VAM) fungi and other beneficial microbial agents for effective nutrient use efficiency. Good water management using well-designed systems is critical for sustaining production and quality of produce, more specifically for horticultural crops. Among others, drip irrigation has proved successful in exhibiting high water productivity by saving irrigation water by 60% in various orchard crops and vegetables, with 10–60% increases in yield compared to conventional methods of irrigation. Fertigation has become the 'state of

the art' in orchard crops and vegetables because nutrients can be applied to plants in the correct dosages and at the appropriate time for the specific stage of plant growth. Due to changing dietary habits coupled with health-consciousness, demand for organic food is on the increase these days (GoI, 2011). Protocol for organic production using resistant varieties, management of soil vermicompost and biofertilizer/biopesticides for managing diseases and pests have been developed. Farming system and cropping system approaches have been successfully demonstrated in perennial horticulture. Various farming system models have been developed and suitable crops in the early years of tree plantation to maximize output in different agroclimatic conditions have been selected. The elephant foot yam is widely grown as an intercrop in lychee, coconut and banana orchards. Spices like black pepper, ginger, turmeric, vanilla, nutmeg, clove and some medicinal plants are the ideal intercrops for coconut.

Most of the horticultural operations in India are done manually or with animal power. Wherever farming operations are mechanized, crop productivity is high. Several machines and tools have been developed to enhance the efficiency of the farm operation. In fruit crops, tractor-operated pit-hole diggers and bucket excavators have been developed and need adoption. Hi-tech horticulture has become the order of the day; it encompasses a variety of interventions such as micro-irrigation, fertigation, protected/greenhouse cultivation, soil and leaf nutrient-based fertilizer management, mulching for *in situ* moisture conservation, micro-propagation, biotechnology for germplasm, genetically modified crops, use of biofertilizers, vermiculture, high-density planting, hi-tech mechanization, soil-less culture and biological control (Singh *et al.*, 2015). Precision farming calls for efficient management of resources through location-specific hi-tech interventions. Activities like greenhouse construction, mulching, shade net and plastic tunnels are also being promoted. The crops where some of the components of precision farming have been practised are banana, grape, pomegranate, capsicum, tomato, chilli, cashew and selected flowers. The chemical control measures for various pests and diseases, such as fruit fly, stem and fruit borer, leaf gall midge, aphids, mites and moths, and diseases like scab, powdery

mildew, leaf spot, brown spot, gummosis and canker have been worked out, but there is a need for ecofriendly practices. Plant health management in horticultural crops involves not only pre-harvest but also post-harvest health management strategies such as production of pest- and disease-free planting materials, use of bio-inoculants and other growth-enhancing soil amendments, indexing for major pathogens and certification of planting materials, seed plot technique and mother garden technique, and other measures. Several biocontrol agents have been identified for various fruit crops but new biocontrol agents from the native zone are required to be identified. Disease forecasting models are useful in determining the role of climatic factors in disease appearance and progression and in devising suitable management strategies (Singh, 2009).

Post-harvest Technology

Production and consumption of fruits and vegetables can help in achieving nutritional security. However, consumption of fruits and vegetables is low in most of the developing world due to lack of buying power. In developing countries, an estimated 25–30% of produce is lost on account of post-harvest events. Increased investment in post-harvest research is therefore justified. Post-harvest losses invariably occur due to improper ripeness, poor initial quality, mechanical damage, inadequate sanitation, inadequate drying, decay, improper temperature and delays between harvesting and market. Some key interventions to reduce losses include selecting varieties with good shelf life, harvesting at proper maturity, avoiding sun exposure to reduce water loss and temperature gain after harvest, cooling (or drying) quickly to the lowest safe temperature, protecting from physical damage, maintaining the cold chain (or dry chain), and expediting marketing whenever possible. For example, using storage containers like reusable plastic containers and plastic bags can improve the quality of produce over time by protecting it from damage and also by serving as a moisture barrier to reduce water loss. Many post-harvest fruits and flowers are regulated by ethylene, a major cause of post-harvest loss, and modulation by ethylene

synthesis can improve the life of horticultural crops (GoI, 2011).

In order to make horticulture a viable enterprise, value addition is essential. Harvest indices, grading, packaging and storage techniques have to be standardized for major horticultural crops. Value addition through dehydration of fruits and vegetables, including freeze-drying, dried and processed fruits, vegetables, spices and fermented products has also been developed. Potato chips, spice flakes, fingers and French fries are becoming popular as fast foods. New products like juice punches, banana chips and fingers, mango nectar, fruit wines, dehydrated products made from grape, value-added coconut products like snowball tender coconut, coconut milk powder and pouched tender coconut water (Cocojal) etc. are also becoming popular these days. Improved blending/packaging of tea and coffee has opened new markets. New products such as tetra-pack-filled fruit juices are now household items. As food consumption patterns are changing towards more convenient foods, the demand for products like pre-packed salads, packed mushrooms and baby corn, and frozen vegetables is increasing and these items are sold in shopping malls. Consumer-friendly products like frozen peas, ready to use salad mixes, vegetable sprouts and ready-to-cook fresh-cut vegetables are major retail items, which have already started appearing in retailers' windows. In order to reduce dependence on refrigerated storage, low-cost, eco-friendly cool chambers for on-farm storage of fruits and vegetables have lately been developed. Standardization of modified atmosphere packaging and storage systems with greater emphasis on safety (pesticide-free), nutrition and quality are also becoming important.

Supply Chain Management

In India, most exporters still rely on the traditional wholesale market to procure fruits and vegetables. The marketing channels for fruits and vegetables vary considerably by commodity and state, but they are generally very long and fragmented. The majority of domestic fruit and vegetable production is transacted through wholesale markets, depending on the state and the commodity. Farmers may sell to traders directly

at the farm gate or at village markets, or directly to processors, cooperatives and others. In the majority of states, the Agricultural Produce Marketing Committee (APMC) regulations have also prevented the private sector from investing in wholesale markets and marketing infrastructure. As a result, most markets have rudimentary infrastructure, particularly for storing and handling perishable products. Recently, there has been an emergence of more coordinated supply chains for fruits and vegetables in India catering to the export market and to the high-end domestic market. The coordinated supply chain involves structured relationships among producers, traders, processors and buyers, whereby detailed specifications are provided as to what and how much to produce, time of delivery, quality and safety conditions, and price. The coordinated supply chains fit well with the logistical requirements of modern food markets, especially for fresh and processed perishable foods. These chains can be used for process control of safety and quality and are more effective and efficient than control only at the end of the supply chain. Recently, a terminal market for fruits and vegetables has been set up in Bangalore known as SAFAL. The market receives sorted, graded and packaged produce from these associations and centres, which is then auctioned at the market. SAFAL has also forward linkages to a number of retail outlets. The market has modern infrastructure including temperature-controlled storage facilities and ripening chambers. Normally, small and marginal fruit and vegetable growers are unable to get an acceptable price for their produce due to small marketable surplus and the highly perishable nature of the produce. Support for such farmers is the need of the hour. For this they should form fruit and vegetable cooperative societies/self-help groups (Singh, 2009).

With a view to establishing a complete supply chain, from farm to market, the infrastructure facilities will have to be created at the following levels: (i) small pre-cooling units and/or evaporative cooled chambers in the production areas where the field heat of the produce is to be removed at a fast rate to bring down the temperature of the produce to the desired level before putting the product into cold storages. The refrigerated transport units, from farm to cold storage, are also utilized as mobile pre-cooling units for this purpose; (ii) collection centres near to the farms; (iii) medium or small cold storages having multi-product, multi-chamber facilities, which are the most popular segment where horticulture produce is stored in transit godown; (iv) specialized cold storages with the facility of built-in, pre-cooling, high humidity and controlled/modified atmosphere are required for storage of the produce for a longer period. These specialized storages are essential for extending the shelf life of the produce, and without these facilities proper storage of the produce to meet demand in the off season is not feasible; (v) other components like ripening chambers close to the markets and display cabinets at retail outlets; and (vi) linkages for conversion of fresh produce into other marketable forms.

The Way Forward

In order to achieve household nutritional security, the role of horticulture will be important. The progress in the last 50 years has been phenomenal. For the last four continuous years, horticultural production has surpassed foodgrain production in the country, and during 2017–18, the production of horticultural crops (299.85 million t) has exceeded that of foodgrain crops (305.4 million t) considerably. For an accelerating expected growth rate of 5–6%, a national strategy is needed for both horizontal and vertical expansion of ecoregion-specific horticultural crops. Greater thrust is needed to produce and make available disease-free saplings and seeds to farmers. For this, the role of the private sector will be crucial. On one side, area expansion programmes did not have proper back-up of seed and planting materials; infrastructure created in the form of plant disease-forecasting labs, tissue analysis labs, biocontrol labs etc. are yet to be created and used fully for the benefit of production and plant protection purposes. Rejuvenation of old and senile plantations also has not picked up; and private sector investment in post-harvest management and marketing infrastructure has not materialized to the desired extent. On the other side, various schemes dealing with horticulture development need integration. Moreover, in order to achieve the desired growth rate in a sustainable manner, there is a need for prioritization in a rational manner and to set targeted

productivity for major crops according to production clusters/states. Prioritization may vary from state to state and from one production cluster to another. The crops of significance for achieving higher growth rates are known as 'focus crops'. However, crops having special significance in terms of the local economy (e.g. lychee, saffron, strawberry, passion fruit, dates etc.), important for processing (peas, white onion etc.) or for export (grapes, white onion, walnut, almond etc.) are also to be included in this category. Banana, mango, citrus, papaya, guava, apple, pineapple, sapota, grapes, pomegranate and lychee are considered as focus fruit crops at the national level. Stone fruits, too, have a significant role in export. Likewise, potato, tomato, onion, brinjal, cabbage, cauliflower, okra and green peas are priority vegetable crops. Drumstick, white onion and curry leaves are also included, keeping in view their export demand. Coconut, cashew nut and areca nut are the focus crops among plantation crops (NAAS, 2009).

For an integrated development of the horticulture industry, and also to achieve targets for feeding the population as well as meeting the requirements of the processing industry and exports, emphasis on quality production needs to be integrated with post-harvest management of the highly perishable horticultural crops. Considering the role horticulture has to play and the constraints on its development, as well as the mandate to double food production and reduce the gap between requirement and availability, the following priority areas (Singh, 2009; Singh et al., 2016) are identified to be given due consideration for better post-harvest infrastructure: (i) technical/financial support for all round development of the horticulture sector; (ii) emphasis on increasing production with an objective of achieving complete nutritional security; (iii) adoption of appropriate post-harvest management technologies for maximizing return to the farmers/growers; (iv) feasibility studies for setting up the marketing, processing plants, cold storage, transportation systems for raw and processed perishable horticultural products, and undertaking designing, planning and execution of projects on that basis; (v) promotional activities to boost the process of employment generation, to increase the income of small and marginal farmers and the involvement of women and backward communities in the horticulture development process;

(vi) encouraging the shifting of food habits from quantity to quality food through increased availability and mass-media promotion of health-oriented benefits of fruits and vegetables; and (vii) stimulating private investment, particularly in the fields of infrastructure, marketing and R&D with particular emphasis on the special needs of the processing industry and exports (Singh et al., 2016).

The horticulture sector in India is suffering due to natural, manmade, technical and economic reasons. Some of the reasons behind the crippling growth that are obstructing the growth of the horticultural sector include: (i) inadequate availability of disease-free, high-quality planting material; (ii) slow dissemination and adaptability of high-yielding cultivars/hybrids; (iii) lack of post-harvest management technology and infrastructure; (iv) weak database and poor market intelligence; (v) instability of prices, with no support price mechanism; (vi) inadequate technical human resources in the farming system; (vii) poor credit supply and high rates of interest, coupled with inadequate crop insurance schemes; (viii) poor linkage between R&D, industry and farming communities; (ix) late implementation of government policies and schemes; and (x) absence of a horticultural crop suitability map of India based on agroclimatic conditions and depicting the most suitable areas for optimum productivity of a particular crop (Singh, 2009) .

Thrusts and strategies (Singh, 2009; Singh et al., 2010) needed to address the above problems include: (i) improving production and productivity; (ii) reducing cost of production; (iii) improving quality of products for export; (iv) value addition, marketing and export; (v) price stabilization; (vi) strengthening of organizational support; (vii) availability of adequate human resources; and (viii) addressing relevant policy issues. Besides these improvement programmes, the government has to concentrate on the WTO issue with regard to marketing and trade-related affairs. For sustainable development of horticulture, the prerequisite infrastructure should be made available by the authorities concerned. The standard of horticulture produce should be maintained in line with quantity and quality approaches to capture markets and fulfil the nutritional standards. The establishment of an organizational framework for horticulture will lead to an organized and coordinated dispersal of functions.

References

FAO (2011) *The State of Food and Agriculture: Bridging the GAP*. Food and Agriculture Organization of the United Nations, Rome.

FAO (2013) *Statistical Yearbook 2013: World Food and Agriculture*. Food and Agriculture Organization of the United Nations, Rome.

GoI (2011) *Report of the Task Force Committee on Horticulture of the Planning Commission for XII Plan Period (2012–2017)*. Government of India, New Delhi.

MoA and FW (1991–92 to 2016–17) *Annual Report (1991–92 to 2016–17)*. Ministry of Agriculture, Cooperation and Farmers' Welfare, Government of India, New Delhi.

NAAS (2009) *State of Indian Agriculture*. National Academy of Agricultural Sciences, New Delhi.

Peter, K.V. (2015) *Horticulture for Nutrition Security*. Daya Publishing House, a division of Astral International Pvt. Ltd, New Delhi.

Singh, A.K., Chadha, K.L. and Patel, V.B. (2010) *Recent Initiatives in Horticulture* (e-publication). Westville Publishing House. New Delhi.

Singh, A.K., Chaurasiya, A.K. and Mitra, S. (2015) *Role of Horticulture in Agriculture Development and Food Security in India*. Available at: https://www.researchgate.net/publication/282185551 (accessed 29 May 2018).

Singh, H.P. (2009) Triggering agricultural development through horticultural crops. *Indian Journal of Agricultural Economics* 64(1), 15–39.

Singh, S.K., Behera, T.K., Prakash, J. and Prasad, K.V. (eds) (2016) Souvenir-cum-invited paper abstracts book, 7th Indian Horticulture Congress, an international meeting for doubling farmers' income through horticulture. ICAR-Indian Agricultural Research Institute, Pusa Campus, New Delhi, 15–18 November.

12

Strategies for Enhancing Oilseed Production

_____

Preamble

India is among the largest producers and consumers of vegetable oils in the world. Oilseeds have been the 'backbone' of the agricultural economy of India. The Indian vegetable oil economy is the fourth largest in the world next to the USA, China and Brazil. Oilseed crops are the second most important in the Indian agricultural economy next to foodgrains in terms of area and production. At present, more than 27 million ha of land are under oilseed cultivation. The area under oilseeds has been increasing over time and the production has registered a many-fold increase, but its productivity is still low compared to other oilseed-producing countries. Low and fluctuating productivity of oilseeds is primarily because cultivation of oilseed crops is mostly done on marginal lands, which are lacking irrigation and have low levels of inputs. To improve the situation of oilseeds in the country, the government has been pursuing several development programmes: the Oilseed Growers Cooperative Project (OGCP), the National Oilseed and Development Project (NODP), the Technology Mission on Oilseeds (TMO) and the Integrated Scheme of Oilseeds, Pulses, Oilpalm and Maize (ISOPOM). The concerted efforts of these development programmes/schemes register significant improvement in annual growth of yield and area under oilseed crops. However, India still imports a significant proportion of its requirement of edible oil. The Technology Mission on Oilseeds adopted a four-pronged strategy in order to harness the best of production processing and storage technologies for attaining self-reliance in vegetable oils. The mission initiated a corporatization and modernization process in the oilseed sector. Area expansion, technology improvement, expansion of irrigation facilities and transfer of new technologies helped a great deal in achieving significantly increased production.

The oilseed scenario in India has been the subject of review in view of its importance in the recent past (Sharma, 2014; Singh *et al.*, 2017). India occupies a prominent place in the global oilseed scenario with 12–15% of area and 6–7% of vegetable oil production. The oilseed sector has remained vibrant, globally, with 4.1% growth p.a. in the last three decades. In India, oilseeds account for nearly 3% of GDP and 5.98% of the value of all agricultural products. India has a rich diversity of annual oilseed crops on account of diverse agro-ecological conditions. Nine annual oilseeds, which include seven edible oilseeds – groundnut, rapeseed-mustard, soybean, sunflower, sesame, safflower and niger – and two non-edible crops – castor and linseed – are grown in the country. In addition, tree-borne oilseeds (TBOs) of over 125 species are cultivated/grown in the country under different agroclimatic conditions in a scattered form in forest and non-forest areas as well as in wastelands/deserts/hilly areas. Some of the important TBOs (Mukta and

Varaprasad, 2013) are mahua (*Madhuca indica*), neem (*Azadirachta indica*), simarouba (*Simarouba glauca*), karanja (*Pongamia pinnata*), olive (*Olea europaea)*, ratanjyot *(Jatropha curcas)*, jojoba (*Simmondsia chinensis*), cheura (*Diploknema butyracea*), kokum (*Garcinia indica*), wild apricot (*Prunus armeniaca*), wild walnut (*Aleuritesmo lucana*), kusum (*Schleichera oleosa*) and tung (*Vernicia fordii*).

India has the largest area under oilseeds in the world (26.11 million ha for the triennium ending 2016–17). However, it is the fourth-largest oilseed-producing country after the USA, China and Brazil, and the second-largest importer in the world after China at a cost of Rs 957.5 billion in 2016–17 (DVVOF, 2017). Furthermore, India is the third-largest consumer of edible oils at 18.13 kg/capita/annum in 2015–16 (FAO, 2016).

Production Scenario

The average productivity of all these major oilseed crops ranges from 1160 kg/ha to 1420 kg/ha, much below the world average and incomparable with the highest productivity recorded in some countries (Table 12.1).

India produces 30.3 million t of oilseeds annually from an area of 26.88 million ha, of which 64% is rainfed. The cultivation of oilseeds in the rainfed areas, in varying agroclimatic regions, with uncertain returns on investment, is the major factor for low productivity. The production scenario of the vegetable oilseed sector in the country can be categorized into four periods: (i) post-independence (1950–1966); (ii) coordinated research programme (1967–1985);

(iii) technology mission (1986–1996); and (iv) post-mission (1996–97 to present).

Post-independence Period

This period witnessed an area expansion by 5 million ha, from 10.73 million ha in 1950–51 to 15.25 million ha in 1965–66, with production of 5.16 and 6.40 million t, respectively, for the aforementioned periods. The area increased by 32% while production increased by 34% with negligible gain in productivity levels, although this period witnessed the release of 40 improved cultivars. The compound growth rate for the period from 1950–51 to 1965–66 indicated that area increased by 2.38%, whereas production increased by 2.46%. The growth rate of productivity was a meagre 0.07% (Fig. 12.1).

Coordinated Research Programme Period

Research on oilseeds got an impetus after the establishment of the All India Coordinated Research Project on Oilseeds (AICORPO) in 1967. This project strengthened the base for the development, verification and adoption of location-specific technologies for increased productivity, especially the new varieties and hybrids. It witnessed massive structural reforms in the national network in oilseeds. Area, production and productivity increased by 18.41% and 19%, respectively, for the quinquennium ending 1971–72, as against 1985–86. The annual growth rates registered

Table 12.1. Average productivity of major oilseed crops. (From: FAOSTAT, 2016)

Crop	World (kg/ha)	India (kg/ha)	Country with highest yield* (kg/ha)
Soybean	2755	1218	4322 (Turkey)
Rapeseed-mustard	2042	1179	3974 (Chile)
Groundnut	1590	1182	4408 (Nicaragua)
Sesame	577	419	2133 (Afghanistan)
Sunflower	1806	830	3101 (Serbia)
Safflower	832	505	1851 (Mexico)
Linseed	1058	426	1941 (France)
Castor	1414	1786	1786 (India)

*From among the countries having not less than 10,000 ha of cultivation

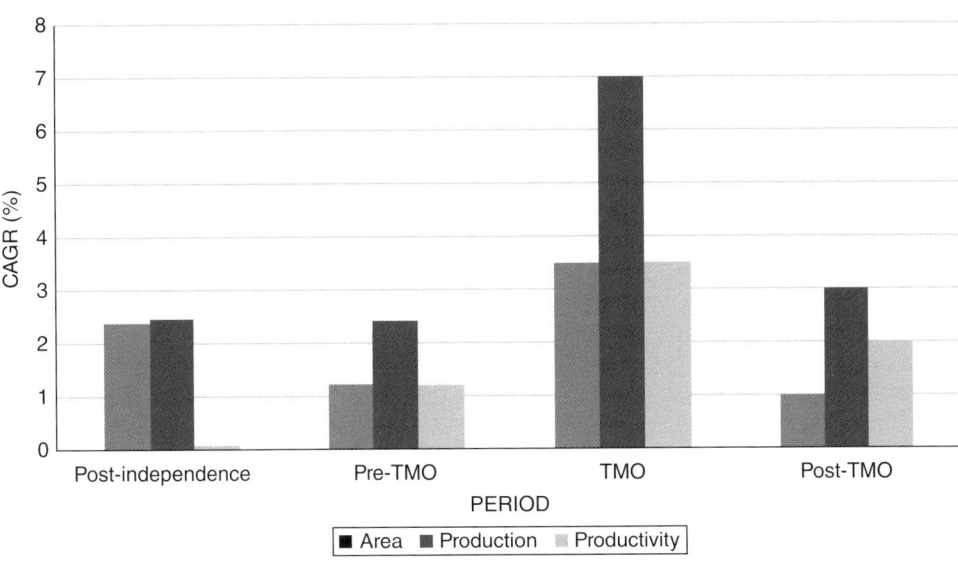

Fig. 12.1. Compound growth rates of annual oilseeds in India. (From: Sarada *et al.*, 2015)

during the period 1967–68 to 1986–87 were 1.21%, 2.41% and 1.19% for area, production and productivity, respectively (Fig. 12.1). The period also witnessed the release of 153 varieties of oilseeds.

Technology Mission Period

The Technology Mission on Oilseeds was initiated by the late prime minister Shri Rajiv Gandhi in May 1986, with very ambitious objectives: (i) self-reliance in edible oils by 1990; (ii) reduction in imports to almost zero by 1990; and (iii) raised oilseed production to 18 million t by 1989–90, 26 million t of oilseeds and 8 million t of vegetable oil by 2000. Thrust was given to the main oilseed crops in 180 selected districts in 17 states, which contributed the maximum quantity of oilseeds to the nation. The scope of the mission and strategies to be adopted to achieve the objectives were set well before the onset of the mission in February 1986, which is elaborated in the excerpts from the prime minister's speech:

> One of our biggest problems today in the agricultural sector is oilseeds. We are setting up a thrust Mission for oilseeds production. When we talk of the Mission, we mean an exercise starting from the engineering of the seeds and

> finishing with the finished products of the vegetable oil (and the byproducts like oil meal) which could be delivered to the consumer. We would like to put one person in charge of such a Mission with full funding, with no restrictions on him, whether bureaucratic or otherwise. The only limits will be certain achievements which must come within a certain time frame. This will cut across a number of ministries.

The mission started functioning as a consortium of concerned government departments – Agricultural Research and Education (DARE), Agriculture and Cooperation (DoAC), Civil Supplies (DoCS), Commerce (DoC), Science and Technology (DST), Biotechnology (DBT), Planning, Health, Irrigation and Economic Affairs. The mission adopted a four-pronged strategy through four mini-missions: (i) improvement of production and protection technologies; (ii) improvement of processing and post-harvest technology; (iii) strengthening the input support system; and (iv) improvement of post-harvest operations.

The constitution of the TMO in 1986 resulted in the country's oilseed production surpassing the target of 18 million t, fixed for the 7th Five Year Plan with an impressive annual growth rate of nearly 6% in the short term. Hence, India achieved near self-sufficiency in edible oils during the early 1990s, which was popularly referred to as the 'Yellow Revolution'. As a result of concerted efforts under TMO, a quantum jump in oilseed production from 10.83 million t (1985–86) to 24.75 million t (1998–99) was made possible through effective coordination of different ministries, departments and organizations like ICAR and the SAUs. The area under oilseed cultivation increased from 19 million ha (1985–86) to 26 million ha (1996–97) and production increased from 10.83 million to 24.38 million t during just one decade, registering an increase of 36% in area and 125% in production. Similarly, productivity of all the annual oilseed crops, on average, increased from 570 to 926 kg/ha, an increase of 62% during this period. This golden era witnessed the release of 200 varieties and hybrids, and performance of improved crop technologies. As a result, India achieved the status of a 'self-sufficient and net exporter' during the early nineties, rising from being a net-importer state. At the same time, imports declined from Rs 7 billion in 1985–86 to Rs 3 billion in 1995–96.

Post-mission Period

The other dominant feature that has had significant impact on the present status of edible oilseeds/oil industry has been the policy of open general licensing (OGL). Controls and regulations became relaxed, resulting in a highly competitive market. At the same time, the increasing per capita income led to enhanced consumption of edible oils. The gap between domestic production and the requirement became widened at an alarming rate. This completely eroded the gains that the country had achieved during the TMO period (Fig. 12.1). Despite the above developments, performance of oilseeds on the domestic front during the last two decades has been commendable. The growth rate of nine edible oils between 2000–01 and 2016–17 vis-à-vis 1990–91 to 1999–2000 provided a fillip for consolidation and revitalization of the oilseed economy (Fig. 12.2). The enhanced growth rate of 3.15% of oilseed production could not match the rate of growth of imports of edible oils, which was 6.99%. The per capita consumption of edible oils grew at a rate of 5.15%. Growth analysis of individual oilseed crops during the decade 2000–01 to 2016–17 suggests that there had been an acceleration in area under soybean, rapeseed-mustard and sesame while stagnation/deceleration had been observed in groundnut, sunflower, niger, safflower and linseed. The growth in area under castor crop, although marginal, resulted

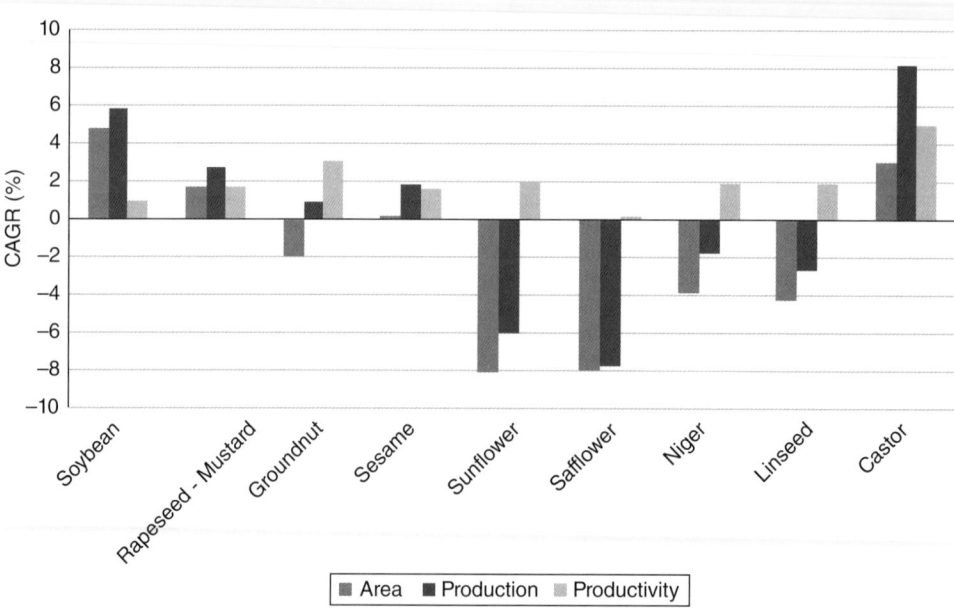

Fig. 12.2. Compound growth rates of annual oilseed crops in India (2000–01 to 2016–17). (From: DAC, 2017)

in production enhancement through considerable productivity improvements.

Demand, Import and Export Scenario

Demand projections

Annual demand is increasing at the rate of 6% while domestic output has been increasing at just about 2%. The average per capita consumption of edible oils was 3.66 kg p.a., which was much less than the norms prescribed by the Indian Council of Medical Research (ICMR)/ World Health Organization (WHO), i.e. a growth rate of 5.01%. The import bill began to increase at an alarming rate of almost 25%, from Rs 313 million to Rs 6.92 billion for the triennium ending 1985–86. In the export scenario of edible oilseeds and products, the rate of growth was a meagre 4% from 1970/71 to 1985/86, with export earnings being Rs 1.01 billion and 1.38 billion for the trienniums ending 1973–74 and 1985–86, respectively. In future, vegetable oils are likely to retain and indeed strengthen their primacy as major contributors to further increases in food consumption of the developing countries. Three decades ago, 136 kcal/person/ day or 6.5% of the total availability of 2110 calories were contributed by oil products (Fig. 12.3). Oil consumption per capita had grown to 10.4 kg by the year 2000, contributing 272 kcal to total food supplies, or 10% of the total 2650 kcal consumed. Average per capita consumption of edible oils for the period 2002 to 2012 rose to 29.4 g/day (10.7 kg from edible oils and 1.2 kg from *vanaspati*). Increase in average per capita consumption of edible oils was 5.25% during the period. The consumption levels of edible oils are beginning to increase to alarming levels as against the recommended 30 g/day to meet average physiological needs. The demand for vegetable oil increases with increase in population, increase in standard of living (income) and increase in use of oil for industrial, pharmaceutical, nutraceutical and cosmetic purposes (Fig. 12.4).

Taking into consideration a host of factors, the projections are based on the assumptions that per capita consumption would be increasing annually at 3% until 2015, followed by an increase at a rate of 2.5% from 2015 declining to 1.75% in 2020, with a further decline in the incremental consumption to negligible levels by the year 2050. The estimated per capita consumption is accordingly placed at 16.43, 17.52, 18.62 and

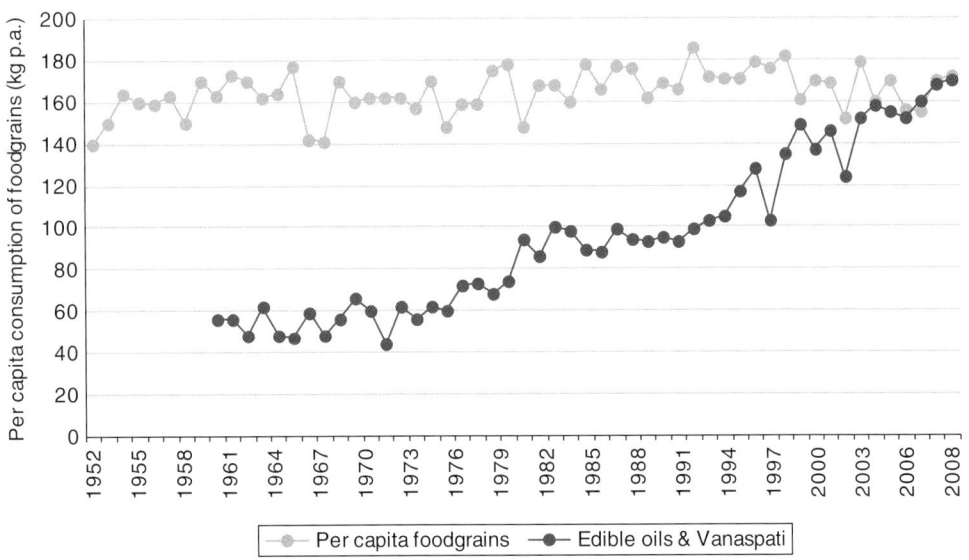

Fig. 12.3. Per capita consumption of foodgrains and edible oils. (From various issues of five-year-plan documents, Ministry of Finance, DGCIS and Economic Survey, Government of India)

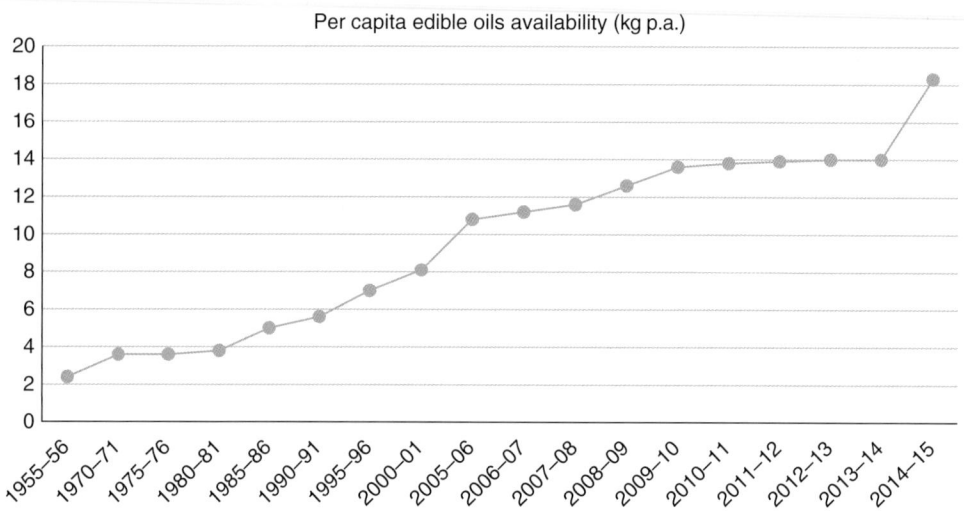

Fig. 12.4. Per capita availability of edible oils. (From: FAO, 2016)

19.16 kg/annum for the years 2020, 2030, 2040, and 2050, respectively (DOR, 2013). However, according to FAO estimates, per capita edible oil consumption was 18.13 kg in 2016–17, which will increase to 24 kg by 2025 (FAO, 2016). A newer dimension of vegetable oil requirement for industrial use is estimated to grow by 15% in 2020, 20% in 2030 and 25% post-2040, thus requiring around 3.57, 6.34, 9.69 and 10.61 million t in 2020, 2030, 2040 and 2050, respectively (Table 12.2). The Indian trade industry predicts much greater expansion. The total vegetable oil requirement is estimated at 25.26, 29.47, 34.27 and 35.90 million t during 2020, 2030, 2040 and 2050, respectively, which is a gigantic task for the country. The contribution of vegetable oil derived from secondary sources including arboreal tree species (20%) is estimated at 5.05, 5.89, 6.85 and 7.18 million t during 2020, 2030, 2040 and 2050, respectively. Thus the total domestic vegetable oilseed requirement from nine annual oilseed crops is estimated at 67.37, 71.45, 80.65 and 82.06 million t by 2020, 2030, 2040 and 2050, respectively.

Import situation

The success of the Technology Mission on Oilseeds was evident from the drastic doubling of oilseed production and the reduction in imports during the triennium ending 1993–94. To aid the process, and for consumer protection against price rises, in the year 2005 import duty was raised on crude palm oil/crude palmolein from 65% to 80% and on refined palm oil/RBD palmolein from 75% to 90%. Subsequently, in August 2006, the import duty was reduced on crude palm oil/crude palmolein from 80% to 70% and on refined palm oil/RBD palmolein from 90% to 80%. In 2007, the custom duty on crude and refined palm oil/palmolein was further reduced to 45% and 52.5%, respectively. The custom duty on crude as well as refined sunflower oil was further reduced to 40% and 50%, respectively. In 2008, the custom duty on all major crude and refined oils was reduced to nil and 7.5%, respectively (Fig. 12.5). In 2008/09, the government introduced a scheme of distribution of up to 1 million t of imported edible oils. Four public sector undertakings (PSUs), namely Projects Equipment Corporation (PEC), Minerals and Metals Trading Corporation (MMTC), State Trading Corporation (STC) and National Agricultural Corporation Marketing Federation (NAFED), were entrusted with the tasks of import, refining, packing and distribution of subsidized edible oils to the states. The scheme, with a subsidy of Rs 15 per kg of oil imported helped to soften the price of edible oils in the domestic market, this providing some relief for consumers, but raised the import bill significantly, which now touches almost Rs 560 billion annually. The globalization/WTO era

Table 12.2. Demand projections of vegetable oils in India. (From: Rabobank, 2011)

	2020	2030	2040	2050
Projected population (billion)	1.32	1.43	1.55	1.68
Per capita consumption considering 50%, 60%, 70% and 75% above the prescribed consumption levels during 2020, 2030, 2040, 2050, respectively				
Per capita consumption (kg/annum)	16.43	17.52	18.62	19.16
Vegetable oil requirement for direct consumption (million t)	21.69	23.13	24.58	25.29
Vegetable oil requirement for non-industrial use (million t)	3.57	6.34	9.69	10.61
Total vegetable oil requirement (million t)	25.26	29.47	34.27	35.90
Vegetable oil availability from secondary source (million t)	5.05	5.89	6.85	7.18
Total vegetable oil requirement from annual oilseed crops (million t)	20.21	23.58	27.42	28.72
Total vegetable oilseed requirement from nine annual oilseed crops (million t)	67.37	71.45	80.65	82.06

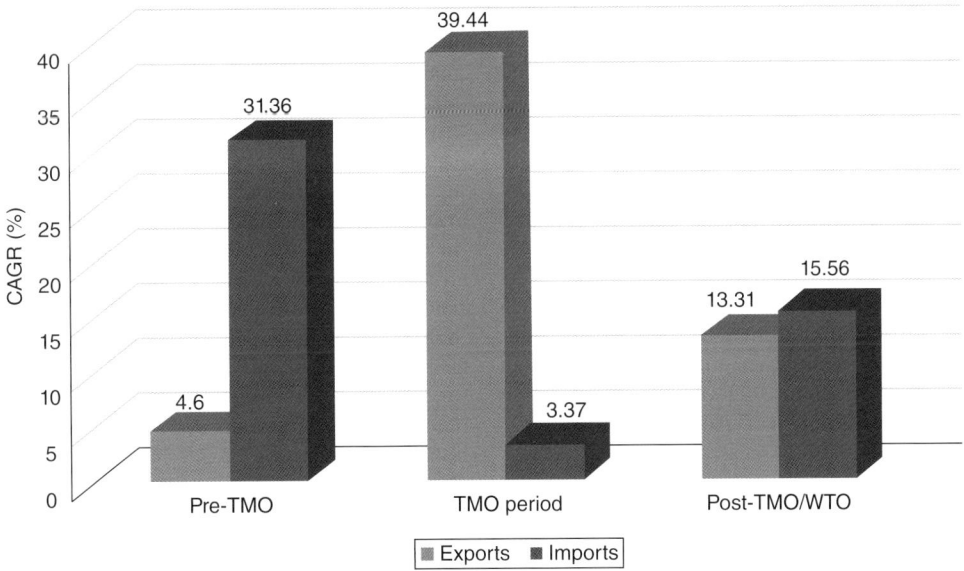

Fig. 12.5. Exports and imports. (Sarada *et al.*, 2015)

failed to consolidate the gains achieved during the TMO period (Fig. 12.5) due to operationalization of market/non-market forces in addition to biotic and abiotic problems. As stated above, the country is now meeting more than 50% of its oil requirement through imports, resulting in a huge drain on its foreign exchange (Fig. 12.6). The current import bill is around Rs 560 billion annually, which is much higher than in the past.

Export trend

India made excellent inroads through export of oil meals and castor oil to the tune of Rs 230 billion, thus plugging almost 50% of its import bill. The advantage of exports can further be consolidated with proper policy back-up and value addition. The overall trend of export of vegetable oilseeds and their products, as well as imports, has been increasing since 1987, except for a brief period in the mid-1990s (Fig. 12.7).

The government, in compliance with the requirements of WTO agreements, and to meet the domestic demands for vegetable oils, took certain decisions during 1994–2000, opening the floodgates to oil imports, which were available at cheaper prices. Edible oils, except coconut oil and palm oil, were placed under open general licence (OGL). The import duty was reduced in steps from 65% to 15%. The resulting heavy imports of edible oils led not only to vast drainage of foreign exchange but also produced a cascading

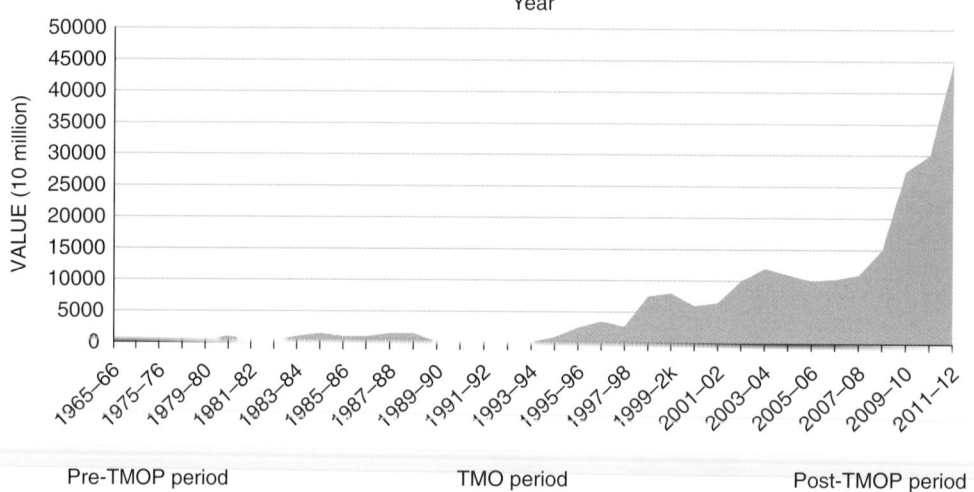

Fig. 12.6. Imports of oilseeds and oilseed products (Rs 10 million). (From: various issues of economic surveys, Government of India)

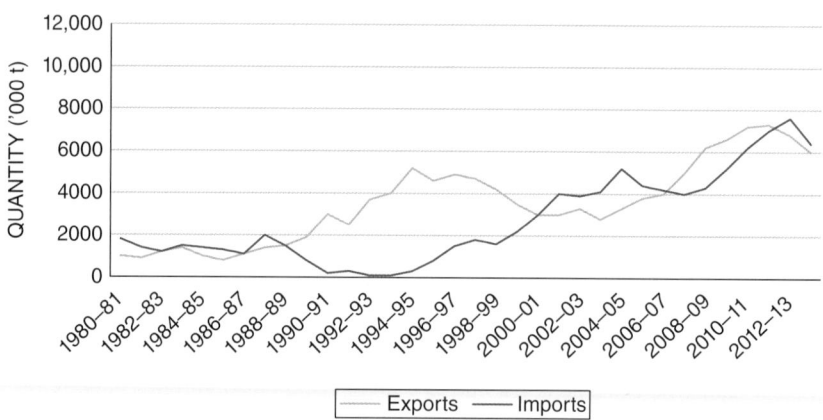

Fig. 12.7. Trends in quantities of exports and imports of edible oilseeds and their products. (From various issues of Five Year Plan documents, Ministry of Finance, DGCIS and Economic Survey, Government of India)

effect on the Indian oil economy. There was a crash in domestic prices causing serious impact on the Indian oilseed industry with considerable disincentive to the oilseed farmers. It is obvious that we need forward-looking policies on mitigation of various risks in oilseed production efficiency and profitability to ensure a healthy oilseed economy. All options for risk mitigation, like timely availability of inputs and credit, MSP and procurement of crop insurance, linking farmers to market, buffer stock options, and other commodity price stabilization schemes need to be put in place for the oilseed sector as a matter of priority.

Trade-related Policy Initiatives

In pursuance of the policy of liberalization, edible oils, which were on the negative list of imports, were first dechannelled, partially, in April 1994, with permission to import edible vegetable palmolein under OGL at 65% duty. This was followed by enlarging the basket of oils under OGL import in March 1995, when all edible oils (except coconut oil, palm kernel oil, RBD palm oil and RBD palm steering) were brought under OGL import at 30% duty. The duty was then further reduced to 20% plus a 2% surcharge in the regular budget for the year 1996–97. India was pursuing a policy of Import Substitution Industrialization (ISI) strategy until 1994–95, under which the oilseed/edible oil sector was protected through quantity restrictions. All imports of edible oils and oil meals were channelled through the State Trading Corporation (STC) and the Hindustan Vegetable Oils Corporation (HVOC), which remained limited to the packaging of oils and channelling to the state governments for sale through the Public Distribution System (PDS). It may be recognized that the ISI strategy pursued until 1994–95 delivered significant benefits to the Indian economy. India was able to transform itself from a deficit to a virtually self-sufficient state in edible oils by the early 1990s. The exports of oil meals increased substantially, both in volume and share, because of the increasing demand for Indian oil meals in the world market, which is mainly flooded with oil meals of GM oilseeds. Indian oil meals command a premium because of their non-GM nature. Soymeal exports in 2017–18 were 1.2 million t,

worth US$492 million. It is worth mentioning that the growth in the livestock industry will be a major force driving future demand for oil cakes with high-income elasticity of demand for milk and milk products, meat, eggs and fish.

Support price

Under the harsh growing conditions faced by Indian agriculture, oilseeds have a clear edge over many minor millets and pulses in terms of higher productivity. However, increase in cultivable area under oilseeds largely depends on higher profitability. Unfortunately, the support price declared each year by the government clearly favours other crops, mainly on account of food security considerations (Fig. 12.8). Similar consideration for oilseeds is, therefore, warranted.

The minimum support price (MSP) index analysis clearly indicated that it mainly favoured wheat and paddy against pulses, coarse cereals and oilseeds. Over and above the relative discrimination in MSP for oilseeds, there was no mechanism for implementation of the MSP without assured procurement. Hence, most of the time the wholesale prices were much lower than the MSP. During the TMO period, there was effective implementation of the MSP through the National Dairy Development Board (NDDB), which gave confidence to farmers about the minimum expected returns. It therefore warrants revival of an institutional mechanism to implement the MSP effectively for oilseed growers to reap the benefits.

Need for Institutional Linkages

The research, development and technology dissemination infrastructure existing in India for oilseed crops is the legacy of past policies and interplay of public and private interests in the sector. Apart from the institutions as such, some institutional support programmes – the National Dairy Development Board (NDDB), the National Agricultural Marketing Federation (NAFED), the flagship programme of the government in the oilseed sector, and the Integrated Scheme on Oilseeds, Pulses, Oilpalm and Maize (ISOPOM) have been tried in the past. These programmes need to be studied to understand their significance and

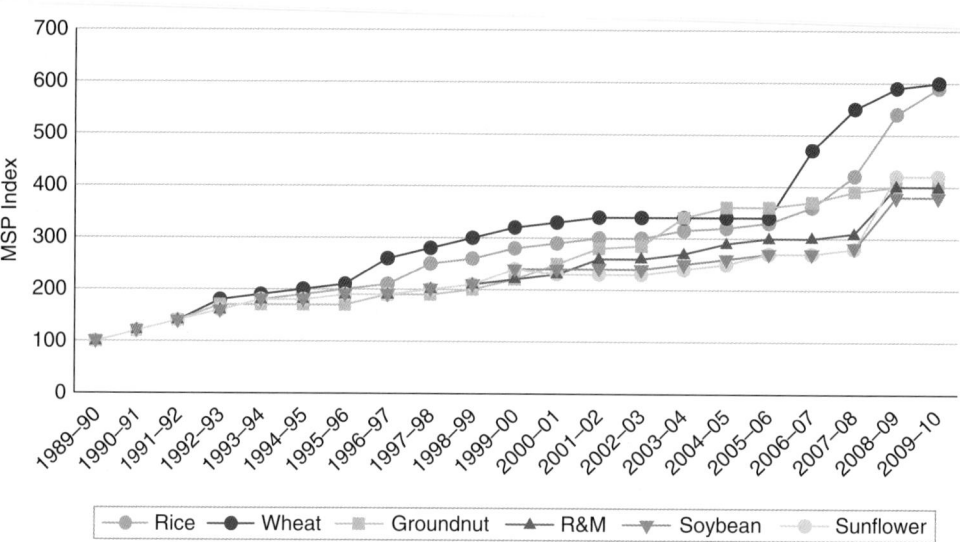

Fig. 12.8. Minimum support price of different crops in different years. R&M = Rapeseed and Mustard (From: Commissioner of Agricultural Costs and Prices, Government of India)

impact so that efficient and functional institutional support is provided in future for the required growth of the oilseeds sector. It must be recognized that the core strength for the success of the technology mission for oilseeds was due to effective dovetailing and coordination among institutions linked with production, processing, input supplies, trade and pricing. Some systems need to be revisited to give a much-needed push for the oilseeds sector.

Ecoregional Approach for Productivity Enhancement

The concept of an ecoregional approach can effectively be utilized for oilseed crops. It refers to the practice of delineating efficient zones for specific crops for realizing potential yields with high input-use efficiency. Supporting services like input supply, marketing and processing have to be linked to these ecological zones besides strengthening research and extension systems and infrastructural facilities. The importance of crop ecological zoning in oilseeds is evidently based on the following facts: 4 districts contribute 33% of groundnut area; 4 districts contribute 37% of sunflower area; 9 districts contribute 31% of mustard area; 12 districts contribute 41% of

soybean area. Concerted efforts on two categories of crop-wise ecoregions, i.e. high area, low productivity and low area, high productivity zones will enhance efficiency in efforts to increase production and productivity of oilseed crops. The classic examples of high productivity of spring-season sunflower in the Indo-Gangetic Region of Punjab, Haryana, western Uttar Pradesh and Bihar; high productivity of safflower in the Malwa region (Madhya Pradesh) and Gujarat; high productivity of sesame in West Bengal in the summer season; and high productivity potential of soybean in Punjab, Haryana and eastern Uttar Pradesh are mainly due to the optimum ecological conditions that are beyond input and management. Hence, providing necessary input supply, technology, market and extraction facilities in these areas can help realize a quantum jump in productivity.

Natural Resource Management

With the current practices of crop cultivation under sub-optimal management, especially without nutrient application, significant soil nutrient mining is occurring. Correcting the present limitation and imbalance in soil nutrients can provide rich dividends. Declining per capita arable land

and increasing oilseed cultivation to poor and marginal soils result in low productivity. Moreover, productivity of oilseed crops is limited owing to their cultivation under rainfed conditions. Currently, only 28% of area is irrigated under oilseeds. Water requirement in oilseeds is, therefore, a key factor for ensuring higher yields. With dwindling water resources, both in quantity and quality, water for irrigation will be costly and face severe competition from different enterprises within the agriculture sector. Castor in Gujarat and Rajasthan is cultivated under irrigation while in Andhra Pradesh it is mainly cultivated under rainfed conditions. Safflower cultivation is limited to vertisols and rabi season under receding soil moisture conditions. Sunflower is cultivated in all seasons and all soil types. A total of 40% area under kharif sunflower is rainfed. Watershed management with appropriate rainwater harvesting, both *in situ* and with proper disposal and storage farm ponds, provides excellent opportunities to mitigate the expected dual problems of long droughts and floods. Site-specific land configuration and management for effective soil and moisture conservation and its economic use can operationalize the drought-mitigation strategy. Enhancing drought tolerance in oilseed crops is, therefore, a priority with associated practices to improve profitability through achieving 'more crop (oil) per drop' of water, resource-use efficiency and a preferential edge over other competing crops. Besides, due to low fertilizer-use efficiency, investments are not remunerative. Improving nutrient-use efficiency of fertilizers through better product development and methods of application should now be a priority to achieve profitable oilseed production. Improving soil fertility to reduce external applications is an achievable solution through site-specific management. Exploiting nutrient interactions as per the soil test and crop response results in higher efficiency and reduced cost. Organic manures are central in the INM of oilseeds under rainfed situation, along with other components such as secondary and micro-nutrients like sulphur bioinoculants, crop residues etc. Precision crop management with conservation agricultural practices and customized fertilizer application schedules would usher in higher efficiency and profitability. Therefore, greater emphasis on integrated natural resource management in oilseeds should be a high priority.

Crop Improvement Strategy

The gains in productivity of oilseed crops have been achieved primarily through exploitation of available genetic variability. Conventional breeding coupled with modern tools such as biotechnology should now be the primary focus in crop improvement programmes. Heterosis breeding should be the major focus in crops like sunflower, castor, rapeseed-mustard, safflower and sesame. To facilitate better exploitation of the available gene pools and overcome the production constraints, research emphasis needs to be given to: (i) augmentation/identification of trait-specific germplasm; (ii) pre-breeding and genetic enhancement; (iii) allele mining; (iv) functional genomics, proteomics, metabolomics and interactomics; (v) marker-assisted breeding and gene pyramiding; and (vi) trait improvement through genetic engineering.

Role of biotechnology

The two main options of biotechnological approaches for crop improvement include molecular marker-based selection and transgene manipulations. Both these approaches have distinct niches with respect to their role in crop improvement. At present, biotechnological research on minor oilseed crops (safflower, castor, niger, sesame, linseed and sunflower) is in its infancy but concerted efforts are needed to address crop-specific problems such as pests like budfly in linseed; antigastra and phyllody in sesame; necrosis; leaf spot and powdery mildew in sunflower; wilt and *Alternaria* in safflower; and botrytis and lepidopteran pests in castor. Quality aspects such as presence of anti-nutritional compounds (oxalic acid and phytates) in sesame; oil quality in mustard; toxic proteins (ricin and *Ricinus communis* agglutin) in castor; and herbicide tolerance in soybean also need to be addressed. Apart from these crop-specific issues, there are research areas of a generic nature such as abiotic stress (drought, salinity, cold) tolerance, increased oil content, and altered fatty acid profiles to suit different industrial and human consumption requirements. The success of 'doubled haploids' in developing superior inbreds is a potential area for immediate gain in oilseed crops limited by availability of superior inbred development. The required infrastructure and support needs to be ensured for operationalization.

Transgenic approach

Transgenic technology has removed the phylogenetic barriers for transfer of useful genes across organisms. Modifying the fatty acid profile of the oil to suit industrial, pharmaceutical, nutritional and cosmetological requirements using genetic engineering approaches has been a priority in application of biotechnology in oilseed crops. Similarly, imparting biotic and abiotic stress tolerance and improved resource-use

efficiencies through transgenic approaches have been areas of focused attention. Transgenic technology is facing stiff resistance from a section of society. It is the responsibility of scientists, policy makers and industry to allay the fears of the public through scientific knowledge and empirical evidence regarding the safety of GM crops. Once the biosafety of transgenic plants is established, they should be treated as any other variety or hybrid. During the last decade there has been considerable progress towards harnessing transgenic technology for oilseed improvement in India (Table 12.3). At the National Research Centre on Plant Biotechnology (NRCPB), *Alternaria*-resistant transgenic mustard expressing either glucanase gene from tomato or other antifungal genes such as annexin and osmotin, and aphid-tolerant plants expressing either wheatgerm

Table 12.3. Biotechnological interventions in oilseed crops in India.

Crop	Institute	Genes being used	Trait
Groundnut	DGR and UoH	Annexin	Leaf spot resistance
	ICRISAT	Rice *Chitinase*	Leaf spot and rust resistance, aflotoxin reduction
	ICRISAT	*Coat protein and replicase*	BND, clump virus resistance
	ICRISAT	*DREB IA from rice*	Drought tolerance
	UAS, Bangalore	*Cry IX*	*Spodoptera* and pod borer resistance
	UAS, Bangalore	*EPSPS*	Glyphosate resistance
	UAS, Bangalore	*Tabacco I, 3 beta glucanase*	Leaf spot, *Aspergillus* resistance
	NABI, Manali	*Cry IEC*	*Spodoptera* resistance
Mustard	DU	*Barnase-Barstar*	Male sterility
	NRCPB	Rice *Chitinase*	*Alternaria* resistance
	Bose institute	Lectin (*ASALI*)	Aphid resistance
	BARC	Synthetic *Cry IAc*	Diamond back moth resistance
	NRCPB	*Glucanase*	Alternaria *resistance*
	NRCPB	*Wheat germ agglutinin*	Aphid resistance
	NRCPB	*Snow drop lectin*	Aphid resistance
Soybean	DSR, Indore	Marker and reporter genes	Transformation protocol development
	MKU	Full-length and truncated movement protein	Viral resistance (Development stage)
	Bharatidasan University	*Alpha tocopehrol methyl transferase*	Vitamin E in oil
Castor	IIOR	*Cry 1Aa, Cry1Ec, Cry1abcf*	Lepidoptern defoliators
	IIOR	Multiple genes (*ERFI, EBPI, BIKI, Chitinase, RsAFP2, AcAMPI*)	Botrytis tolerance
	IIOR	Silencing of ricin and RCA	Reduction of endosperm toxins
Sunflower	IIOR	TSY-Coat protein gene	SND resistance
Sunflower	IIOR	*OrfH522, u-nad3*	Male sterility
	IIOR	*DAGATI and GPAT9*	Increased oil content

agglutinin or snowdrop lectin gene, have been developed. Similarly, aphid-tolerant mustard transgenic plants have been developed at the Bose Institute, Calcutta, using the lectin gene from garlic. At the University of Delhi, a transgenic male sterility system has been developed in Indian mustard using the popular barnasebarstar system and the experimental hybrids obtained through this technology have already been tested in multi-location trials.

At the Directorate of Groundnut Research (DGR), Junagadh, transgenic groundnut plants have been developed with coat protein 25; also genes for incorporating resistance to peanut bud necrosis disease (PBND) and peanut stem necrosis disease (PSND), currently being evaluated in glasshouses; mtlD gene for enhancing tolerance to drought and salinity; defensin gene for enhancing resistance to fungal diseases; and annexin and PR 10 genes for enhancing tolerance to abiotic stresses. Also, transgenic groundnut plant expressing annexin gene has been developed at the University of Hyderabad as well as at the Directorate of Groundnut Research (DGR), Junagadh, to fight against leaf spot disease. Transgenic groundnut plants with resistance to Spodoptera, fungal disease and glyphosate tolerance have been developed at the University of Agricultural Sciences (UAS), Bangalore. At the National Agri-Food Biotechnology Institute (NABI), Mohali, Spodoptera-resistant plants have been developed by deploying cry1EC gene. At the International Crop Research Institute for the Semi-Arid Tropics (ICRISAT), transgenic groundnut lines with aflatoxin resistance conferred by rice chitinase gene, and bud necrosis virus resistance imparted by expressing viral coat protein gene have been developed and characterized. Similarly, improved drought tolerance has been achieved by deploying DREB gene. Limited field trials are being carried out to select the lines for further studies and commercialization. In soybean, concerted efforts are going on to develop good transformation protocols for the Indian genotypes at the Directorate of Soybean Research (DSR), Indore. At Bharatidasan University, Tiruchirapalli, Tamil Nadu, transgenic soybean lines with enhanced vitamin E in the oil are being developed, while at Madurai Kamaraj University, viral resistance is being achieved by expressing the movement protein of the virus. At the Indian Institute of Oilseed Research (IIOR), Hyderabad,

transgenic castor lines with resistance to defoliators have been developed by deploying different cry genes with specificities against target pests (semilooper and Spodoptera). Limited field trials have been conducted with this material to select lines with higher pest mortality potential and the selected events will be further tested in confined field trials. Also, multigene constructs have been developed to counter Botrytis disease and these constructs are being validated now for their efficiency in controlling necrotrophic fungi. Similarly, RNAi gene constructs developed for suppressing genes encoding ricin and RCA are being validated using a model plant system. In safflower, attempts are being made to develop transgenic male sterility and fertility restoration using orfH522 and u-nad3 genes. Over-expression of the rate-limiting enzymes (DAGAT and GPAT9) in a seed-specific manner is being attempted to increase the oil content in safflower, which currently is about 30% in the cultivated safflower varieties or hybrids. The problem of sunflower necrosis disease, which once threatened the very cultivation of sunflower crops, is also being tackled through transgenic approaches using coat protein in sense and anti-sense orientations. Therefore, concerted efforts should be made to develop transformation protocols in all oilseed crops. Efforts are needed to develop methods and strategies to have the transgenes inserted in targeted regions of the genome to avoid positional effects as well as insertional inactivation of unintended genes. Also, it is envisaged that technologies for cis-genesis, intra-genesis, gene-stacking, marker-free transgenesis, zinc-finger nucleases, ribonucleic acid (RNA) dependent DNA methylation etc., once perfected in model crops, should be adopted for the improvement of oilseed crops. This information could be used in developing and realizing designer transgenic oilseed crops to meet pharmaceutical, nutraceutical and industrial demands. The first GM mustard in India is ready and waiting for official release for cultivation in the country. In spite of the progress made in the use of transgenic technology in oilseed crops, current policy controversy is a setback, delaying the fruit of the research reaching stakeholders. The GM mustard developed by Delhi University scientists is held up for sociopolitical consensus while the country is importing GM soya oil (2.5 million t) and GM canola (1–1.5 million t) and consuming domestically produced

GM cotton seed oil (1.5 million t), constituting a total GM oil component of nearly 5 million t out of 21 million t of annual edible oil consumption.

Exploring frontier sciences

Significant innovations in frontier science and technologies such as nanotechnology, genetic engineering and biotechnology, synthetic lipid science and technology, information science and modelling, simulation and forecasting; and the recent developments in related sciences such as hydroponics, vertical farming, protected agriculture, precision agriculture systems, biosecurity and biodiversity management provide unlimited opportunities for supporting higher production and product development to meet changing requirements through precision farming and protection/conservation practices. Post-production developments and dynamic integration of production, processing and quality with global trade would make production of vegetable oils profitable and competitive. Oilseed production will also benefit from innovations in the industrial sector for oilseed processing and small-farm mechanization and *ex ante* approaches for quantification of economic output. These frontier sciences will have to be harnessed and integrated into ongoing research programmes for productivity improvement – increasing resource-use efficiency; improving processing; value addition; diversified uses; improved access to stakeholders through ICT; enhanced delivery systems; and better targeting of technologies for yet better production and marketing environments, including supply chain mechanisms. Traditional knowledge should also be valued for its wisdom in technology generation, refinement and adoption.

Public–private Partnership and Linkages

Oilseeds, unlike other food crops, depend on other enterprises for their ultimate use/consumption. The necessity of extraction of oil from seeds enables interdependence of industry and oilseed producers and consumers, thus making success in the vegetable oil production business interdependent at each stage of production, processing and pricing. The potential of PPP through linkages in all aspects of oilseed production and marketing needs to be harnessed for a win–win situation. The grey areas for PPP in oilseeds include incentives for seed production, forward/backward linkages for processing, value addition, contract research in niche areas, contract farming, and joint ventures for higher-order derivatives and speciality products. The edible oil industry is largely dominated by the bulk segment, which creates an opportunity for the agri-business sector. The unbranded segment accounts for anywhere between 80 and 90% of the total consumption, which can be targeted for better value addition and thus minimize health hazards that otherwise would occur on account of adulteration of edible oils. The share of raw oil, refined oil and *vanaspati* in the total edible oil market is estimated at 35%, 55% and 10%, respectively. The former group is a viable agri-business venture. The shift in consumer preference for branded edible oils has resulted in the corporate sector targeting the packaged edible oil segment in the last few years. Hence, PPP for R&D efforts towards value addition emerges as a new priority in the future.

Diversification and Value Addition

Profitability of oilseeds solely from primary products like seed and oil will not be sustainable. Besides the primary product oil, oilseed crops provide immense scope for diversified uses with high-value speciality products and derivatives. From the vegetable oil consumption point of view, either for edible or for fuel purposes, the situation is envisaged of valuing oil for its intrinsic value for calories or desired fatty acids. Designer oils with requisite blends can meet that expectation, and to that extent the individual oilseed crop's potential would be its yield of oil or desired fatty acid and not as oil from a specific crop. Thus the present wide diversity of oilseed crops may narrow down to a few high-oil-yielding crops. As for unique non-oil value aspects, for specific aroma or non-oil uses (medicinal, ornamental or other uses), the individual oilseed crops would be grown for speciality purposes irrespective of productivity level. Major opportunities for oilseed crop diversification and value addition include its introduction as a catch crop in paddy fallows to utilize residual moisture and fertility; a component crop in major wide-spaced field crops such as sugarcane, pigeon pea, cotton and maize for sunflower; as a main crop with groundnut, soybean, finger millet, pigeon pea, cluster bean and short-duration pulses for castor and sunflower; with chickpea and coriander for safflower; rabi castor under limited irrigation protection; and sunflower for the Indo-Gangetic plains of Punjab, Haryana, western Uttar Pradesh in spring, and Bihar, Odisha and West Bengal in rabi/summer. Soybean also offers opportunity for rice-wheat cropping systems in the north.

Adaptation to Climate Change

The low productivity and uncertain production of oilseeds is mainly due to their cultivation under rainfed conditions (about 70%). The inherent tolerance of oilseeds to drought and other edaphic stresses is construed as though they are low-input crops. On the contrary, oilseeds need higher inputs for increasing productivity. Adaptation strategy for drought, high temperature and rainfall variations must, therefore, be put in place as a matter of priority. Oilseed production is constrained by several biotic stresses like insect pests and diseases that are being further aggravated by changing climatic conditions. *Botrytis*, root rot and capsule borer have emerged as major threats to castor production. Sunflower production is limited by diseases like *Alternaria* leaf blight, sunflower necrosis, downy mildew and powdery mildew, while mealy bug is an emerging pest. The foliar diseases *Alternaria* and *Cercospora* leaf spots and *Macrophomina* root rot are becoming increasingly important while wilt and aphid continue to challenge safflower production. Global warming-induced climate change is expected to trigger major changes in population dynamics of pests, their biotypes and activity, abundance of natural enemies and efficacy of crop-protection technologies. Studies on the epidemiology of plant diseases including variation in pathogen population in the light of climatic change are necessary to develop integrated disease management (IDM) modules. Studies on wilt disease aetiology in the context of reniform nematode in castor and sunflower, and root-knot nematode in sunflower,

coupled with identification of sources of resistance, deserve attention. There is a need to generate information on the likely effects of climate change on pests so as to develop robust technologies that will be effective. The approach to pest management has seen a significant change over the years from chemical control to IPM, with emphasis currently on biointensive IPM involving use of pest-resistant varieties, bioagents, biopesticides and natural products like botanical pesticides and pheromones. Several ecofriendly products of biological origin have been developed at the Indian Institute of Oilseed Research, Hyderabad, for the management of important pests of oilseed crops like castor semi-looper, sunflower head borer and tobacco caterpillar as well as wilt of castor and safflower. However, the relative efficacy of many of these pest-control measures is likely to change as a result of global warming, necessitating identification of temperature-tolerant strains.

Transfer of Technology

There are several technologies available for enhancement of oilseed production (Akter *et al.*, 2012; Varaprasad and SudhakaraBabu, 2015). Concerted efforts are urgently needed for the dissemination of technologies and new approaches in a participatory mode to be strengthened for effective delivery by showcasing the potential technologies/products. The Farmer-Institution-Industry linkage mechanism should be strengthened besides the existing formal delivery mechanisms so that the gap between the potentially attainable yield and the yield realized on the farmers' fields is reduced, which will make the industry more vibrant and profitable on account of assured quality supply, reduced obstacles in the supply chain, enhanced capacity utilization and increased economic surplus, which will benefit both the producer and the consumer. The potential ICT tools should be harnessed on a dynamic and interactive mode. This can minimize dissemination loss while sharing information and provide benefits to all the stakeholders involved in oilseeds. Also, a dedicated TV channel on agriculture will help in faster dissemination of knowledge. Creation of agri-clinics with provision of outsourcing through the involvement of a new breed of young, well-trained technology agents would go a long way in outscaling innovation for greater impact.

Exploring New Crops for Industrial Use

Efforts need to be intensified to explore native as well as exotic crops to find new options for industrial use as has been done in the European Union (Zanetti *et al.*, 2013). Such efforts are likely to prevent diversion of edible oils for industrial use. Ethiopian mustard, meadowfoam, honesty and pennycress are likely to contain similar or higher amounts of long-chain fatty acids (FAs) than are found in rapeseed. Cuphea FAs profile is unique, characterized by high content of capric and lauric acids. Worldwide, the major source of hydroxylated FAs is castor seed. One alternative source is lesquerella. Current domination by India and China for castor cultivation may change if an alternative to ricinolic acid is found. Oilseeds naturally rich in epoxidized FAs include euphorbia, vernonia and stokes aster. Calendula does not contain epoxidized FAs but does have conjugated double-bond FAs that can be used in green industry. Flax and camelina seed oils are rich in alpha linolenic acid. Camelina is also rich in eicosenoic acid, with unique industrial applications (e.g. polyamide-11, a polyamide bio-plastic). Other minor oilseeds rich in polyunsaturated fatty acids (PUFAs) are echium, evening primrose and borage. The relatively high content of unusual FAs such as stearidonic acid in echium and hemp seed oils opens up attractive perspectives for the pharmaceutical, nutraceutical and cosmetic industries. Sunflower is by far the most important source of oleic acid, globally, for industrial applications. Among the alternative monounsaturated fatty acids (MUFA) crops, high oleic types exist in rapeseed, safflower and cardoon. Recently, the Indian Institute of Oilseed Research (IIOR) developed high-oleic safflower varieties on demand from private industry. There is interest in coriander cultivation as a feedstock for the chemical industry. Coriander seed oil content is approximately 25% and can be used for the production of surfactant, soap and detergent due to high content of petroselinic acid (65–74%) in the seed oil, which is also suitable for the production of unique derivatives such as lauric acid (C12:0). Coriander seed oil methyl esters have excellent fuel properties as a result of their unique fatty acid composition. Coriander, now used largely as herb and seed spice, needs re-evaluation from an industrial application angle.

Strategies for Enhancing Oilseed Production

Following research, development and policy strategies would be needed to increase both oilseed production and vegetable oil availability in the country.

Research

- Greater emphasis needs to be given to innovation to achieve a quantum jump in productivity using new science and translational research.
- There is a need for integration and effective coordination of all oilseed research institutes under NARS for a holistic research approach.
- Short-duration and high-yielding varieties need to be developed for better adaptation to climate change through integration of modern biotechnological tools like marker-assisted selection (MAS) and transgenic breeding, supplementary to conventional breeding, and development of cultivars with in-built resistance to biotic and abiotic (especially drought and heat) stresses. In this context, greater use of germplasm through pre-breeding will be highly desirable.
- There is an urgent need to develop small-farm machinery for different operations specific to each crop so as to ensure timely farm operations and efficient use of costly inputs.
- Increased emphasis needs to be given to post-harvest technology and value addition for diversified uses in order to ensure higher profitability.

Development

- Strong links need to be developed for successful operation of the 'seed village' concept with producers, technocrats, certifying agencies and concerned state departments of agriculture for timely procurement and distribution to ensure higher seed replacement under improved varieties/hybrids.

- Oilseed cultivation needs to be promoted in new and non-traditional areas and seasons to ensure crop diversification and additional areas for expansion. The eastern region offers options for potential area expansion, especially in paddy fallows. Similarly, soybean offers great opportunities for diversification of the rice-wheat cropping system in northern India.
- Location-specific, efficient dry-farming technologies need to be adopted for sustainable oilseed production under drought situations. There is a need to integrate oilseed production with watershed programmes for holistic development and to ensure life-support irrigation for assured harvests.
- There is a need to increase the area under protective irrigation and to promote efficient irrigation methods, especially micro-irrigation, for achieving higher production and stability.
- Greater attention needs to be given to promote adequate and balanced fertilizer use with emphasis on the use of sulphur and limiting micronutrients through proper soil amendments, based on soil testing.
- Effective transfer of technology with assured input, market and technological backstopping needs to be done by both public and private sector agencies.
- There is a need to promote intercropping systems involving oilseeds for achieving higher efficiency of resources, profitability and risk minimization.
- Needs-based plant protection measures need to be adopted through effective and biointensive IPM.
- Large-scale production of promising small-farm equipment, with the involvement of state governments, will help improve efficiency of farm operations. Also, provision needs to be made for credit and incentives for manufacturing of small-farm equipment and machinery by small-scale industries and promotion of custom hiring to ensure resilience in farming.
- Greater emphasis on the use of soybean as food rather than just as oil and feed will help the nation address current concerns about protein malnourishment, while ensuring household nutritional security.
- There is a need to exploit additional features of crops like high-value safflower petals and fibre from linseed for realizing additional profits. Also, there is a need to accelerate area expansion of oil palm plantations and to ensure irrigation, power, local processing facilities and competitive prices in order to realize higher production of vegetable oil/ unit area/unit time.
- Use of rice bran directly as feed should be avoided in order to promote greater extraction and use of rice bran oil.
- There is a need to promote scientific processing of cotton seed for higher oil recovery and to get high protein retention (42%) compared to traditional processing practices (22%).
- The efficiency of extraction of oil should be improved through solvent extraction for hard seeds.

Policies

- There is a need for revival of an oilseed mission, with focus on the five Ps: priorities, policies, productivity, profitability and private-sector participation, with emphasis on increased oilseed production in the country.
- Greater emphasis needs to be given to public awareness concerning rationalization of vegetable oil consumption for proper health.
- Increased investment on research, development and extension to promote production of oilseed crops and their products is urgently needed.

Conclusion

The success of the Yellow Revolution, achieved through the Technology Mission on Oilseeds and Pulses (TMOP) during the late 1980s fully justifies the revival of an oilseeds mission, with the greater commitment of all stakeholders to ride out the present crisis of large-scale import of edible oils. There must be a clear national policy of bridging the yield gaps and increased oilseed production, with the specific aim of reducing vegetable oil imports, as was achieved during the TMOP period. No doubt, to achieve this, a clear policy direction and a missionary zeal on the part of all concerned will be required. The development and implementation of appropriate strategies along with enabling policies favouring domestic production, rather than dependence on import, will certainly help in meeting the challenges and harnessing the opportunities available to enhance production of oilseeds and their products. This will help the majority of smallholder farmers engaged in the cultivation of oilseed crops in India. Enhanced allocation for R&D will reduce the burden on large-scale imports of oilseeds from abroad and the nation will save almost Rs 700 billion crores on its annual import bill. With clear strategy and an implementation plan supported by policy makers, self-sufficiency in oilseeds is achievable.

References

Akter, N., Varaprasad, K.S. and Kalam, A.A. (eds) (2012) *Enhancing Oilseeds Production through Improved Technology in SAARC Countries*. SAARC Agriculture Centre, Dhaka.

DAC (2017) Directorate of Economics and Statistics (DES), Department of Agriculture and Cooperation (DAC) and Department of Commerce (DoC) Commodity Profiles.

DOR (2013) *Vision 2050*. ICAR-Indian Institute of Oilseeds Research (IIOR), Andhra Pradesh, India.

DVVOF (2017) *Commodity Profile of Edible Oil for March–2017*. Directorate of Vanaspati, Vegetable Oil and Fats (DOVVF) and Department of Commerce.

FAO (2016) *OECD-FAO Agricultural Outlook 2016–2025*. OECD, Paris.

FAOSTAT (2016) Crops. Available at: http://www.fao.org/faostat/en/#data/QC (accessed 30 May 2018).

Mukta, N. and Varaprasad, K.S. (2013) Tree borne oilseeds for agroforestry. In: Rao, G.R., Kumar, N.R., Prasad, J.V.N.S., Prabhakar, M., Venkatesh, G., Srinivas, I., Ramana, D.B.V. and Venkateswarlu, B. (eds) *Agroforestry as a Strategy for Adaptation and Mitigation of Climate Change in Rainfed Areas*. Central Research Institute for Dryland Agriculture, Hyderabad, India, pp. 102–110.

Rabobank (2011) In 2030: Four Future Scenarios in Businesses. Available at: https://economics.rabobank.com/PageFiles/3444/OutlookIN2030ENG_tcm64-132400.pdf (accessed 30 May 2018).

Sarada, C., Alivelu, K., Sambasiva Rao, V., SudhakaraBabu, S.N. and Varaprasad, K.S. (2015) *Oilseeds Statistics – A Compendium 2015*. ICAR-Indian Institute of Oilseeds Research, Hyderabad, India.

Sharma, V.P. (2014) Problems and Prospects of Oilseeds Production in India. Final Report submitted to DAC, Ministry of Agriculture, Government of India. Centre for Management in Agriculture, Indian Institute of Management, Ahmedabad. Available at: https://www.iima.ac.in/c/document_library/get_file?uuid=981f4ee1-2595-4090-a563-73fc703e5118&groupId=62390 (accessed 30 May 2018).

Singh, A.K., Singh, A.K., Choudhary, A.K., Kumari, A. and Kumar, R. (2017) Towards oilseeds sufficiency in India: present status and way forward. *Journal of AgriSearch* 4(2), 80–84.

Varaprasad, K.S. and SudhakaraBabu, S.N. (2015) Technology for increasing oilseeds production. National Seminar on Technologies for Enhancing Oilseeds Production through NMOOP. January 18–19. Directorate of Oilseeds Rajendranagar, Hyderabad, India.

Zanetti, F., Monti, A. and Berti, M.T. (2013) Challenges and opportunities for new industrial oilseed crops in EU-27: a review. *Industrial Crops and Products* 50, 580–595.

13

Accelerating Forage Crop Production

Introduction

The livestock sector contributes almost 30% to India's agricultural GDP and plays a crucial role in national food and nutritional security. The sustainability and viability of livestock production depends on the availability of affordable fodder and feed resources, as they constitute almost 60% of the total expenditure in dairy farming. Current estimates of fodder crop cultivation, though, may not be accurate, and are not more than 4–5% of the total cultivated area. Punjab, Haryana and western Uttar Pradesh have a higher share (7–10%). The importance of the silvipastoral farming system is well recognized in the arid regions. In India, rich genetic diversity, institutional infrastructure and competent human resources, besides policy support for linking smallholder farmers to markets, resulted in a 'White Revolution'. At present, India is the world's largest milk producer (producing more than 155 million t p.a.). In spite of all these achievements, dairy farmers are facing challenges of high cost of fodder and feed, non-remunerative price for milk, lesser incentives for value addition and export, lack of credit and insurance, and the adverse impact of climate change.

Forage crop production and development have not been given due importance. Production and availability of their improved seeds is still a grey area. The main challenge is of the ownership by both agriculture and animal husbandry departments. Dairy farmers need new knowledge on innovations that can enable them to produce more with less input and help them improve their livelihood. The majority of livestock farmers are smallholders and landless, and require both technical and financial support. Significant research has been carried out to identify suitable forage crops, domesticate them, breed new varieties, develop cultivation and pasture management practices, rehabilitate degraded and waste land through the introduction of suitable grasses/legumes, manage silvi/hortipastures, improve forage utilization and develop efficient techniques for seed production. However, there has not been any significant change in the status of forage supply in the country (Table 13.1) because these research findings/technologies have been adopted to a limited extent. Therefore, a strategic road map is urgently required for the production and development of forage crops.

Strengthening Forage Breeding

Breeding of forage crops needs to be strengthened considerably. Rich genetic diversity forms the basis for effective breeding programmes aimed at developing high-yielding nutritive and widely adapted varieties. The Indian subcontinent enjoys a very rich genetic diversity in different native grasses and legumes of forage potential. According to an estimate, 21 genera and 139 species are endemic out of 245 genera and 1256 species of Gramineae and one third of Indian grasses are

Table 13.1. Demand and supply estimates of dry and green forages (million t). (From: Dikshit and Birthal, 2010; IGFRI, 2014)

Year	Demand		Supply		Deficit (%)	
	Dry	Green	Dry	Green	Dry	Green
2017	526.9	845.5	469.6	572.6	11.41	33.10
2020	530.5	851.3	467.6	590.4	11.81	30.64
2025	549.3	881.5	483.8	638.9	11.92	27.52
2030	568.10	911.69	500.03	667.70	11.98	26.76

of forage value. There is also a need to develop varieties with better plant types suitable for mixed cropping/farming systems. Therefore, plant genetic resources (PGR) activities in fodder crops, including range grasses and legumes should be given high priority along with their effective management and characterization. The ICAR-Indian Grassland and Fodder Research Institute, Jhansi, is maintaining diverse germplasm of different forage crops including range grasses and legumes for fodder improvement. Diversity of a few important crops – maize, pearl millet, sorghum and other crops of fodder value – have been characterized and maintained as core and mini-core collections by different institutes under the Consultative Group on International Agricultural Research (CGIAR). These genetic resources must be utilized in forage breeding to develop diverse varieties/lines suitable for various agro-ecological niches. The important forage breeding objectives are: high green and dry matter yield, nutrition quality parameters – crude protein (%), *in vitro* dry matter digestibility (IVDMD) along with low percentage of neutral detergent fibre (NDF), toxicity in some crops, tolerance to adverse soil and weather conditions like high rainfall/low moisture condition, acidic/saline soils, good response to agronomic inputs, resistance/tolerance to important diseases and insect pests, and higher regeneration ability in multi-cut perennial legumes and grasses. Concerted efforts are needed through a pre-breeding approach to search and transfer useful genes from wild to cultivated species. The desirable genetic stocks can be further used in crop improvement activities along with the creation of wide-ranging variability aimed at identification of better plant types and development of new varieties suitable for mixed cropping/farming systems. Development of interspecific hybrids, i.e. pearl millet, napier hybrids, trispecific hybrids (pearl millet–Napier-*P.*

squamulatum) comprising desirable traits, i.e. nutritional quality, perenniality, hardiness, resistance to biotic and abiotic stresses and better adaptability, is needed. Modern techniques like embryo-rescue and micro-propagation have been used for the production of hybrids in *Lolium-Festuca* complex and can be extended to other interspecific combinations. Use of brown midrib (bmr) mutant for enhancing quality traits along with stay-green trait is an essential component for future forage breeding programmes. For popularizing dual-purpose fodder crops, it is essential to transfer stay-green trait in popular varieties. Combined efforts are needed to incorporate these traits through molecular as well as conventional breeding approaches. Molecular mapping and tagging the gene of interest in forage crops have not been explored very much. Gene pyramiding and incorporation of the gene(s)/quantitative trait loci (QTLs) of desirable traits, i.e. high tillering, perenniality, ability to withstand adverse conditions and tolerance/resistance to multiple stresses (both biotic and abiotic), is another aspect of great importance for developing improved varieties of fodder crops that needs to be given greater attention. It is always advisable to include cereal and legume components for a balanced diet. The development of intercropping-compatible new plant types of forage cereals and legumes will be of great importance in achieving the production of a balanced diet. Concerted efforts are needed to develop short-duration, multi-cut, stay-green and dual-purpose varieties to fit into diverse cropping systems to serve the specific needs of farmers. Apomixis in range grasses also needs to be exploited in order to develop new improved strains/varieties and fixation of heterosis. Most of the tropical range grasses are apomictic in nature. Apomixis facilitates the maintenance of varieties due to their true-to-type nature. However, the development of variation for the selection process under breeding is severely hampered resulting in lack of hybrids as well as high-yielding varieties with good heterozygosity in most of the grass species. The presence of sexual lines in some species, i.e. *Cenchrus ciliaris*, *Brachiaria* species, helped in the development of hybrids in these crops. The identification of sexual lines and their use in the development of hybrids shall be prioritized in tropical forages of India. Important varieties of fodder crops are listed in Table 13.2.

Practising Food-fodder Mixed Farming

A food-fodder cropping system needs to be promoted on a large scale to provide balanced nutrition to livestock. In mixed farming systems, the following measures should be adopted: (i) enhanced nutritional quality of crop residues through proper storage and value addition for livestock; (ii) emphasis on green forage-based feeding system; (iii) use of legumes and top feeds; and (iv) minimization of methane emissions. Efforts should be made to improve forage production in a farming system's mode for efficient utilization of available land.

Since farmers and livestock keepers require fodder all-year-round, they need to adopt an overlapping cropping system. This system consists of raising berseem, interplanted with hybrid napier in spring and intercropping the inter-row spaces of the grass with cowpea during summer after the final harvest of berseem. Under assured irrigation, cropping systems like guinea grass + (cowpea – berseem), hybrid napier + (cowpea – berseem), hybrid napier (IGFRI 3) + leucaena (K8), fodder sorghum + cowpea – berseem – maize + cowpea, MP chari + cowpea – berseem – cowpea, MP chari + cowpea – berseem + mustard – maize + cowpea, and MP chari + cowpea – berseem + mustard have been recommended for round-the-year fodder production and availability. Among these systems, hybrid napier + (cowpea – berseem + mustard) had the highest biomass production potential (273.1 t/ha green and 44.3 t/ha dry fodder). Region–specific, high-intensity forage production systems like maize + cowpea – lucerne + oats – mustard, NB hybrid + berseem – cowpea, NB hybrid + lucerne, and

sorghum + cowpea – berseem + mustard – maize + cowpea may also be considered for hill, tarai, semi-arid and sub-humid regions, respectively, with green fodder production potential ranging from 85 to 255 t/ha p.a.

The cultivation of fast-growing forage crops during the gap periods is also expected to provide around 30–35 t/ha green fodder. Fodder production systems like MP chari + cowpea, bajra + cowpea, and maize + cowpea for lean periods in a rice-wheat cropping system (April–June) and fodder turnip, carrot, Japan rape and Chinese cabbage after harvesting of early rice/sorghum are most suitable. These systems are of particular interest to farmers having smallholdings with assured irrigation who cannot divert land exclusively for fodder production. This system has specific benefits of efficient utilization of the gap period, additional yield from the system and enhanced farm income. There is even a need to integrate forage crops in the existing food-based cropping systems to enhance forage availability without compromising yield of food crops. Cowpea-wheat sequence was found to be more remunerative (38.8 t/ha green fodder, 4.63 t/ha wheat grain and 5.02 t/ha wheat straw) than grain sorghum-wheat cropping systems with enriched soil fertility and economized nitrogen fertilizer to the extent of 40 kg/ha. In food-fodder production systems, fodder cowpea (1.36 t/ha dry fodder) + annual *Sesbania* (1.89 t/ha dry fodder) + grain sorghum was found to be the best crop combination.

Rehabilitation/Development of Grasslands and Wastelands

Rehabilitation of degraded grasslands for livelihood support especially in hill, semi-arid and arid regions, and also utilization of wastelands with emphasis on range grasses and legumes, should be taken on a participatory basis. Since area expansion in cultivated lands under forage production is a remote possibility, concerted efforts need to be made to make best use of available grazing resources (Table 13.3), including wasteland along railway tracks and on the roadside, by planting better varieties of range grasses and range legumes and by managing and using these areas for livestock grazing.

Table 13.2. Important varieties of cultivated fodder crops and grasses/legumes. (From: Hazra, 1989, 1995)

Crop	Varieties	Production potential (green forage t/ha)	Adaptable region	Suitable for existing production systems
Sorghum	Pusa Chari 1, CO27, SSG 59-3 (Meethi Sudan), CSH 20 MF (UPMCH 1101), PAC 981, CSV15	35–45	Whole country except temperate hills	Food-forage cropping system/sole forage
Bajra	Avika Bajra Chari (AVKB19), Raj Bajra Chari 2, CO 8, APFB 2, PCB 164	30–40	Whole country except temperate hills	Food-forage cropping system/sole forage
Maize	Pratap Makka Chari 6	40–50	Whole country	Sole forage/silage (milkshed areas)
Sudan grass	Meethi Sudan, Sweet Sudan Grass, Punjab Sudex Chari-1 (LY 250)	45–65	Whole country except temperate hills	Sole forage/silage (milkshed areas)
Oat	FOS1/29, Bundel Jai 822, Bundel Jai 992 (JHO 99 2), JHO 2009-1	35–45	North, central & north-western, hill region	Sole forage/silage (milkshed areas)
Cowpea	Bundel Lobia1, Bundel Lobia 2 , S 450	20–25	Whole country	Food-forage cropping system/sole forage
Guar	Durgajay, Durgapura Safed, HFG 119, Bundel Guar 1, Bundel Guar2, Bundel Guar 3	20–25	Whole country except temperate hills	Food-forage cropping system/sole forage
Sem	Bundel Sem1	20–22	Whole country except temperate hills	Food-forage cropping system/sole forage
NB hybrid	CO 1, NB 37	150–180	Whole country except temperate hills	Round-the-year forage system, on farm boundaries, hortipasture
Guinea grass	Bundel Guinea 1 (JHGG 96-5), Bundel Guinea 2 (JHGG 04 01)	120–150	Whole country except temperate hills (very high altitude)	Round the year forage system, on farm boundaries, hortipasture
Dinanath grass	Bundel 1, Bundel 2, COD 1	35–45	Whole country except temperate hills	Silvipasture, forest fringes, degraded lands/ watersheds, community lands
Anjan grass	Bundel Anjan 1, CO 1, Bundel Anjan 3	25–35	Whole country	Silvipasture/horti-pasture, forest fringes, degraded lands/watersheds, community lands
Motha dhaman grass	CAZRI 76, Marwar Dhaman (CAZRI 175)	25–35	Central, western & dry arid regions	Silvipasture, forest fringes, degraded lands/ watersheds, community lands
Black spear grass	Bundel Lampa Ghas 1	25–30	Whole country (arid & semi-arid regions)	Silvipasture, forest fringes, degraded lands/ watersheds, community lands
Stylosanthes sp.	Stylosanthes hamata, S. seabrana, S. scabra	25–30	Whole country (semi-arid regions)	Silvipasture/horti-pasture, forest fringes, degraded lands/watersheds, community lands
Clovers	White and red clovers	20–25	Temperate/sub-temperate regions	Hortipasture/Silvi-pasture, forest fringes, degraded lands/watersheds, community lands

Overgrazing is usually considered the major cause of grassland degradation/deterioration. There are a number of proven management options to improve grassland productivity including its quality (Table 13.4). The principles of grassland management are based on ecological succession, assisted ecological succession and intensive management. It is usually achieved through offering protection for vegetation recovery, soil and water conservation, management of bushes, reseeding of better-quality and more productive grasses/legumes, fertilizer application, cutting and grazing management. The gains as a result of judicious management are quite significant. Such options may be used in isolation or in a combination of two or more, depending on the local situation, feasibility and resource availability (Roy, 2017). Different grass/legume species can also be cultivated or introduced on specified lands like forest lands, permanent pastures and other grazing areas. This will provide additional quality fodder resource for the animals as well as reduce the pressure on natural grasslands, facilitating their restoration.

During the recent past, silvipasture systems/models are also becoming popular, which optimize production rather than maximization from the same unit of land through integrated management. Silvipasture is an agroforestry practice for the production of trees, tree products, forage and livestock and has a significant role in semi-arid and arid regions. Trees and bushes provide green fodder during lean periods when grasses are not available. These systems are for different types of degraded lands and suitable tree/crop species are introduced based on prevalent climate and kind of soil (Table 13.5). A comparative assessment of the production scenario of various systems indicated that the production of forage can be raised by three times and its quality by seven times. The degraded wastelands (shallow, red, gravelly soil), under semi-arid condition, of Jhansi, which were producing 1 t/ha p.a. earlier, produced 10 t/ha p.a. at ten years' rotation through silvipastoral systems. Besides improved

Table 13.3. Grazing resources in India. (From: ICAR, 2009)

Resources	Area (million ha)	%
Forests	69.41	22.7
Permanent pastures, grazing lands	10.9	3.6
Cultivable wasteland	13.66	4.5
Fallow land	24.99	8.1
Fallow land other than current fallows	10.19	3.3
Barren uncultivable wastelands	19.26	6.3
Total common property resources other than forests	54.01	17.7

Table 13.4. Management options for rehabilitation of degraded grasslands. (From: Roy, 2017)

Management options		Gains (%)	
		Forage yield	Crude protein
Protection of area from grazing	Two-year closure	158	15.98
	Five-year closure	233	17.24
	Long-term closure	238	24.45
Soil and water conservation techniques	Natural grasslands	83	–
	Reseeded grasslands	58	–
Management of shrub density	Canopy (18%)	18	–
	Canopy (14%)	27	–
	Canopy (11%)	31	–
Fertilizer application	*Sehima nervosum* grassland (@ 60 kg N/ha)	83	37.23
	Heteropogon contortus grassland (@ 40 kg N/ha)	60	23.17
	Iseleima laxum grassland (@ 40 kg N/ha)	42	24.67

Other management options to be considered: reseeding at periodical intervals, burning at periodical intervals, harvest management, grazing management.

Table 13.5. Silvipasture models for degraded lands in semi-arid regions. (From: Roy *et al.*, 2005)

| Tree | Pasture | |
	Grass	Legume
Acacia nilotica	*Cenchrus ciliaris*	*Stylosanthes hamata, S. scabra*
	Dichanthium annulatum	*S. hamata*
Acacia tortilis	*Cenchrus ciliaris, C. setigerus*	*S. hamata, S. scabra*
	Dichanthium annulatum	*S. hamata*
Azadiracta indica	*Cenchrus ciliaris*	*S. hamata*
	Dichanthium annulatum	*S. hamata*
Albizia amara	*Cenchrus ciliaris, C. setigerus*	*Macroptilium atropurpurreum, S. hamata*
	Dichanthium annulatum	*S. hamata*
	Panicum maximum	*S. hamata*
Albizia lebbeck	*Cenchrus ciliaris, C. setigerus*	*Clitoria ternatea, S. hamata*
	Chrysopogon fulvus	*Clitoria ternatea, S. hamata*
	Sehima nervosum	*Clitoria ternatea, S. hamata*
Albizia procera	*Cenchrus ciliaris, C. setigerus*	*Clitoria ternatea, S. hamata*
Dalbergia sissoo	*Cenchrus ciliaris*	*S. hamata, S. scabra*
	Dichanthium annulatum	*S. hamata, S. scabra*
Hardwickia binata	*Cenchrus ciliaris + Chrysopogon fulvus*	*Macroptilium atropurpureum, S. hamata*
	Sehima nervosum	*Clitoria ternatea, S. hamata*
	Panicum maximum	*S. hamata*
Leucaena leucocephala	*Cenchrus ciliaris, C. setigerus*	*S. hamata, S. scabra*
	Dichanthium annulatum	*S. hamata, S. scabra*
	Panicum maximum	*S. hamata*
Zizyphus mauritiana	*Cenchrus ciliaris*	*S. hamata*

yield improvement by eight to ten times, the quality of mixed forage was also improved by six to seven times. A study on forage or grass production in a 14–18-year-old plantation of four desert trees revealed that dry matter yield was the maximum under *Prosopis cineraria* (1.54 t/ha) and minimum under *Acacia senegal* (0.69 t/ha). The increase in forage yield was attributed to an increase in organic matter and increased availability of nutrients under a *Prosopis cineraria* tree (Ahuja *et al.*, 1978).

Establishing Forage Banks

There is a need to promote a forage bank concept for preserving surplus production from rangelands during the rainy season to be used in lean periods by transporting economically baled and nutritionally enriched dry fodder from surplus-production areas to low-production areas. There is a need to explore the possibility of interstate transport of crop residues for fodder and feed at the time of harvesting of paddy and also wheat straw. The facility may be strengthened to promote

commodity forage banks at *Tehsil/Taluka* level, where surplus fodder can be stored as hay/silage/fodder blocks for use during the scarcity periods.

Forage Seed Availability and Supply

Availability of quality forage seed is crucial in realizing the enhancement of forage production. With the increased preference for stall feeding due to reduced grazing areas and degraded common property resources, the inclination of farmers towards fodder production is to be capitalized upon. Seed availability is crucial under these circumstances. As per an estimate, the availability of quality seed is less than 25% of the required quantity (Table 13.6). Even though breeder seed is being produced as per the requirement, the absence of further multiplication leads to severe shortages of seed to the end-users. The breakage of the seed chain at foundation and certified levels due to lack of involvement by production organizations coupled with marketing has aggravated the situation.

Table 13.6. Estimated seed requirements of cultivated fodders. (From: Hazra, 1989, 1995)

Crop	Area (million ha)	Average seed rate (kg/ha)	Breeder seed (t)	Foundation seed (t)	Certified seed (t)
Maize	0.9	20	1.8	180	18,000
Jowar	2.6	10	2.6	260	26,000
Bajra	0.9	10	1.4	112	9,000
Oats	0.2	75	46.9	937	18,700
Berseem	2.0	20	64.0	1600	40,000
Lucerne	1.0	15	21.6	562	15,000
Cowpea	0.3	20	6.7	200	6,000
Guar	0.2	20	2.0	89	4,000
Total	–	–	**147**	**3941**	**136,700**

Policy-level constraints have crippled forage development. The non-availability of authenticated data, lack of coordination for indent and supply, and stringent certification requirements have invariably affected seed production adversely. The increase in the seed replacement rate of fodder crops from the existing 2–3% to almost 20% shall be achieved with a 2% increase every year. The utilization of uncultivable lands for forage purposes along with revitalization of denuded grasslands and rangelands is the best possible solution for reducing the shortage of fodder. The large-scale forage production activities will yield significant results by following proper management as discussed above. The assured purchase of seed will ensure development of the forage seed sector as well as pave the way for PPP, provided proper policy support is extended. The lack of an assured market accounts for the lack of enthusiasm of the private sector for forage crops. Development programmes involving the public and private sectors for seed production with technical backstopping of research institutes like the Indian Grassland and Fodder Research Institute (IGFRI), with certification for seed quality, will have great impact on the whole scenario of fodder shortage. There is even a justification to get away with certification of seeds when the produce is entirely used for fodder.

The establishment of seed banks should be initiated on the pattern of the seed village on a pilot basis by participatory forage seed production, by organizing forage seed markets and by organizing farmers' training programmes, especially for forage seed production. Production of seeds of range grasses and legumes needs to be encouraged for the spread of these grasses even inside the forest areas. Common property resources (CPRs) should also be re-vegetated with range grasses and legumes. Farmers should be encouraged to maintain grasses like napier × bajra (NB) hybrid, guinea grass (*Panicum maximum*), and anjan grass (*Cenchrus* species) for supply of rooted slips to meet local demands. The revived CPRs and forest wastelands will also enhance biodiversity and help in ecological restoration. Grasses being the most climate-resilient species, their spread in the natural habitat will result in development of sustainable forage production niches under a climate-change scenario.

The disconnect of the seed chain between foundation to certified seed class is one of the major reasons for quality seed shortage in cultivated fodder crops. In Uttar Pradesh during 2011–12, only 26–30% multiplication was observed from foundation to certified seed in cowpea and berseem. The other reason for seed shortage is lack of sufficient breeder seed indent by various production and multiplication agencies. This will ultimately result in less production of certified seed even if the seed chain is properly followed (Table 13.7). The percentage deficit in the last column clearly indicates that the breeder seed was not indented as per the certified seed requirement.

In the case of cultivated fodder crops, seed availability can be increased by strengthening the seed chain (breeder seed/foundation/certified/TFL) and entrusting the responsibility to different agencies with regular monitoring of seed multiplication. The production of breeders' seed shall be entrusted to ICAR institutes like the IGFRI and the SAUs as they are the primary institutions engaged in the development of improved varieties. The seed multiplication phases containing foundation and certified seed production

Table 13.7. Expected and estimated certified seed quantities of different forage crops. (From: Hazra, 1989, 1995)

Crop	Breeder seed production 2011–12 (t)	Expected foundation seed 2012–13 (t)	Expected certified seed 2013–14 (t)	Estimated certified seed requirement (t)	Percentage excess/deficit availability
Maize	10.925	1092.5	109,250	18,000	+507
Jowar	7.364	736.4	73,640	26,000	+183
Bajra	0.922	73.7	5901	9000	−34
Oats	61.130	1222.6	24,452	18,700	+31
Berseem	7.668	191.7	4793	40,000	−88
Lucerne	0.660	17.1	446	15,000	−97
Cowpea	1.372	41.1	1235	6000	−79
Guar	0.540	24.3	1094	4000	−73

shall be strictly entrusted to Regional Stations for Forage Production and Demonstration (RSFPD) of the Department of Animal Husbandry, Dairying and Fisheries (DAHDF), the National Seed Corporation (NSC), the State Seed Corporations (SSCs) and the milk cooperatives under the National Dairy Development Board (NDDB). The presence of small-scale dairy units with approximately 97% of farms having one to three cattle resulted in a very fragmented forage market. The lack of awareness about the importance of green fodder and the unwillingness to invest for fodder resulted in the absence of an assured market for forage. Thus, even though there is a huge requirement for green fodder, it is not being converted into market demand. This has a direct implication on the demand for fodder seeds per se. Necessary measures in terms of creating awareness, marketing facilities and incentives for forage seed production similar to food crops will help increase forage production significantly.

The supply of forage seeds by the public sector lacks a proper marketing structure with a network of dealers and distributors. The private sector involving reputed firms are very few in the forage arena. The major market share by the informal seed sector is also one of the obstacles for forage business growth. The lack of a quality assurance system in the informal system results in reduced production and profitability, thereby diminishing the interest and importance shown by farmers in these crops. The encouragement of the private sector, with assured purchases and link-up with government development programmes, including development of mining rehabilitation sites, highways, riverbanks and forest and railway wasteland, will augment seed and forgage production.

Improvement in Forage Quality and Value Addition

Concerted efforts should be made by the plant breeders with the help of animal nutritionists to develop forage crop varieties with improved quality traits. Development of grasses and forage crops with higher water-soluble nutrients, enhanced digestibility and more organic acids would open new avenues for forage-based, ecofriendly livestock production with reduction of methane emission while rearing bovine livestock. Cereal crop residues are again important in Indian agriculture as they are the staple feed for livestock. But earlier, under the breeding programmes of cereal food crops, efforts were made to maximize the yield of grains, neglecting the straws/stovers component both quantitatively and qualitatively. This calls for urgent attention to the screening of sole/dual-type cereal crop varieties for better quality and higher yields of both grains and straws/stovers.

The major constraints associated with the use of these cereal crop residues as feeds are low nutrient density and poor digestibility, which lead to low nutrient intake and reduced animal performance. The low nitrogen and mineral profile, along with the high lignin and silica levels associated with these dry forages leads to lower nutrient digestibility resulting in hardly meeting the maintenance needs of even non-productive animals on straw-based rations. Hence, they need to be processed or supplemented if they are meant for production in terms of milk, meat or draught purposes. These crop residues are subjected to various processing methods like chaffing, grinding, pelleting, and chemical

Table 13.8. Different processing methods for cereal crop residues.

Physical	Chemical	Physico-chemical	Biological
Soaking	Sodium hydroxide	NaOH/pelleting	Addition of enzymes
Chopping/grinding	Calcium hydroxide	Urea/pelleting	White rot fungi
Pelleting	Urea-ammoniation	Lime/pelleting	Mushroom cultivation
Boiling	Ammonium hydroxide	Chemicals/steaming	
Steaming under pressure	Anhydrous ammonia	NaOH/temperature	
Gamma irradiation	Sodium carbonate		
	Sodium chlorite		
	Chlorine gas		
	Sulphur dioxide		

and biological treatments, especially for their incorporation into complete feeds with the aim of increasing palatability, intake, nutrient utilization and improved animal performance (Krishna, 2003). Many methods of treatment have been exhaustively studied before accepting urea ammoniation as the most potential method for field-scale application in most of the Asian countries including India (Table 13.8). Thus, post-harvest technologies such as processing, enrichment/ fortification of straws/stovers, ensiling, drying, densification etc., alone or in combination, hold the key for improved and efficient utilization of poor-quality forage resources.

Studies have also confirmed that cereal crop residues or dry forages, when fortified or supplemented with concentrates, legume forages, tree leaves and protein supplements, can significantly improve the dry matter digestibility, feed intake and, ultimately, animal performance. Supplementation helps in overcoming the low palatability of these dry forages. However, the level of supplementation plays a vital role in deciding the level of performance. There are different types of supplementation (catalytic, strategic and substitutional), which can be distinguished one from another, and each one serves a particular purpose and fits the need of a particular farming and animal production system.

needs to be adopted for the controlled use of vegetation/grasses available in the forest margin, which is an important source of forage for livestock. Animals should not be blamed solely for deforestation; studies have indicated that animals are not the destroyers of vegetation in forest areas, provided controlled grazing or removal of vegetation/grasses is practised in a scientific manner, which helps in further revival of vegetation. The forest departments have also a real need for seeds of different grasses with high-quality biomass yields that require proper attention. The planting of trees that have fodder value should also be considered under different forestation programmes. Thus the staff of forest departments need to be educated as to suitable grass/tree species, keeping in view the prevalent forest ecosystem. This will not only reduce the gap between demand and supply of forage in normal years, but will also act as a live fodder bank for livestock during periods of drought and flood. However, grazing should be based on scientific carrying capacity of the system. Indeed, an appropriate system for livestock management needs to be developed through interdepartmental programmes, and stall feeding of animals needs to be encouraged. But, usually, it is not being practised by the livestock farmers and overgrazing is rife, leading to degradation of forest forage resources (Anonymous, 2016).

Utilization of Forage Resources from Forests

In India, around 69 million ha of land are under forest cover, which need to be exploited both scientifically and judiciously. A synergistic approach between the forestry and livestock departments

Use of Non-conventional Forage Resources

There are a number of non-conventional feed/ fodder resources that can supplement existing green herbage for ruminants under varied management situations. Efforts should be made to improve the basket of feed resources through evaluation of non-conventional/under-utilized feed resources like azolla (humid and sub-humid conditions), turnip and fodder beets (intensive management system) and cactus (semi-arid and arid condition) for their inclusion and effective utilization in livestock diets (Ghosh *et al.*, 2013). Azolla is highly productive, having the ability to double its weight within a week. It can produce nine tonnes of protein/ha of pond p.a. and is cultivated widely in different countries, mostly along the rice fields. It is rich in crude protein (18–20%) and a potential feed ingredient for livestock. Azolla has the potential to replace expensive concentrate feeds in ruminant diets. Fresh azolla is usually mixed with other feeds in a 1:1 ratio and fed to dairy cattle, sheep and goats. Fodder beet (*Beta vulgaris*) has also the potential to become a good forage crop for livestock due to its drought tolerance, excellent root-keeping qualities, high sugar content, good leaf fodder characteristics, high nutritive value and high yields (95.6 t/ha) compared to other forage crops. However, fodder beets do not form a complete diet (acting more like a concentrate) and so should be supplemented with a high-protein fodder such as berseem or lucerne hay. If fodder beets are fed with grass hay, it is necessary to supplement the feed ration with a high protein additive such as bran or oilseed cakes. In general, the fodder beets are sliced or shredded to prevent animals from choking on large pieces, and mixed with green fodder or concentrate feeds for feeding. Cactus is a succulent plant, well adapted to extreme drought conditions. It grows in areas having 200–250 mm p.a. of rainfall and tolerates high temperatures. It can thrive where common forage species cannot grow. Usually, the yields vary between 30 and 100 t of fresh cladodes and between 2 and 20 t of fruits/ha p.a. Thus, one hectare can produce a consumable biomass yield that exceeds those of many common forage species when cultivated in semi-arid and arid regions. Cactus cladodes are mostly fed fresh to cows, sheep, goats and dromedaries. In order to avoid material loss, it is recommended to cut cladodes into small slices before offering to animals.

Many non-conventional feed resources are considered as waste, which is not true. In fact, such non-conventional feeds can easily be used by the livestock and can become new feed materials of importance. In addition, they can be used to supplement existing feed resources. The demonstration of potential value can thus make any of these waste products new or alternative feeds of value and importance (Amata, 2014). In many Asian countries, including India, there is a continuing shift in the cropping pattern from cereals to more remunerative fruit and horticultural crops. This results in generation of huge quantities of fruit and vegetable residues. Recent investigation on post-harvest studies revealed that 5.8–18% of losses were estimated from fruit and vegetables grown in the country. Presently, such residues are not efficiently used and either composted or dumped in landfill causing environmental problems. But such non-conventional resources can act as an excellent source of nutrients and alternative fodder for ruminant animals (Table 13.9). However, a major issue that must be addressed scientifically involves the nutritional benefits and effects of feeding fruit waste to animals. There is a need to develop suitable methods to convert waste to edible products and to contribute to value-added feed resources.

Table 13.9. Nutritional value (% DM basis) of horticultural wastes/residues. DM = Dry Matter; CP = Crude Protein; NDF = Neutral Detergent Fibre; ADF = Acid Detergent Fibre. (From: Gowda and Vijay Bhasker, 2017)

Waste/residues	Botanical name	DM	Ash	CP	NDF	ADF
Apple pomace	*Malus domestica*	35.9	2.6	7.7	52.5	43.2
Banana peels	*Musa acuminata*	9.4	11.1	8.1	35.8	25.3
Citrus pulp	*Citrus limetta*	9.5	4.5	10.5	26.5	24.5
Grape pomace	*Namily vitaceae*	35.0	7.9	8.9–12.2	51.56	48.4–52.6
Pineapple bran	*Ananas comosus*	9.9	3.5	4.6	73	37
Pineapple fruit residue	*Ananos camosus*	15.0	2.3	7.5	56.04	19.76
Tomato pomace	*Solanum lycopersicum*	25.3	6.0	22.1	63	51

Convergence and Strengthening Linkages

Forage resource development-related activities are required to be tailored in harmony with the policies of central government for developmental and livelihood-supporting projects such as the National Horticulture Mission (NHM), the Mahatma Gandhi National Rural Employment Guarantee Act (MGNREGA) and the National Rural Livelihoods Mission (NRLM). Credit and market linkages to forage-based livestock production needs support from central and state governments enabling livestock keepers to boost their income from dairying/animal husbandry. There is also a need to establish producer companies and market linkages with private sector agencies, whilst involving ICAR institutions, SAUs and farmers. Technologies for forage-based livestock production have percolated at a very slow rate to the end-user. Now strategies need to be changed from simple mini-kit programmes on cultivated fodder from the Department of Animal Husbandry, Dairying and Fisheries (DAHDF) to focused technology demonstrations. Adoption and applicability of lCT to promote forages at field level need to be explored.

Supportive Strong Policies

As is evident from the above, the forage resource development is a more complex issue than is food and commercial crops. Lack of momentum in feed and fodder development in the country is largely the result of poor organizational structure. Some of the prominent aspects related to policy, required to provide a favourable environment for accelerated forage development in India, are: collection of data on fodder production and utilization; investment in forage resource development; credit facilities to enhance forage production; a support price for forage and marketing of seed; non-diversion of edible crop residues for other uses like packaging; a policy on grazing and common property resources; and legal protection of grasslands.

Strengthening of the Livestock Mission

The National Livestock Mission (NLM), which has been created based on the Hooda Committee recommendations (Hooda Committee Report, 2010) as part of the government's 12th Plan, needs to be strengthened for rapid livestock and dairy development with a holistic approach. It should address concerns for an efficient artificial insemination service to cover all breedable bovine females through AI or NS (natural service) using better genetic material; conserving native breeds; and improving local breeds using sexed semen and biotechnology, leading to overall development of livestock species. Similarly, dairy development activities should address issues like farmer to consumer, quality milk production and cold-chain infrastructure, strengthening procurement, processing, marketing, capacity-building of farmers, strengthening dairy cooperatives, technical inputs for scientific feeding and management of animals.

India is blessed with diverse livestock, one of the largest in the world; but the average productivity of these animals is quite low compared to the global average and that of developed countries. The production potential of these animals is not realized fully because of constraints related to feeding, breeding and health management. Proper feeding and nutrition of healthy animals is most important; thus the availability of quality fodder resources becomes critical. With an ever-growing human population, the demand for livestock products will increase substantially, requiring improvement in livestock productivity. Also, additional cropland area for forage production is not likely to be available. Hence, other options will have to be explored and strategies adopted to accelerate forage production and its availability for livestock development.

The Way Forward

Accelerating production of forage crops is more complex than with food and commercial crops production. Earlier, the main emphasis was on foodgrain production to achieve national food security. Now, we need household nutritional security through greater emphasis on livestock development. Fortunately, the increased contribution of livestock would require a structural shift in the agricultural sector. Policy makers must now recognize that the livestock sector is an important engine of agriculture growth, since livestock contributes more than a quarter of GDP and hence deserves adequate technological, institutional and policy support.

The following strategies need to be put in place to accelerate the production and utilization of forage resources, so important for faster growth of livestock sector:

- The forage genetic resource base must be enhanced. Fortunately, very rich genetic diversity exists in forage crops, which needs to be conserved, characterized and utilized in breeding programmes for developing high-yielding, nutritive, dual-purpose and climate-resilient forage crop varieties. Therefore, plant genetic resource activities in fodder crops, including range grasses and legumes, should be given high priority.
- Efforts on pre-breeding will have to be strengthened in order to create wide genetic variability aimed at identification of better-yielding and nutritive plant types. Incorporation of tolerance/resistance to multiple stresses (both biotic and abiotic) will be another objective that needs to be given greater attention. Concerted efforts are also needed to develop short-duration, multi-cut, stay-green and dual-purpose varieties to fit into diverse cropping systems. Interspecific hybridization to transfer desirable genes, including apomixis in range grasses, will help in accelerating the forage breeding process.
- Improved agronomic practices for different fodder crops need to be developed, standardized and disseminated to farmers for enhancing forage production in different cropping/farming systems in different agro-ecoregions. These improved practices need to be developed both for mono-cropping and mixed-cropping systems.
- Addressing availability of good-quality seed of improved varieties of forage crops is an important issue. There is a need to have an effective mechanism to gather authentic data on forage seed production. Some out-of-box solutions like dispensing with a certification system of seeds to be used for forage production, establishing producer companies and market linkages with private sector agencies etc. are required. Greater involvement and effective coordination between agriculture and livestock departments as well as seed-producing agencies and the private sector,

with farmers' participation, will help ensure production of an adequate quantity of forage crop seeds.

- A food-fodder cropping system needs to be promoted to ensure provision of balanced nutrition to livestock. Since the farmers and livestock keepers would require fodder year-round, they should be encouraged to adopt overlapping cropping systems that are suitable for different agroclimatic conditions.
- Growing forages on 'bunds' and along railway tracks following a non-competitive land use approach may be an option. Similarly, there is an urgent need for formulation of a comprehensive strategy for rejuvenation of grazing land and common property, like encouraging the establishment of cooperatives for forages and pasture management and utilizing wasteland for forage production. This may involve a huge seed requirement, which, in turn, may attract the seed industry to meet the additional requirement.
- Non-conventional/under-utilized fodder resources need to be exploited for fodder availability during lean/drought periods. There is a need for identification, evaluation and domestication of forage halophytes, growing of azola as feed, and the utilization of saline water in water-scarce areas for forage production. Similarly, aquatic and water-logged areas need to be exploited for aquatic forage production such as para-grass, coix etc.
- A system of inventory and assessment (covering both yield and quality) of forage resources in major forest types is also required with an in-built system of monitoring and periodically updating. A synergistic approach between the forestry and livestock departments needs to be adopted for the controlled grazing of pasture grasses and legumes in forest margins by livestock farmers.
- Concerted efforts need to be made to establish fodder seed banks with an appropriate network for easy access to information and availability of seeds within a short distance. Fodder banks should also be established near the forest areas for needed rejuvenation and afforestation initiatives.
- Forage resource development and related activities need to be tailored in line with developmental and livelihood-supporting projects as per the policies of central government such as Horti-Mission, MGNREGA,

the National Rural Livelihoods Mission and the National Livestock Mission.

- The Livestock Insurance Scheme needs to be implemented efficiently and effectively in order to make farmers risk-free. Animal insurance should be structured to involve product innovations and effective delivery through farmer organizations. Institutional credit at low interest rates needs to be ensured for dairy farmers.
- Adequate infrastructural development in the livestock sector needs to be carried out

through investment by public and private institutions. Farm income can be enhanced by reducing or sustaining the cost of livestock products, augmenting R&D towards increased milk/meat and fodder productivity, improving quality and adopting innovative farming practices. There is also a need to link the production system to consumer demand and to have a balanced growth in the crop-livestock production system, making livestock farming globally competitive and rewarding for smallholder farmers.

References

Ahuja, L.D., Verma, C.M., Sharma, S.I. and Lamba, T.R. (1978) Range management studies on the contribution of ground storey (grass) in afforested areas in arid region. *Annals of Arid Zone* 17, 304–310.

Amata, I.A. (2014) The use of non-conventional feed resources for livestock feeding in the tropics: a review. *Journal of Global Biosciences* 3, 604–613.

Anonymous (2016) Augmenting forage resources in rural India: policy issues and strategies. Policy Paper No. 80, National Academy of Agricultural Sciences, New Delhi.

Dikshit, A.K. and Birthal, P.S. (2010) India's livestock feed demand: estimates and projections. *Agricultural Economics Research Review* 23, 15–28.

Ghosh, P.K., Singh, J.P., Mahanta, S.K., Rai, A.K., Yadav, V.K. and Das, N. (2013) *Technology Developed for Enhanced Forage Productivity*. Indian Grassland and Fodder Research Institute, Jhansi, India.

Gowda, N.K.S. and Vijay Bhasker, T. (2017) Fruit residues as alternate forage resources for livestock. In: Ghosh, P.K., Mahanta, S.K., Singh, J.B., Vijay, D., Kumar, R.V., Yadav, V.K. and Kumar, S. (eds) *Approaches Towards Fodder Security in India*. Studera Press, New Delhi, pp. 535–550.

Hazra, C.R. (1989) *Technology for Increasing Forage Production in India*. Indian Grassland and Fodder Research Institute, Jhansi, India, pp. 1–36.

Hazra, C.R. (1995) *Advances in Forage Production Technology*. Indian Grassland and Fodder Research Institute, Jhansi, India pp. 1–51

Hooda Committee Report (2010) Report of Working Group on Agriculture Production under the chairmanship of Bhupinder Singh Hooda to recommend strategies and an action plan for increasing agricultural production and productivity, including long-term policies to ensure sustained agricultural growth.

ICAR (2009) *Forage Crops and Grasses: Handbook of Agriculture*. Directorate of Information and Publications of Agriculture, Indian Council of Agricultural Research, New Delhi.

IGFRI (2014) *Vision-2050 Document*. Indian Grassland and Fodder Research Institute, Jhansi, India.

Krishna, N. (2003) Processing of roughages for improved utilization. In: Reddy, G.V.N., Krishna, N., Prasad, V.L.K., Reddy, K.J. and Reddy, Y.R. (eds) *Feed Technology: Present Status and Future Strategies*. ANGRAU, Hyderabad, India, pp. 86–100.

Roy, M.M. (2017) Grassland development on degraded lands. In: Ghosh, P.K., Mahanta, S.K., Singh, J.B., Vijay, D., Kumar, R.V., Yadav, V.K. and Kumar, S. (eds) *Approaches Towards Fodder Security in India*. Studera Press, New Delhi, pp. 267–282.

Roy, M.M., Kushwaha, D., Gupta, S.K. and Singh, M. (2005) Lessons learnt in promoting MPTS on wastelands of farmers in semiarid rainfed regions. *Forests, Trees & Livelihoods* 15, 383–388.

14

Agricultural Biotechnology for Food and Nutritional Security

India, with its 1.34 billion people, is the second most populous country in the world and supports 17.74% of the world's population (7.55 billion) (United Nations, 2017). Around the world, 815 million people, or about one in ten, are undernourished (FAO *et al.*, 2017). Of these, 23.4% (190.7 million) live in India. The country's foodgrain production has increased four-fold over the last five decades; but the yield of major foodgrain crops is reaching a plateau, while the population continues to rise and is predicted to reach 1.7 billion by 2050. Hence, there is a need to increase foodgrain production from the current level of 277.49 million t in 2017–18 to 345 million t by 2030. Among staples, India needs to produce 120 million t of rice, 100 million t of wheat and 32 million t of pulses. Since net sown area has reached an optimum level (140 million ha), the only option available is to go vertical by increasing crop productivity through cultivation of high-yielding, more stress-tolerant varieties, which can give desired yields even under adverse growing conditions. In the past, policy makers and scientists rose admirably to the urgent national need for increasing food and agricultural production. The Green Revolution transformed the country from a highly food-deficient country, critically dependent on imports, to a food-surplus country.

Food and Nutritional Security

The World Food Summit of 1996 defined food security as: '…when all people at all times have access to sufficient, safe, and nutritious food to maintain a healthy and active life'. This means that in order to enjoy food security there must be, on the one hand, provision of safe, nutritious food of quality and in the right quantity, and, on the other, every person must have access to it. Food security thus has three dimensions: (i) it is available in sufficient quantity and quality, supplied through domestic production or imports; (ii) it can be accessed by households and individuals; and (iii) there must be optimal uptake of a sustaining diet, clean water and adequate sanitation, together with proper healthcare. The diet of the poor in the developing countries consists usually of disproportionately high amounts of carbohydrate-rich staples such as maize, wheat and rice, and low amounts of protein and mineral-rich pulses, meat, fish, fruits and vegetables. The consequences for health of such an imbalanced diet are devastating and can result in poor growth, blindness, stunting and death. Malnutrition in India, especially among children and women, is widespread and acute. The Global Hunger Index 2017 (von Grebmer *et al.*, 2017) – based on parameters like undernourished population, children suffering from wasting, stunting and mortality – ranks India at position 100 among 119 countries, excluding industrialized countries. More than 40% of Indian women and young children have serious nutritional deficiencies. The most common mineral and vitamin deficiencies are iron, zinc, calcium, iodine, magnesium, selenium and vitamin A.

Need for Technology Interventions

Conventional and modern biotechnological innovations have helped immensely in enhancing agriculture through: precision plant breeding and animal reproductive technologies incorporating exotic genes for yield; resistance to biotic and abiotic stresses; production of high-quality disease-free planting material; biofertilizers and biopesticides; and disease diagnostic and genetic resource conservation tools (FAO, 2011). Accordingly, there is justification in intensifying research and developing human capacity for introducing cutting-edge science and technology in agriculture. This is all the more necessary in view of changing climate, and pest, disease and abiotic stress dynamics in different ecoregions. It is also clear that increase in agricultural production would help significantly in meeting the main SDGs – good food, nutrition and health.

Non-genetic Modification (Non-GM) Biotechnology

While the term 'biotechnology' is often confined to genetic modification, it, in fact, encompasses a very broad range of tools and techniques and is defined as any technological application using biological systems, living organisms, or derivatives thereof, to make or modify products or processes for a specific use. Tissue culture, induced mutations, fermentation technology, biopesticides, biofertilizers, marker-assisted breeding, assisted reproductive technologies in farm animals, and biochemical and molecular diagnostics are some non-GM biotechnologies. Commercial application of tissue culture multiplication (micro-propagation) technology for quality and mass production of ornamentals is a global industry. Among other agricultural crops, potato, sugarcane, sweet potato, banana, citrus, date palm and several other

horticultural crops are micro-propagated on a commercial scale and grown by farmers in many developing countries (Karihaloo, 2015). India has also benefitted a great deal through better performance of micro-propagation-based planting materials and better quality of produce. Furthermore, tissue culture techniques are being used for developing seedless or sterile fruits and creating polyploids, especially triploids, by embryo culture or regeneration from endosperm. As of now, many seedless fruits have been developed using these methods, including citrus, acacia, kiwi fruit, loquat and passion flower. However, the rampant spread of viral diseases through planting materials has brought an urgency to the need for production of disease-free, quality planting material, in which micro-propagation has a major role.

Conventional breeding programmes of the national agricultural system are being strengthened with the adoption of molecular marker technologies. Molecular breeding or molecular marker-assisted breeding (MAB) is defined as the application of molecular biotechnologies, specifically molecular markers, in combination with linkage maps and genomics. Marker-assisted selection (MAS), marker-assisted backcrossing (MABC), marker-assisted recurrent selection (MARS) and genome-wide selection (GWS) are different molecular breeding strategies used to improve crop-plant traits. Some achievements of MAB are the introduction of blast and blight resistance in rice, rust resistance in wheat, white rust resistance in mustard, downy mildew resistance in pearl millet and tolerance to submergence and salinity in rice (Kadirvel et al., 2015).

GM Technology

There can be limitations to transfer of genes from one plant to another by conventional cross-breeding owing to several factors including cross-incompatibility, absence of viable pollen and eggs, seasonal differences in flowering or absence of requisite traits in crop germplasm. Crops such as banana, cassava, citrus, cucurbits (all gourds such as cucumber, bottle gourd, ridge gourd, ash gourd, bitter gourd etc.), potato, papaya and tomato are severely affected by viruses, fungi and insects, for which most of the available compatible genetic resources do not possess resistance. Heavy use of chemical control measures for some of these pests has only caused residual contamination in the vegetables and increased cultivation expense for the farmers.

GM technology, defined as the alteration of genetic material of living cells or organisms to make them capable of producing new substances or performing new functions, provides a means of overcoming such constraints. Bt cotton is one such GM crop in which a gene from a bacterium (*Bacillus thuringiensis*), already present in Indian soils, has been transferred into cotton plant so that the latter produces a protein otherwise produced by the bacterium that is specifically toxic to bollworms. Hence, the plant becomes resistant to bollworm attack. Incidentally, the bacterium was known to be effective against bollworm quite a long time back and was approved by the Pesticide Registration Committee of the Government of India as a biopesticide. The biopesticide was being sprayed routinely on cotton fields even before Bt GM cotton was developed. But this method of bollworm control is less effective for several reasons and requires large-scale Bt production capacity.

Global Trends in GM Crops

Commercial cultivation of GM crops started in 1996, and by the end of 2017 it had reached 189.8 million ha covering 26 countries, including several developing countries, namely Argentina, Brazil, China, India, Pakistan and The Philippines (ISAAA, 2016; Table 14.1). GM varieties/hybrids have been commercialized in maize, soybean, cotton, canola, sugarbeet, alfalfa, papaya, squash, poplar, tomato, sweet pepper and others, while research and field testing on more than 200 plants including food plants and trees is underway. Among Asian countries, India grows Bt cotton on 10.8 million ha, Pakistan on 2.9 million ha, Myanmar on 0.3 million ha, and China on 2.8 million ha, along with some other GM crops; and the Philippines grows GM maize on 0.8 million ha. Bangladesh is gradually expanding its area under Bt brinjal and is reported to be taking steps towards adopting Bt cotton to enhance cotton production. Countries that have legalized field cultivation of GM crops have benefitted significantly in terms of environmental and economic parameters (Brookes and Barfoot, 2017).

Table 14.1. Ten largest GM crop-growing countries in 2016. (From: ISAAA, 2016)

Countries	Area (million ha)	Crops
USA	72.9	Maize, soybean, cotton, canola, sugar beet, alfalfa, papaya, squash
Brazil	49.1	Soybean, maize, cotton
Argentina	23.8	Soybean, maize, cotton
Canada	11.6	Canola, maize, soybean, sugar beet
India	10.8	Cotton
Paraguay	3.6	Soybean, maize, cotton
Pakistan	2.9	Cotton
China	2.8	Cotton, papaya, poplar, tomato, sweet pepper
South Africa	2.7	Soybean, maize, cotton
Uruguay	1.3	Soybean, maize

GM crops released lately or which are in the process of testing by regulatory authorities, include reduced-lignin alfalfa; potato with reduced bruising, blight-resistance and reduced acrylamide formation when fried; Bt brinjal resistant to fruit and shoot borer; drought-tolerant sugarcane; browning-resistant apple; high-wood eucalyptus; high-oleic soybean; virus-resistant cassava and bean; and β-carotene-fortified rice and banana. Others are being developed for a variety of biotic and abiotic stresses such as heat, salinity, long spells of dryness, submergence, and virus and insect resistance. It is also important to note that the potential of GM crops to improve crop productivity, increase crop adaptation to climatic stresses such as drought, and mitigate greenhouse gas emissions has been recognized by many national and international bodies including the FAO, as per the World Development Reports of 2008 and 2010.

Status of GM Crops in India

Recognizing the potential of biotechnology, India has invested significantly in R&D of GM crops relevant to the country's agriculture. Research work on plant transformation in Indian laboratories started in the 1980s and GM crop plants started being produced in the laboratory in the 1990s. Currently, a large number of crop plants, including brinjal, cabbage, cauliflower, cotton, groundnut, chickpea, maize, mustard, okra, pigeon pea, potato, rice, sorghum, tomato, and wheat, are being targeted for introducing traits like insect resistance, virus resistance, fungal resistance, nutritional enhancement, delayed ripening, reproduction control and abiotic stress tolerance. Insect-resistant and herbicide-tolerant cotton, insect-resistant and herbicide-tolerant maize, insect-resistant chickpea, mustard with pollination control, and rice with salinity tolerance and high agronomic performance, have either been completed for biosafety evaluation or are in the process of completion. However, only insect-resistant cotton has been commercially released thus far. Presently, both public and private sectors are actively engaged in transgenic research.

Bt Cotton Experience in India

Since the introduction of Bt cotton in farmers' fields in 2002, there has been a near doubling of cotton production, from 15.8 million bales in 2001–02 to 35.1 million bales in 2016–17 (Cotton Corporation of India; Fig.14.1). Within a short span of time, the area under Bt cotton has increased to cover 96% of the total cotton area. This increased cotton production is mainly attributable to increases in cotton productivity from 302 kg/ha in 2001–02 to 568 kg/ha in 2016–17 due to the introduction of Bt cotton and improved cultivation technologies. In 2008, India became the fourth-largest adopter of biotech crops in the world, overtaking Canada. With an average holding of 1.5 ha it is the small and medium-sized farmers who are the major beneficiaries of increased production. More importantly, the use of pesticides to kill bollworms has reduced considerably (from about 9410 t in 2002 to just 222 t in 2012), which not only means a monetary saving for the farmer but also an escape from health and environmental hazards. At the national level, the bollworm insecticide market declined from US$160 million in 2004 to US$25 million in 2010. Several reports

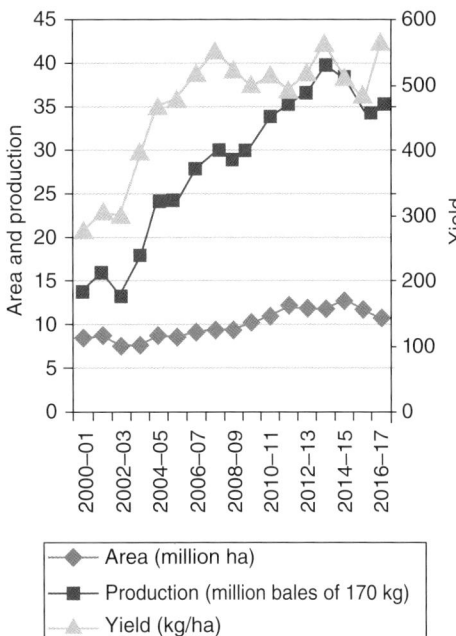

Fig. 14.1. Area, production and yield of cotton in India from 2000 onwards. (From: Cotton Corporation of India National Cotton Scenario (https://www.cotcorp.org.in/))

published in highly reputable journals have confirmed the production, yield, economic, social and environmental benefits of Bt cotton in India (Karihaloo and Kumar, 2009; Kathage and Qaim, 2012). Klumper and Qaim (2014), on analysing 147 original studies on GM soybean, maize and cotton carried out the world over, concluded that, compared to their non-GM counterparts, GM crops gave 21.6% higher yield and consumed 36.9% fewer pesticides.

Agri-food Biotechnology

As pointed out earlier, nutritional deficiency is rampant among the Indian population. Hence, fortification of food is an important mission of the government. Nutritional value of food crops is being enhanced through conventional plant breeding as well as through biotechnological approaches. Yellow corn biofortified through breeding with pro-vitamin A is an important example of biofortification. Golden rice engineered to increase provitamin-A content is the best-known example of GM technology for biofortification of a major staple. Folate (vitamin B9) enhancement is another example of biofortification in rice. Efforts are being made to create provitamin-A-enriched transgenics in wheat, corn, cassava and potato. However, multi-biofortification in many of the crops consumed as staples in different parts of India, along with their popularization, is a must to make an impact on peoples' health and nutrition.

The Way Ahead

Biotechnology has an important place in the country's journey towards food and nutrition security. While conventional crop improvement has served the country quite well, and will continue to do so in future, there are high production and productivity goals where recourse to biotechnology is imperative. Gene manipulation technologies are undergoing revolutionary changes, which could also address concerns expressed by some sections of society regarding GM technology. Gene-editing technologies like CRISPR-Cas 9 enable precise altering of native genomes for desired phenotypes. Such, and other, advanced technologies need to be adopted as a priority to overcome apprehensions about GM crops. Although the end-user decides whether to opt for a transgenic or non-transgenic product in the market, the new-generation biotechnology tools deserve to be appreciated and used judiciously in addressing food and nutrition security problems effectively. However, to fully harness the available benefits of biotechnology, we need to have a clear policy and road map for its implementation. Accordingly, action on the following is desirable:

- The 'Evergreen Revolution' is needed, especially for nutritional security, since India has the maximum concentration of malnourished children and anaemic pregnant women in the world. We need good nutrition rather than food alone. For this, the use of biotechnology is relevant in the present context. This technology offers new options to enhance nutritional security through designer grain, oilseed, pulses, fruit and vegetable crops and to meet challenges of biotic

and abiotic stresses as well as global climate change. Moreover, the poverty of smallholder farmers can be overcome by providing them with new technologies for reducing cost on inputs, building resilience in farming and increasing their income by linking them to markets. In this, a prominent role for biotechnology is envisaged.

- The development and adoption of appropriate biotechnologies are required, for which our public research system has to be strengthened. Along with the public sector, private sector investment in biotechnology needs to be ensured and enhanced, for which an enabling environment is a prerequisite. Appropriate protocols and IPR regimes need to be developed to encourage PPP.

- There is a need for prioritization of crops to use GM technologies efficiently to improve specific traits. To achieve this, a National Mission on GM food crops needs to be constituted jointly by the Department of Biotechnology (DBT) and the ICAR.

- The Biotechnology Regulatory Authority of India (BRAI) Bill, which is already in the Parliament, has to be cleared as soon as possible and a strong message needs to be sent to all policy makers and implementation authorities. The proposed BRAI is also necessary to ensure a single-window system for testing, clearing and monitoring GM plants.

- The biosafety regulatory system, though well defined, needs to be made efficient. A transparent system is needed, for which some reputable laboratories in the public sector, like the National Institute on Nutrition (NIN), the Indian Agricultural Research Institute (IARI), the Indian Veterinary Research Institute (IVRI) and the Central Drug Research Institute (CDRI), having good infrastructure and trained staff, can be accredited at the earliest opportunity.

- India needs to have a well-defined, post-release monitoring system for GM crops, operated jointly by the ICAR and the DBT. Also, a proper survey of farmers' fields is required to assess uptake and impact of GM technologies. Socioeconomic assessment needs to be an integral part of the GM crop-evaluation process. The cost of not adopting the technology should also be part of this assessment.

- Plant breeders and biotechnologists, especially in the public sector, should join hands and work as one to address specific research problems. Their efforts should be synergistic and not competitive. They must adopt a targeted approach with defined expected outputs and outcomes. In this context, an aggressive time-bound approach is necessary. Also, strong PPP from the initial stages of project development would help in better mutual understanding, trust and benefit sharing.
- Public perceptions about biotechnology are often based on hearsay rather than on scientific facts. Information communication systems, including public extension and awareness services, need to be considerably improved to deliver correct and unbiased information to farmers, policy makers and the general public.
- Increased investment is needed in capacity building, especially for biosafety research, regulatory systems (including legal aspects), communication tools and IPR issues, as they are all critical for outscaling innovations for greater impact.
- There must be a defined focus on agri-business and agri-biotechnology. Agri-business platforms and technology parks must be established for building much-needed PPP and for ensuring faster delivery of GM products to farmers and consumers.

Biotechnology offers several opportunities to reduce cost on inputs, ensure planting material and product quality, increase productivity and produce better returns for farmers, including those of resource-poor farmers with small holdings. There is, therefore, a need to reap its benefits, as has been achieved in many countries. To ensure this, there is a need for policy support, robust public research systems and much-desired PPP. Public awareness based on accurate knowledge and information will help dispel concerns and enhance biotechnology adoption for the overall benefit of producers and consumers and the country at large.

References

Brookes, G. and Barfoot, P. (2017) *GM Crops: Global Socio-Economic and Environmental Impacts 1996–2015*. PG Economics Ltd, Dorchester, UK.

FAO (2011) *Biotechnologies for Agricultural Development*. Proceedings of the FAO International Technical Conference on Agricultural Biotechnologies in Developing Countries: Options and Opportunities in Crops, Forestry, Livestock, Fisheries and Agro-industry to Face the Challenges of Food Insecurity and Climate Change. Food and Agriculture Organization of the United Nations, Rome.

FAO, IFAD, UNICEF, WFP and WHO (2017) *The State of Food Security and Nutrition in the World 2017: Building Resilience for Peace and Food Security*. Food and Agriculture Organization of the United Nations, Rome.

ISAAA (2016) *Global Status of Commercialized Biotech/GM Crops: 2016*. ISAAA, Ithaca, New York.

Kadirvel, P., Senthilvel, S., Geethanjali, S., Sujatha, M. and Varaprasad, K.S. (2015) Genetic markers, trait mapping and marker-assisted selection in plant breeding. In: Bahadur, B.,Venkat Rajam, M., Sahijram, L. and Krishnamurthy, K.V. (eds) *Plant Biology and Biotechnology. Vol. II: Plant Genomics and Biotechnology*. Springer India, New Delhi, pp. 61–88.

Karihaloo, J.L. (2015) Promoting safe application of agricultural biotechnologies in Asia and the Pacific. In: Teng P.P.S. (ed.) *Agricultural Biotechnology and Global Competitiveness*. Asian Productivity Organization, Tokyo, pp. 218–232.

Karihaloo, J.L. and Kumar, P.A. (2009) *Bt Cotton in India – A Status Report* (2nd edn). Asia-Pacific Consortium on Agricultural Biotechnology, New Delhi.

Kathage, J. and Qaim, M. (2012) Economic impacts and impact dynamics of Bt (*Bacillus thuringiensis*) cotton in India. *Proceedings of the National Academy of Sciences of the USA* 109, 11652–11656.

Klumper, W. and Qaim, M. (2014) A meta-analysis of the impacts of genetically modified crops. *PLOS ONE* 9, e111629. DOI: 10.1371/journal.pone.0111629.

United Nations (2017) *World Population Prospects: The 2017 Revision, Key Findings and Advance Tables*. Department of Economic and Social Affairs, Population Division, United Nations, New York.

Von Grebmer, K., Bernstein, J., Hossain, N., Brown, T., Parsai, N. *et al.* (2017) *2017 Global Hunger Index: The Inequalities of Hunger*. International Food Policy Research Institute, Washington, DC.

15

The International Treaty – Current Concerns

The International Treaty on Plant Genetic Resources for Food and Agriculture (ITPGRFA) is a legally binding instrument, adopted by the FAO Conference in 2001. It came into force on 29 June 2004 and at present has 134 contracting parties. Member states are obliged to conserve their plant genetic resources for food and agriculture in accordance with the Convention on Biological Diversity (CBD) to ensure their sustainable use and to share benefits arising from their use. The treaty recognizes 'farmers' rights', the traditional rights of farmers as producers, maintainers and developers of agrobiodiversity.

Planning workshops for strengthening national capacities to implement the ITPGRFA was essential to promote participation of countries in the multilateral system of access and benefit sharing and to identify means to improve access to plant genetic resources. For the effective implementation of the multilateral system of access and benefit sharing at country level, there were a number of core requirements to be fulfilled according to the needs of each country. The time has come to move beyond just raising awareness about the ITPGRFA and to develop a road map for its fast and effective implementation.

The institutionalized management of plant genetic resources for food and agriculture (PGRFA) can be viewed from the period leading up to the Earth Summit held in Rio in 1992 to the one that followed it. During the pre-summit period, three important things were taught about genetic resources. The first was that genetic resources are the building blocks for improving productivity using new genes in plant breeding. The second was that they were the common heritage of humankind. The third was that they were freely exchanged for human welfare. Unfortunately, these principles no longer hold, since the global debate on conservation of biodiversity which began in the early nineties (Cooper et al., 1994; Cooper, 2002).

The United Nations Conference on Environment and Development (UNCED) promoted a major paradigm shift in the management of genetic resources, subjecting them to rights of nations, which required them to be protected with proper legal instruments. Furthermore, the sustainable use of their components and the fair and equitable sharing of the benefits arising from the utilization of genetic resources were enshrined in the CBD. The CBD, which was adopted in 1992, also envisioned genetic resources to be conserved for posterity. Ten years later, during the World Summit for Sustainable Development (WSSD) in Johannesburg, it was realized that conservation is not only required for 'posterity' but also for the present. Hence, 'conservation through use' has become a common buzz phrase. Many studies indicate relatively less use of genetic diversity at present than in earlier periods. This has jeopardized the growth of agriculture. The FAO, with support from the

Bill and Melinda Gates Foundation (BMGF), has begun a Global Initiative for Plant Breeding (GIPB) to build the required capacity for enhanced use of genetic resources.

In the past, India had strong national breeding programmes, especially under the All India Coordinated Research Projects (AICRPs), on almost all crops for food and agriculture. Several improved varieties and hybrids were developed under these coordinated projects. At present, we seem to have become complacent and more dependent on pre-breeding materials that are provided by many of the international centres/institutions.

The CBD is an internationally legally binding treaty with three main goals: conservation of biodiversity; sustainable use of biodiversity; and fair and equitable sharing of benefits arising from the use of genetic resources. It relates to all forms of biodiversity. But greater concern was for agricultural commodities, including crops, which were immediately required for the food and nutritional security of humankind. Thus a dialogue was initiated under the auspices of the FAO to revise the international undertaking on plant genetic resources for food and agriculture. The deliberations culminated in the development of the ITPGRFA. During this time, there was a wide range of debates concerning farmers' rights and the revision of the ITPGRFA. There was a general consensus that only plant breeders should have rights, and the definition of farmers' rights was not established. The FAO Working Group on Farmers' Rights took almost two years to arrive at a clear definition of the term. It was then realized that not only plant breeders but also farmers should have rights over their landraces and varieties (FAO, 2009; Paroda, 2012).

Undoubtedly, all these developments have changed the way genetic resources are being managed today. In the process, the free exchange of genetic resources has almost stopped. India was among the first countries to ratify the ITPGRFA in 2002. The ITPGRFA came into force in 2004, and in 2006 its governing body adopted the Standard Material Transfer Agreement (SMTA) as the instrument for carrying out multilateral germplasm exchange under the ITPGRFA. In India, it was envisioned that there would be a bilateral system of germplasm exchange under the CBD, and multilateral exchange under the

umbrella of the ITPGRFA. Although the process had not been easy, and was slow, India has moved forward. The government enacted the Protection of Plant Varieties and Farmers' Rights (PPV&FR) Act in 2001, for the establishment of an effective system for the protection of plant varieties, rights of farmers, plant breeders and researchers, and to encourage the development of new varieties of plants of economic importance. At the same time, the government addressed issues related to the biodiversity fund and access and benefit-sharing mechanisms by enacting the Biological Diversity Act (BDA) in 2002. In many countries, similar laws are yet to be formulated or passed by respective governments (FAO, 2009; Paroda, 2012).

Prior to these international regimes and national laws, genetic resources were exchanged for faster genetic enhancement. Our country's food basket would have been entirely different had we not freely exchanged those genetic resources. There was a lot of debate during negotiations of the ITPGRFA as to why soybean and some important vegetables should not be included in the Annex I list of crops in the multilateral system of access and benefit sharing, as they are important food crops. Somehow, these were excluded because of political rather than scientific considerations. Several other crops were also discussed but not included due to the commercial interest of some countries. Decisions regarding the Annex I list of 64 crops (35 food crops and 29 forage species) were taken after intense debate with the understanding that countries would eventually come forward and decide if the list should be expanded; but, unfortunately, no country was willing to debate and extend the list.

Although the ITPGRFA was ratified almost ten years ago, discussions are still continuing about raising awareness and developing strategies for its implementation. Countries like India, though previously forerunners in using and exchanging PGRFA, have not yet fully implemented multilateral access to those materials under the ITPGRFA, currently under the domain of the FAO and available in the collections of the Consultative Group on International Agricultural Research (CGIAR) Consortium of International Agricultural Research Centres. A large amount of germplasm of Indian origin was acquired by several international gene banks (including

CGIAR gene banks) before CBD ratification (1993). This germplasm is being continuously exchanged, globally, through the ITPGRFA. It is paradoxical that India has yet to agree upon a mechanism under the ITPGRFA to implement multilateral exchange of Annex I crops, when most of the germplasm is already held in the global multilateral domain. There is a general opinion that India and many other countries are not very open with their genetic resources under the obligations of the ITPGRFA. In spite of its great merits, the Standard Material Transfer Agreement (SMTA) has not yet been adopted by many countries, including India. To address these issues, the Asia-Pacific Association of Agricultural Research Institutions (APAARI), along with Bioversity International, has played a significant role in creating awareness about the enhanced use of genetic resources through multilateral exchange using the SMTA, or bilateral exchange systems based on a mutually agreed material transfer agreement. APAARI, Bioversity International, Rural Development Administration (RDA), Republic of Korea and the Global Forum for Agricultural Research (GFAR) jointly organized an international symposium on Sustainable Agricultural Development and Use of Agrobiodiversity in the Asia-Pacific Region at Suwon, Republic of Korea, from 13–15 October 2010, where 84 experts from 32 countries participated. The symposium adopted unanimously the Suwon Agrobiodiversity Framework and provided an opportunity to review and redefine the role and directions of agricultural research and development for conservation and use of agrobiodiversity, for inclusive agricultural growth and development (FAO, 2009; Paroda, 2012). It became quite clear that the current situation calls for a better understanding and urgent implementation by the countries concerned, rather than merely raising awareness on the ITPGRFA.

For the general well-being of humanity, the pre-CBD era was certainly better than the post-CBD era. As a result of sovereign rights of nations over their genetic resources in the post-CBD era, several legal and policy dimensions have been added to PGRFA. A Global Plan of Action was adopted in 1996 and has 20 priority activities to address various aspects of the conservation and use of PGRFA (FAO, 1996). The Second Report on the State of the World's Plant Genetic Resources for Food and Agriculture, published in 2010, reviewed and assessed the situation of PGRFA and reflected on many interesting lessons learned (FAO, 2010). The report revealed that great effort towards strengthening capacity building and partnerships to fulfil legal obligations was needed, and to refrain from putting more hurdles in the way of implementing India's obligations under the ITPGRFA. In fact, with all these negotiations, exchange of genetic resources, which was previously the domain of scientists, now to be carried out with the involvement of bureaucrats, legal experts and farming communities, has become more difficult. Thus, in all these developments, issues and concerns need to be looked into more seriously in the context of the rights of beneficiaries as well as the expected benefits to society.

An oft-raised question is, 'What benefits can be obtained from access and benefit-sharing laws?' Access itself is important, and the ITPGRFA recognized this in the multilateral system of access and benefit sharing. Benefit sharing has long been an unresolved issue. While debating in India with private seed sector organizations, a general agreement was reached whereby the organizations agreed to share approximately 5% of the sale proceeds from publicly bred varieties and hybrids. Although the seed industry in India made significant progress with the efforts of the public and private sectors, private sector organizations expressed concern that they were not getting enough genetic resources for crop improvement. With the advent of plant breeders' rights, and the application of IPR in agriculture, there was a hesitation in sharing germplasm with the private Indian seed industry for fear of loss of ownership and biopiracy. Probably trust was shaken as the private sector always desired to acquire genetic resources but was reluctant to share the material. In fact, now even sharing information on the availability of material is also avoided, which is a matter of great concern. Hence, there is an urgent need to initiate a process to build trust among the various actors to develop an appropriate mechanism to facilitate sharing of germplasm between the private sector and the national system (Paroda, 2012).

Farmers are the custodians of many traditional varieties and landraces. Currently, their rights are being protected through the PPV&FR Act. However, there is a need to see what benefits have gone to farmers so far. A suitable practical mechanism needs to be developed so that farmers can benefit directly from the invaluable services they provide to humankind in protecting rich genetic resources in different hotspots and agro-ecological conditions. The PPV&FRA, in collaboration with some well-recognized and responsible non-governmental organizations, may help in this activity.

The current state of affairs in the international arena is due to the fact that those who have not yet accepted and ratified the ITPGRFA are the most vocal people during debates in international meetings. Mostly, those debating are either lawyers or bureaucrats, not serious technocrats. In the debate on farmers' rights, there were so many 'clauses' and 'sub-clauses', making things more complicated and less clear; it was an arduous task to get even one line cleared as there were more than ten legal experts sitting with the delegations of developed nations; whereas developing countries, from where most genetic resources originate, were represented by only a bureaucrat/scientist, and sometimes by no-one at all.

The Nagoya Protocol on access to genetic resources and the fair and equitable sharing of benefits arising from their utilization, under the CBD, was open for ratification. India signed it on 11 May 2011 and ratified it on 19 October 2012. It was a matter of great concern that, on the one hand, the international protocols and treaties were signed by the countries, while on the other hand, their direct benefits did not reach society. To harness the benefits of these protocols and treaties, a national strategy was urgently needed for the convergence and coordination of all relevant issues/legal requirements (FAO, 2011).

The Biological Diversity Act is broad and encompasses all forms of biodiversity in nature, including agrobiodiversity. In India, the Department of Agriculture and Cooperation (DoAC) of the Ministry of Agriculture and Farmers' Welfare (MoA & FW) is the nodal agency for agrobiodiversity, while the technical aspects are handled by the Indian Council of Agricultural Research (ICAR), which is another wing of the Ministry of Agriculture and Farmers' Welfare. Often, the ICAR is not invited to participate in these international debates where its role is most pertinent. The ICAR formed a National Advisory Board on Management of Genetic Resources. The Advisory Board realized the need to discuss many policy issues in its meetings, and in view of this it resolved to move forward with urgency and made several important decisions. It was decided that the Department of Agricultural Research and Education (DARE), in tandem with the National Biodiversity Authority (NBA) may take immediate steps towards providing access to germplasm of crops listed in Annex I, as per the provisions of the ITPGRFA. It was also decided that information on genetic resources must be made available in the public domain for the purpose of openness in information sharing. The availability of information about germplasm would not only be useful to share and enhance utilization but would negate the belief that a gene bank is merely a 'black box'. It is indeed extremely important for researchers to know what is available in the gene banks, otherwise they would remain black boxes and would not serve any useful purpose. Detailed information on germplasm held in gene banks needs to be made available and also accessible under required legal instruments in order for it to be used judiciously for the benefit of humankind. Furthermore, there is a need for harmonization of different protocols/treaties. This would require better understanding, facilitated by the organization(s) for in-depth discussions at national, regional and global level. To generate awareness at all levels, all stakeholders, including researchers, breeders, policy makers, NGOs and farmers involved in conservation through use, need to be included in initiatives on capacity development. In the absence of all relevant players, the deliberations and discussions of such important meetings would be of little purpose. If we really mean business, we should do something that is well-planned and more tangible to address the issue of conservation and utilization of plant genetic resources such as access and benefit sharing (Paroda, 2012).

It is hoped that the scope of benefit sharing would increase in the future once the government supports creation and strengthening of the gene fund with US$8 million. Private sector organizations and associations can also be approached and asked to contribute to the gene fund. This would be the best step forward to show their solidarity with the national approach. Even the private sector is not sharing germplasm and is not willing to keep it in the national gene bank. This one-way process will not work, and hence a conclusive dialogue with the private sector is very much needed.

There is an urgent need for partnerships among all stakeholders, including the public and private sectors, NGOs and farmers. If the people who conserve the precious germplasm in the interests of the nation are not encouraged through appropriate incentives and rewards, the tribal communities would not protect genetic resources for the benefit of the rest of us whilst living at subsistence level. These are issues and concerns that require serious deliberations and urgent action.

India is richly endowed with a wealth of genetic resources, which have been nurtured from time immemorial. We have been debating and making a good case for effective and urgent implementation of farmers' rights and benefit sharing with local communities. This process has to be initiated without further delay. That is the way the national plan of action during the National Agricultural Technology Project (NATP) was built in 1998 and implemented. Under the plan, a national germplasm collection programme was launched. Prior to the launching of the collection programme, there were 200,000 accessions in the national gene bank at the National Bureau of Plant Genetic Resources (NBPGR). As of 31 March 2017, there are 432,490 accessions of which 200,000 were collected in just five years under the National Agriculture Technology Project (NATP) project. This was achieved through a participatory approach, by involving all stakeholders. This germplasm needs to be systematically characterized, evaluated and made serviceable for its effective use. An institute like the National Bureau of Plant Genetic Resources (NBPGR) cannot do this alone, but it can be achieved through partnership, by a national network programme on collection, characterization, evaluation and supply of genetic resources.

All the above issues and concerns are to be addressed jointly by all stakeholders, especially those working directly with plant genetic resources. Germplasm must be shared more freely in India through the multilateral system, under the ITPGRFA, using the SMTA. This could serve as a good example for other countries in the Asia-Pacific region. There are serious challenges before us. All our joint energy and actions are needed to design a clear road map to address both national and international concerns more effectively.

References

Cooper, D.H. (2002) The international treaty on plant genetic resources for food and agriculture. *RECIEL* 11(1), 125.

Cooper, D.H., Engels, J. and Frison, E. (1994) A multilateral system for plant genetic resources: imperatives, achievements and challenges. In: *Issues in Genetic Resources No. 2*, IPGRI, Rome.

FAO (1996) *Global Plan of Action for the Conservation and Sustainable Use of Plant Genetic Resources for Food and Agriculture*. Food and Agriculture Organization of the United Nations, Rome.

FAO (2009) *International Treaty on Plant Genetic Resources for Food and Agriculture*. Food and Agriculture Organization of the United Nations, Rome.

FAO (2010) *The Second Report on the State of the World's Plant Genetic Resources for Food and Agriculture*. Commission on Genetic Resources for Food and Agriculture, Food and Agriculture Organization of the United Nations, Rome.

FAO (2011) *Nagoya Protocol on Access to Genetic Resources and the Fair and Equitable Sharing of Benefits Arising from Their Utilization to the Convention on Biological Diversity*. Text and Annex. Secretariat of the Convention on Biological Diversity, United Nations Environmental Programme, Quebec, Canada.

Paroda, R.S. (2012) Implementing the International Treaty to address current concerns about managing our plant genetic resources. Strategy paper. Trust for Advancement of Agricultural Sciences, Pusa Campus, New Delhi. Available at: http://www.cbd.int/convention/text (accessed 2 June 2018).

16

Agrobiodiversity: Dynamic Change Management

Introduction

Agrobiodiversity has developed as a result of natural selection and human intervention. Its conservation and sustainable use is essential for the survival of humankind. Besides its supporting role in the risk management of millions of smallholder farmers around the globe by assuring their survival and livelihood, it is an important key for adaptation of agriculture to a future changing environment, especially in terms of climate change and diseases. During the past few decades, agrobiodiversity has decreased at an alarming rate and losses are increasing rapidly in the areas where it has often been very rich. Modernization and intensification, mechanization and monoculture, lack of knowledge and incentives for conservation have reduced access to genetic resources and their free use; and other processes of social and economic change have affected agrobiodiversity. As the world is dynamic, the need for diversity is continuous and increasing owing to the increasing number of people to be fed, kept warm, housed and cured.

The world landscape and biodiversity profiles are changing fast with forest shrinkage, agricultural lands shadowed by ambitious urban and peri-urban developments, genetic vulnerability of crops, genetic erosion on account of greater spread of high-yielding varieties and threat of climate change. In three centuries (1700–2000) there has been more than a 500% increase in the area under agriculture, with corresponding global forest reduction of over 20%. Almost a 100 million ha increase in agricultural land has occurred in just two decades (1980–2000), of which around 55% has been added from forest cover (FAO, 1967, 1998, 2010).

Exponential population increase and demand for more food, feed and fibre have been the main causes of over-exploitation of natural resources. Every morning the world wakes up to a demand for food for an additional 200,000 people. Globally, half of all food produced comes from 1.5 billion smallholder farmers. Subsistence farmers depend mainly on landraces in their cropping systems and they use nearly 60% of the total agricultural land. Hence, making smallholder farmers aware of conservation and rational use of agrobiodiversity is a critical prerequisite for global sustainable development. In fact, time is running out, and 'business as usual' will not suffice to salvage the rich genetic diversity that is being eroded due to human intervention and climate change. In the process of development, as well as depletion of natural resources, we are on the verge of losing, or have actually lost, valuable agrobiodiversity in different regions. Unfortunately, such realization often comes too late to reverse the process. We have hardly done what is needed for both conservation and replenishment of natural resources (Paroda, 2016).

During the UN Conference on Environment and Development (UNCED) in 1972 in Stockholm, various nations started thinking of inter alia care to protect agrobiodiversity landscapes and to conserve and use dynamic gene pools of agricultural species and their wild relatives for overall sustainable development. In 1983, the UN FAO provided a non-legally binding platform as per the International Undertaking on Plant Genetic Resources (IUPGR) to act locally on the principle of 'germplasm is the common heritage of humankind', to maximize international free flow of germplasm and its use in crop improvement. Subsequently, the Convention on Biological Diversity (CBD) in 1992, the International Treaty on Plant Genetic Resources for Food and Agriculture (ITPGRFA) in 2001 and the Nagoya Protocol on Access and Benefit Sharing (NP-ABS) in 2010 set legal standards for facilitating access and benefit sharing. This arrangement, although ratified by many countries, has halted the use of genetic resources to a great extent (FAO, 1998, 2010, 2011).

However, the responsibility of implementing conservation, and access and benefit sharing (ABS), is broadly left to countries without much international commitment. Nevertheless, actions at the local and regional level are crucial to harness desired genes/attributes for better adaptability, fitness and higher source-sink relationship from available gene-rich agrobiodiversity. Inescapable interdependence of countries and people around the world in terms of meeting one another's needs, preferences and tastes has significantly changed our food baskets. Therefore, it is important and urgent to manage and maintain at least the current level of genetic resources, and for this, new, innovative approaches, ways and means need to be properly planned (FAO, 1983; Paroda, 2016).

Agrobiodiversity Conservation

Agrobiodiversity dates back to the settlement and domestication era, over ten millennia BC, whereas the centres of origin and diversity of crop plants were conceptualized only in the past century, mainly in the late 1920s. The knowledge regarding centres of origin and richness of available genetic resources and evolution of technological approaches for germplasm use over the past nearly six decades did have a paradigm change for agricultural research from direct selection to systematic plant breeding and use of biotechnology. Again, it is well argued that while the importance of agrobiodiversity and genetic resources has to be understood globally, the specific actions to conserve and protect them for posterity will also have to be managed locally. Hence it is critical to make efforts to minimize the gap between needs and developments relating to agrobiodiversity management at global, regional and national levels. Bigger countries with diverse agro-ecologies have to focus greater attention on the zonal level also (Scarascia Mugnozza, 1995; Bala Ravi et al., 2010).

Global initiatives

The widespread genetic resource collection efforts in the 1950s across the world, followed by the establishment of national gene banks for ex situ conservation of seeds of plant germplasm, led to the beginning of the plant genetic resource (PGR) conservation movement in the 1960s (FAO, 1967). The change management around genetic resources for food and agriculture (GRFA), as contemplated by the international agencies, especially the Food and Agriculture Organization of the United Nations (FAO), helped investment in PGR management and institutionalizing processes for collecting available diversity to be conserved and used by farmers in the future and to meet existing crop breeding needs across the world. It is also appropriate to pay tribute to the international community of farmers who shaped agrobiodiversity through their conscious selections and subconscious interventions over generations; and also to pioneers like Albert and Gabriella Howard, N.I. Vavilov, B.P. Pal, Sir Otto H. Frankel, E. Bennett, R.O. Whyte, J.G. Hawkes, J.H.W. Holden, J.T. Williams, and many others who shaped or joined the global germplasm movement for agrobiodiversity augmentation, conservation and use (Frankel and Bennet, 1970; Frankel, 1975; Harlan, 1992; Paroda, 2016).

The global mechanisms led by the FAO and the Consultative Group on International Agricultural Research (CGIAR) successfully catalysed and nurtured the germplasm conservation movement. In the process, the CG centres, including Bioversity International (BI), which emerged from the erstwhile International Plant

Genetic Resources Institute (IPGRI), the earlier International Board for Plant Genetic Resources (IBPGR) and other CG centres like the International Rice Research Institute (IRRI), International Maize and Wheat Improvement Center (CIMMYT), International Crop Research Institute for the Semi-Arid Tropics (ICRISAT), International Center for Agricultural Research in the Dry Areas (ICARDA), International Centre for Tropical Agriculture (CIAT), International Food Policy Research Institute (IFPRI), International Livestock Research Institute (ILRI), International Institute of Tropical Agriculture (IITA) and International Potato Center (CIP) laid considerable emphasis on sharing and using available agrobiodiversity resources for much-needed genetic enhancement for yield, stress tolerance and quality. In the process, both *ex situ* and *in situ* on-farm approaches were aggressively promoted (FAO, 1983; Paroda, 2016).

As regards sustainable use of plant genetic resources (PGRs), held *ex situ* in CG gene banks and *in situ*/on-farm in farmers' fields in diversity-rich areas, the process for multilateral access of the GRFA with equal emphasis on Farmers' Rights was triggered through the International Treaty on Plant Genetic Resources for Food and Agriculture (ITPGRFA). The FAO Working Group defined and adopted a definition of farmers' rights, agreed to the creation of a gene fund and finalized the list of 64 crops (incorporated as Annex 1 of the treaty). Also, an important initiative was taken to establish a Global Crop Diversity Trust (GCDT). During the first global conference, organized by the Global Forum on Agricultural Research (GFAR), the forum came out with a Dresden Declaration on Biotechnology and Management of Agrobiodiversity. All these initiatives culminated in a long-term, safe collection of genetic diversity available with all CG centres, and the FAO designated collections for multilateral access under the ITPGRFA, conserved in a permafrost facility at Svalbard, created by the GCDT (Gepts, 2004; Paroda, 2012, 2016).

The FAO initiatives, through first and second technical reports on the 'State of the World's Plant Genetic Resources', and subsequently the 'State of the World's Animal Genetic Resources', also catalysed different National Agricultural Research Systems (NARS) in many parts of the world to initiate national action plans for managing genetic resources. At the same time, the

World Information and Early Warning Systems (WIEWS) and many CG centre gene banks organized and provided valuable information on the GRFA relating to plants, animals, fish, insects and microbes, which drew needed attention for scientific evaluation, conservation and use.

While initiatives in the context of agrobiodiversity conservation, management and use were significant, some other major initiatives taken worldwide by different nations included putting in place their plant variety protection or *sui generis* systems to harmonize with a global WTO Trade-Related Aspects of Intellectual Property Rights (TRIPS) regime. In the process, germplasm exchange at the national, regional and global levels received a setback, mainly owing to uncertainties arising from issues like access and benefit sharing (ABS), determination of mutually agreed terms (MATs) and material transfer agreements (MTAs), ensuring effective protection and enforcement of IPR. Public awareness of these issues and institutional capacity to handle them became more prominent concerns. In many cases, the processes for sharing of genetic resources, even for research or direct human welfare, got complicated, with more and petty legal issues concerning IPR-ABS domains being flagged every now and then (Gepts, 2004; Paroda, 2016).

The institutions responsible for implementing biodiversity and *sui generis* IPR laws resorted to more and more awareness-creating activities, and consequently finding solutions to conflicts on a real terms basis did not get much priority. As a result, the ongoing processes and speed of germplasm exchange and benefit sharing were affected adversely. This trend for relatively slow or even no movement of germplasm, otherwise so critical for genetic advancements in various crop plants and animal species, needs to be reversed with a determined international push and appropriate financial commitment for germplasm enhancement and use at local levels.

Regional initiatives

Major regional initiatives on germplasm exchange and use were undertaken by most of the CG centres, located in different regions, mainly in view of their core activity as well as their major mandate for accelerated pre-breeding activities in different mandated crops/species. In the process, significant progress took place, including realization of the Green Revolution in South Asia. These developments over the last 50 years catalysed faster agricultural growth and ensured poverty reduction and increased food security in all regions. The CG centres also facilitated networks of germplasm exchange, benefitting mainly weaker NARS, thus ensuring availability of international public goods in the form of new high-yielding varieties and hybrids. In the process, some regional associations like the Asia-Pacific Association of Agricultural Research Institutions (APAARI), the Association of Agricultural Research Institutions in the Near East and North Africa (AARINENA), the Forum for Agricultural Research in Africa (FARA) and the Central Asia and Caucasus Association of Agricultural Research Institutions (CACAARI) were actively involved in facilitating such networks in partnership with various NARS of respective regions. APAARI worked closely with Bioversity International (earlier IPGRI) in establishing three sub-regional networks on genetic resources – in south Asia, south-east Asia and the Pacific. APAARI and CACAARI accelerated the process of germplasm exchange and strengthened national gene banks in developing countries' NARS, including adoption of the Suwon Declaration on Agrobiodiversity Management in the Asia-Pacific region, which laid a clear road map for action by all stakeholders. Unfortunately, most of these networks are currently non-functional, mainly due to funding constraints and lack of commitment by the CG centres. Also, pre-breeding initiatives seem to have become casualties at most of the crop research centres in the process of funding through CGIAR Research Programmes and lack of core funding under windows 1 and 2. Even a scoping study of BI to focus mainly on *in situ* conservation strategies seems not to have received due priority for a stand-alone CRP by the CGIAR. Thus, the whole process of agrobiodiversity management has received a setback during the past two decades. It is high time that the whole issue of retarded exchange and use of agrobiodiversity across the regions is revisited, to correct priorities and encourage changes to management of valuable agrobiodiversity, so critical for the future sustainability of global agriculture (Williams and Holden, 1984; Paroda, 2016).

National initiatives

In India, the process of identifying, assessing and augmenting landrace diversity of amber wheat and other crops, especially in the Indo-Gangetic plains (Howard and Howard, 1911), had started a century ago, mainly by economic botanists, resulting in single plant selections of best amber grain quality Pusa wheat varieties (Pusa 4, 6, 12 etc.), which served a useful cause subsequently as donors for grain quality. For example, use of Type 8A and 9D selections in the early 20th century resulted in breeding popular varieties like C591 in the 1930s and C306 in the 1960s. During the 1940s, plant breeding efforts were accelerated through the rational use of agrobiodiversity by B.P. Pal, as narrated in his famous article 'The search for new genes' (Pal, 1937, 1942). Subsequently, the dedicated efforts of pioneers like Harbhajan Singh, M.S. Swaminathan, A.B. Joshi, K.L. Mehra, R.S. Paroda, R.K. Arora, R.S. Rana and K.P.S. Chandel led to the establishment of a unique institutional mechanism by the ICAR called the National Bureau of Plant Genetic Resources (NBPGR), which is one of the most prominent national genetic resource management systems in the world, with the most modern gene bank facility inaugurated in 1996 and built with funding support from the USAID. Having a current holding of 0.43 million seed accessions and a genebank capacity of 1 million for long-term storage, the NBPGR genebank also has cryo- and *in vitro* gene banks and a network of active collections in medium-term storage modules and field collections in different parts of the country. Lately, the genomic resources have been augmented, and the system is supported by ICAR's world-class computing system to develop the field of bioinformatics. Later, a network of national-level genetic resource bureaux for animals, fish, agriculturally important micro-organisms and insects was created. Concurrently, to provide the opportunity for agricultural scientists and other stakeholders to come together from across the globe and deliberate about agricultural research for development (AR4D) under one roof, the National Agricultural Science Centre (NASC) complex was built in New Delhi. This complex in Pusa Campus also houses regional or Indian offices of various CG centres, the Centre for Agriculture and Biosciences International

(CABI) and important national authorities – the National Rainfed Area Authority (NRAA), the Protection of Plant Varieties and Farmers Rights Authority (PPV&FRA), the National Academy of Agricultural Sciences (NAAS) and the National Agricultural Science Museum. The complex houses an international guest house, an auditorium, boardrooms, meeting halls and lecture halls, and boasts splendid lawns (NRC, 1972; IBPGR, 1986; Paroda, 2016).

The national germplasm management system was started as a Plant Introduction Unit of the IARI, headed by H.B. Singh, which gradually expanded its activities under the guidance of M.S. Swaminathan and A.B. Joshi and developed into an ICAR institute, the NBPGR (NBPGR, 2000). The private seed industry was given an initial boost through these germplasm introduction efforts since the 1950s, and public policy support under the New Policy on Seed Development in 1987, which provided access to breeder seed of varieties and hybrids, developed by the public research system – ICAR institutes and SAUs.

Regulatory domain

In the post-CBD era, the ITPGRFA holds the key to the system of multilateral access and benefit sharing (MLS) for crop germplasm of identified food, forage and other agricultural species among contracting parties. However, the national biodiversity laws were mainly enacted in harmony with the CBD. Thus, currently there are many grey areas in these laws, particularly with regard to prescribed procedures and processes for germplasm flow of food and agricultural commodities, which require further attention by legislators, regulatory agencies and the executive machinery of governments. Not only corrective steps and their simplification are needed, but also expansion of the scope of the ITPGRFA beyond the Annex 1 coverage at national level, depending upon their strengths in other commodities, trade and commerce, or for sustainable food and nutritional security. Thus, further means and processes to improve access to PGRFA for use in breeding, research and training must be addressed by the national systems. For this, national institutional capacities will have to be developed further, national focal points elaborated and explained to stakeholders,

and promotional activities supported and financed as a matter of national priority. For effective implementation of the multilateral system of access and benefit sharing at country level, regulatory authorities must publish literature with explanatory notes to help candidate beneficiaries and potential stakeholders in meeting a number of core and/or supplementary requirements. The elaboration of such requirements needs to be clearly spelled out in the prescribed procedures and processes under the treaty as well as laid out according to the needs of the respective countries. In India, some steps have already been taken in this direction. The government has notified that there is no need to seek prior permission from the NBA by foreign applicants accessing germplasm materials covered under the ITPGRFA. Similarly, guidelines for the implementation of the provisions of access and benefit sharing under the Nagoya Protocol have been notified to implement the Protocol in letter and spirit. However, the implications of the relevant gazette notifications by the Ministry of Environment, Forests and Climate Change must be clearly understood in conjunction with other relevant laws in a case-specific manner to avoid undue litigation or public mistrust (Gepts, 2004; Paroda, 2016).

To encourage countries to adopt measures for the smooth implementation of the ITPGRFA, the governing body of the treaty should also recommend to the FAO that it recognizes and highlights/documents the specific national contributions of countries in the international collections already in circulation by the international centres and national systems for public and private use, through multilateral exchange under the Standard Material Transfer Agreement (SMTA). This arrangement may eventually help in developing and adopting some good, long-term benefit-sharing arrangements among gene-rich and technology-rich countries as well as farmers and innovators.

Understanding dynamism

In 1993, the Indian Society of Plant Genetic Resources (ISPGR) Dialogue for National PGR Policy Options was held in New Delhi, which commemorated the establishment of the CBD. It emerged that the CBD had drawn as much attention to the need to institutionalize ABS as had the con-servation of biodiversity through *in situ* means for sustainable use. The *ex situ* collections were being mainly covered by the FAO global system under the International Undertaking on Plant Genetic Resources (IUPGR), which subsequently was governed under the ITPGRFA. Thus for ABS matters, it would be critical for countries to link these with both the CBD and the ITPGRFA do-mains (FAO, 1986; Paroda, 2016).

Now that the global debate on conserva-tion and sustainable use of PGRFA has partly settled under the ITPGRFA, it is high time that the regulatory regimes at national level are vol-untarily reviewed in the interests of peace and prosperity through agriculture. At this stage, the greatest tributes go to Nobel Peace Laureate Dr Norman E. Borlaug, due to whom agricul-ture was seen as the most potent sector respon-sible for sustainable peace on earth. His critical breeding and selection strategy in wheat had led to the greatest ever Green Revolution. Indians and agricultural scientists in particular will al-ways feel proud to portray him as a member of their fraternity and his name is associated with countless commemorations in south Asia (FAO, 1986; Paroda, 2016).

In the past, the Indian National Agricultur-al Research System had strong national breed-ing programmes in many crops, which included national crossing blocks, regional cooperative trials by the ICAR institutes and SAUs and/or multi-location testing for identification of supe-rior varieties under the All India Coordinated Research Projects (AICRPs), on almost all crops for food and agriculture. Several improved varie-ties and hybrids were developed under these pro-jects using native or exotic germplasm without restriction. At present, most of the national pro-grammes seem to have become dependent on the pre-breeding materials provided and/or in-ternational nursery trials constituted by many of the international centres/institutions. To rebuild national capacities in enhancement of GRFA and to enhance probabilities of generating more diverse international commons through collaborative research, more participatory activ-ities within and across regions are to be organ-ized and financially supported. Innovation must be encouraged and rewarded in the first place to push the global AR4D agenda for farmers' welfare–development paradigm (Paroda, 2016).

The CBD relates to all components of bio-logical diversity that broadly concern all sectors (health, industry, agriculture, rural develop-ment etc.), but PGRFA are of immediate necessi-ty for food and nutritional security and the well-being of humankind. These must receive priority and fast-track handling by regulatory bodies. The issue of farmers' rights raised under the IUPGR is equally important. Both these con-cerns remained 'outstanding issues' at the time of finalizing the text of the CBD. Accordingly, these two issues had to be renegotiated, which took over seven years to settle in the form of the ITPGRFA. There was no consensus for the defini-tion of farmers' rights and other definitions pro-vided under the treaty. In the FAO Working Group on Farmers' Rights, it emerged that not only plant breeders should have rights over new varieties developed by them but also farmers over the varieties evolved and perpetuated by them. Eventually, farmers' rights became part of the international law under ITPGRFA and India became the first country to internalize it in its legal and policy systems. Many other developing countries look to the Indian experience in this regard with a view to develop their own national systems. The International Agrobiodiversity Congress (IAC) 2016 has provided the opportu-nity for other countries to share information arising from such developments in the agrobio-diversity domain (Paroda, 2016).

Change we must for facilitated ABS

Studies clearly show how nations have histori-cally been dependent on one another for diversi-fication of their food baskets or meeting their needs for genetic resources for the increased productivity of agricultural commodities. This dependence is predicted (Galluzzi *et al.*, 2016) to increase more in the future, given the current trends of climate change and the need for an ex-panding food basket with consumers' preferences for more healthy foods. Therefore, future deter-minations about how access is to be provided, and what benefit sharing will be agreed upon, will hold the key to sustaining interdependence; and a judicious interpretation of international and national legal obligations and processes un-der which exchange is to be governed will dictate terms. In this context, administrative, structural and political compulsions are not uniform across countries; and this has rendered the exchange of agro-bioresources/PGRFA much more complex and sometimes uncertain.

Instead of easing the process of facilitation, the ITPGRFA has indirectly led to reduced exchange of germplasm between nations, despite clear recognition of a multilateral system for exchange. Now the Nagoya Protocol of the CBD has increased the complexity in handling situations. Experience shows that the MLS has not functioned at the anticipated level, nor has it helped in generating financial benefit through the proposed international Benefit Sharing Fund (BSF). In India, there is still debate concerning exchange of germplasm, even with local private seed sector organizations engaged in plant breeding. Even SMTA has not yet been put into practice for want of procedural clearance and understanding. ICAR as a policy allowed free access to parental lines of hybrids bred by the public system since the mid-1980s, recognizing that seeds of these hybrids would otherwise not reach end-users, i.e. smallholder farmers. This policy decision accelerated the coverage of hybrid seeds resulting in increased crop productivity and helped strengthen the existing Indian private seed sector. With the pronouncement of plant variety protection and the rise of IPR regimes in agriculture and biotechnology, there is hesitation in the developing countries to share their germplasm accessions due to uncertainties and fears over possible effects of ABS and IPR. There is a definite lack of much-needed trust and partnership. A kind of fatigue has jeopardized agricultural growth. This will require an enabling policy environment to foster sharing of germplasm as well as information between the public and private sectors (Swaminathan, 2002; Paroda, 2016).

In many cases, farmers are custodians of traditional varieties in different diversity-rich regions. In India, their rights are now being protected under the *sui generis* PPV&FR Act. The system of genome saviour awards and recognition has evolved considerably with government funding, and farmers are being made aware through ICAR's *Krishi Vigyan Kendras* (Farm Science Centres) and extension units of SAUs. The PPV&FR Authority needs to be commended for implementing farmers' rights and creating awareness. The Authority has been assured of

government support to build an Indian gene fund of Rs 500 million (around US$7.5 million) to ensure long-term recognitions, rewards and incentives to farming communities engaged in conserving valuable genetic resources. It is also expected that the evolution of benefit-sharing mechanisms along with funding support from the seed sector will help in building up the gene fund to around US$20 million in the future. Simultaneously, there is a need to develop a clear mechanism to benefit farmers directly for their invaluable service regarding PGRFA to society (FAO, 1997).

Turning Youth into Catalysts of Change

Rural youth are undoubtedly the key to food security, agricultural sustainability and innovation in farming. Yet few youths in villages see a future for themselves in agrobiodiversity management, agriculture or farm enterprise. The declining interest among rural youth in agriculture is directly related to existing poor physical amenities, socioeconomic conditions and lack of an enabling environment. Economic factors like low-paid employment and inadequate credit facilities discourage them from remaining in agriculture. It is clear that for sustainable rural development to occur, young farmers, especially smallholders, must be at the centre of all policy decisions, which can be handled at local level. Funding from central government should be made available to reach rural youth with the right message and viable options. Concerted efforts are needed to equip them with technology, innovation and market-linked facilities.

Some imperatives for sustainable agrobiodiversity management in gene-rich rural areas are: safeguarding of available natural resources; sharing available knowledge (both traditional and formal); building local PGRFA inventories, local access processes and capacity to harmonize with existing policies and laws; promoting conservation for sustainable PGRFA use; and developing links and partnerships at local, district, state and country levels (Paroda, 2016). Rural youth must be sensitized, trained and supported to the level of sustainable self-dependence to manage dynamic agrobiodiversity.

References

Bala Ravi, S., Rani, M.G. and Swaminathan, M.S. (2010) Conservation of plant genetic resources at the Scarascia Mugnozza. In *Memorie di Scienze Fisiche e Naturali*. Aracne Editrice, Rome, pp. 47–58.

FAO (1967) *The State of Food and Agriculture 1967*. Food and Agriculture Organization of the United Nations, Rome.

FAO (1983) Report of the Conference of FAO. In *Proceedings of the Twenty-Second Session*, Rome, 5–23 November 1983.

FAO (1986) *Global Plan of Action for the Conservation and Sustainable Use of Plant Genetic Resources for Food and Agriculture*. Food and Agriculture Organization of United Nations, Rome.

FAO (1997) Paraguay – financial and economic implications of no-tillage and crop rotations compared to conventional cropping systems. TCI Occasional Paper Series No. 9. Investment Centre Division, FAO, Rome.

FAO (1998) *Report on the State of the World's Plant Genetic Resources*. Food and Agriculture Organization of the United Nations, Rome.

FAO (2010) *The Second Report on the State of the World's Plant Genetic Resources*. Food and Agriculture Organization of the United Nations, Rome.

FAO (2011) Nagoya Protocol on Access to Genetic Resources and the Fair and Equitable Sharing of Benefits Arising from their Utilization to the Convention on Biological Diversity. Secretariat of the Convention on Biological Diversity. United Nations Environmental Programme, Montreal, Canada.

Frankel, O.H. (1975) Conservation of crop genetic resources and their wild relatives: an overview. In: Frankel, O.H. and Hawkes, J.G. (eds) *Crop Genetic Resources for Today and Tomorrow*. Cambridge University Press, Cambridge, UK.

Frankel, O.H. and Bennet, E. (1970) *Genetic Resources in Plants: Their Exploration and Conservation*. Blackwell Scientific, Oxford, UK.

Galluzzi, V., Guzzetta, L., Mancinelli, P., Giacomini, L., Ferranti, L., Massironi, M., Palumbo, P., Pauselli, C. and Rothery, D.A. (2016) 47th LPSC – Poster 404. Available at: https://www.researchgate.net/publication/299477871 (accessed 4 June 2018).

Gepts, P. (2004) Who owns biodiversity, and how should the owners be compensated? *Plant Physiology* 134, 1295–1307.

Harlan, J.R. (1992) *Crops and Man*. American Society of Agronomy, Madison, Wisconsin.

Howard, A. and Howard, G.L.C. (1911) *Wheat in India, Its Production, Varieties and Improvement*. Thacker Spinn & Company, Calcutta, India.

IBPGR (1986) *Programme and Structure of the International Board for Plant Genetic Resources*. International Board for Plant Genetic Resources, Rome.

NBPGR (2000) *20 Glorious Years of NBPGR (1976–1996)*. National Bureau of Plant Genetic Resources, New Delhi.

NRC (1972) *Genetic Vulnerability of Major Crops, National Research Council*. National Academy of Sciences, Washington, DC.

Pal, B.P. (1937) The search for new genes. *Agriculture & Livestock in India* 7, 573–578.

Pal, B.P. (1942) Genetic nature of self- and cross-incompatibility in potatoes. *Nature* 149(3774), 246–247.

Paroda, R.S. (2012) Implementing the International Treaty to address current concerns about managing our plant genetic resources. Strategy paper. Trust for Advancement of Agricultural Sciences, Pusa Campus, New Delhi. Available at: http://www.cbd.int/convention/text (accessed 4 June 2018).

Paroda, R.S. (2016) Agrobiodiversity Needs Dynamic Change Management. 1st International Agrobiodiversity Congress: Science, Technology, Policy and Partnership. Indian Society of Plant Genetic Resources & Bioversity International, New Delhi, 6–9 November, pp. 1–11.

Scarascia Mugnozza, G.T. (1995) *The Protection of Biodiversity and the Conservation and Use of Genetic Resources for Food and Agriculture: Potential and Perspectives*. Food and Agriculture Organization of the United Nations, Rome.

Swaminathan, M.S. (2002) The past, present and future contributions of farmers to the conservation and development of genetic diversity. In: Engels, J.M.M., Rao, V.R., Brown, A. and Jackson, M.T. (eds) *Managing Plant Genetic Diversity*. CAB International, Wallingford, UK, pp. 23–32.

Williams, J.T. and Holden, J.H. (eds) (1984) *Crop Genetic Resources: Conservation & Evaluation*. Allen & Unwin, Winchester, Massachusetts.

17

Managing Agrobiodiversity through Use: Changing Paradigms

Scientific and conscious management, use and conservation of agrobiodiversity have undergone many paradigm shifts in the past few centuries. Management of agrobiodiversity, including its conservation and use, is extremely important for the welfare of society at large. Some suggestions are given below to ensure its effective and long-term management.

Biodiversity under Domestication

In nature, all organisms have been living in harmony for millions of years. Humans (nomadic and forest tribes) have been highly dependent on the endless diversity among and within species along with their habitats and ecosystems. When humans transited from being nomadic hunter-gatherers to having a more settled lifestyle, due to the adoption of agriculture some 12,000 years ago, they started searching for such bioresources that could provide them with food, feed, fodder, fibre and improved livelihood. The intervention of humans by way of domestication and farming affected the pattern of evolution, diverting selection from 'fitness' to 'human preference'. The available diversity of domesticated species, which is the basis for the quality, range and extent of choices available to humankind, is the result of such evolution, influenced by frequent human interventions, especially farm women, over millennia.

First, a clear understanding between biodiversity and agrobiodiversity needs to be grasped. Biodiversity is essential for food security and nutrition. Thousands of interconnected species make up a vital web of biodiversity within the ecosystems upon which global food production depends. With the erosion of biodiversity, humankind loses the potential to adapt ecosystems to new challenges such as population growth and climate change. Achieving food security for all is intrinsically linked to the maintenance of biodiversity. Agrobiodiversity is the result of the interaction between the environment, genetic resources and management systems and practices used by culturally diverse people, and therefore land and water resources are used for production in different ways. Thus, agrobiodiversity encompasses the variety and variability of animals, plants and micro-organisms that are necessary for sustaining key functions of the agro-ecosystem, including its structure and processes for, and in support of, food production and food security (FAO, 1997). Local knowledge and culture can, therefore, be considered as integral parts of agrobiodiversity, because it is the human activity of agriculture that shapes and conserves this biodiversity. Many people's food and livelihood security depends on the sustained management of various biological resources that are important for food and agriculture. Agricultural biodiversity, also known as agrobiodiversity, or the genetic resources for food and agriculture, includes: harvested crop varieties, livestock breeds, fish species and non-domesticated (wild) resources within field, forest and rangeland including: tree products; wild animals hunted for food and in aquatic ecosystems (e.g. wild fish); non-harvested species in production ecosystems that support food provision, including soil micro-biota, pollinators and other insects such as bees, butterflies, earthworms and greenflies; and non-harvested species in the wider environment that support food production ecosystems (agricultural, pastoral, forest and aquatic ecosystems). Had we not judiciously used agrobiodiversity, our food basket may not have been what it is today. Yet there is a need to diversify it further to meet increasing demands for food and nutrition (Paroda and Agrawal, 2017).

By 2050, we will be requiring 70% additional foodgrains. To ensure this happens, use of available genetic resources needs to be more effective and efficient. If these resources had not been protected properly by tribals living at subsistence level, vital resources would possibly not have been saved. Further, the number of species existing on the earth is enormous, but research conducted on them so far has been limited. Unfortunately, in the past, research was mainly on crops that were of direct use to humans. The whole world is dependent for 60% of its energy and food requirements on three crops – wheat, rice and maize (Swaminathan, 2011; 2016).

Origin and Ownership of Genetic Resources

Earlier, two common principles were posited for the development of genetic resources and their use – genetic resources were considered a common heritage of humankind; and they were freely exchanged. If this were not to have happened, many of the daily food staples like maize, potato and tomato, having their centres of origin in South America, would not have come to India, and many crops like sugarcane, pulses and eggplant would not have reached other countries. It is well known that rice in south-east Asia, wheat in west Asia and north Africa, maize in central America and potato in Latin America and parts of Europe, emerging as staple foods, proliferated

and subsequently dominated the world food bowl along with millets, pulses, oilseeds, saccharum, cucurbits, citrus, forage crops and many other species. Among non-food crops, cotton, jute and bamboo are worth mentioning, having originated from the Indian sub-continent. Vavilov travelled the whole world and collected a large number of seeds and plants, which enabled him to understand and suggest the concept of centres of origin of crop plants. Today, these are fondly called Vavilovian centres of origin (FAO, 1999; Frison *et al.*, 2011; Paroda and Agrawal, 2017).

The collection, evaluation, exchange and utilization of genetic resources in exotic areas accelerated during the second half of the 20th century. This enabled the whole world to boost food production while keeping pace with an ever-increasing population. India looks back with amazement at its degree of dependency on the genetic diversity that came from outside. India, and many other countries, capitalized on agrobiodiversity resources, though there was not a single plant that originated from within the country. This dependency is predicted to increase more in future, given the current trends of climate change, an expanding food basket and changing consumer preferences towards healthy and nutritious food (Mohapatra, 2016; Upadhyay, 2016).

As stated earlier, until the late 20th century, genetic resources were exchanged freely, not only among farmers but also between plant breeders and researchers within and outside the country. In the late 1980s, this perception changed, with biodiversity being regarded as a treasure and the subject of national sovereignty. This paradigm shift from free flow of genetic resources to a restricted exchange emerged as a reality soon after the Convention on Biological Diversity (CBD) came into force in December 1993. Thus, germplasm exchange became operational under the legal instrument or *sui generis* system, as per the guidelines of international treaties. The underlying idea was that if genetic resources are used to develop commercial products such as new plant varieties, then the subsequent benefits must be shared with the provider(s) of the genetic resource. Hence, the concept of free exchange of agrobiodiversity was changed to protect the rights of the owners of the germplasm. Thus, new legal issues, which prominently emerged, restricted the flow of germplasm. This is an obvious challenge that must be addressed jointly by all (Paroda and Agrawal, 2017).

Humans – the Catalysts of Change

The obvious change in public perception towards genetic resources has also been due to an alarming rise in the world's population. For thousands of years, the population grew rather slowly, but in the last century it jumped dramatically. Between 1900 and 2000, the increase in the world's population was three times greater than in the entire history of humanity – an increase from 1.5 billion to 6.1 billion. There are more than 7.5 billion humans living on earth today, whereas 200 years ago, this number was less than a billion. It is expected that at this rate the world will have around 9.7 billion people by 2050 (Swaminathan, 2016). Hence, the balance of nature has been massively disturbed. Mahatma Gandhi, the Father of the Nation, had said that 'nature has provided for everyone's need but not for everyone's greed'.

Unfortunately, human greed has disturbed the equilibrium. Geologists have begun to predict that almost 12,000 years of the Holocene have come to an end. Why? Because it is human beings that have adopted a path of destruction of the life-support system. A new epoch is said to have begun around 1950, when radioactive elements from nuclear testing were spread over the globe, and has been characterized by extinctions, plastic pollution and a spike in carbon emissions in the atmosphere. It is now said that we are entering an era called 'Anthropocene', wherein anthropogenic activities are reshaping the earth's land, oceans, air and biodiversity. Consequently, biological diversity has significantly reduced, the earth has become warmer, and all over the world we are facing greater incidences of natural catastrophic events (Paroda and Agrawal, 2017).

A recent study has shown that about 58% of the world's land surface, and 9 out of 14 of the world's terrestrial biomes, have fallen below the 'safe threshold' of biodiversity, impacting a wide range of services provided by biodiversity, including crop pollination, waste decomposition, regulation of the global carbon cycle and sociocultural services critical for human well-being. Another study has shown that over the past 500 years, the rate of extinction of vertebrates is a clear signal of elevated species loss, which has markedly accelerated over the past hundred years or so. In fact, these rates are so high that life on the earth is embarking on its sixth greatest extinction event in its 3.5 billion-year history. In the Anthropocene, humanity faces the question of how to transform agriculture to enable it to feed its population, eradicate poverty and contribute to a stable planet. Most importantly, it has been said that averting a dramatic decay in biodiversity and subsequent loss of ecosystem services is still possible through intensified conservation efforts, but this window of opportunity is rapidly closing. This cannot be allowed to happen, and strenuous efforts need to be made to ensure that it does not.

Global Outlook towards Agrobiodiversity

Global thinking and intergovernmental approaches to managing genetic resources to improve food and nutritional security within this changing scenario have witnessed many developments since the late 20th century. They started with the UN Conference on Human Environment held in Stockholm in 1972, with its emphasis on population, agriculture and environment (Philippe, 1972). Later, world leaders congregated at the UN Conference on Environment and Development (UNCED), also known as the Earth Summit at Rio de Janeiro in 1992. One of the major outcomes of this was the adoption of the Convention on Biological Diversity (CBD) in 1993, which directly addressed ways to protect biological resources, being our life-support system, from becoming extinct. A major shift caused by the CBD was to place these resources under the territorial sovereignty of individual nations where they are found or where they originated, with legal rights of the nations to determine their own system of access and benefit sharing (ABS). For addressing trade-related concerns, the World Trade Organization was established, which helped to enact the Agreement on Trade Related Aspects of Intellectual Property Rights (TRIPS), including those related to agriculture (International Union for the Protection of New Varieties of Plants (UPOV) and patents) (FAO, 1999; Frison et al., 2011; Kotschi and Lossau, 2011; Paroda and Agrawal, 2017).

Almost a decade later, discussions around the International Undertaking for Plant Genetic Resources (IUPGR) of the FAO culminated in the adoption of the International Treaty on Plant Genetic Resources for Food and Agriculture (ITPGR-FA) in 2001. The overall objective of the ITPGRFA was to ensure conservation and sustainable use of PGRFA and to have both fair and equitable sharing of benefits derived from their use, in harmony with the CBD, for sustainable agriculture and food security. It also recognized for the first time farmers' rights on GRFA. The centrepiece of the treaty was the multilateral system (MLS) of facilitated access of PGRFA through a Standard Material Transfer Agreement (SMTA), which was freely accessible for breeders and researchers of member countries. The treaty covered a series of crops listed in Annex 1, which included 35 food crops and 29 forages. It also covered *ex situ* collections of those crops held by the CG gene banks. Though these crops accounted for about 80% of the world's food calories from plants, they did not represent all 100 food crops of importance to food security and 18,000 forages of value to food and agriculture. Soybean, groundnut, sugarcane and oil palm were among those that were still not included in Annex 1 even 16 years after the implementation of the treaty.

This treaty, while in harmony with the CBD, created an alternative multilateral ABS regime for the agriculture sector to gain access to, and transfer, those plant genetic resources that were 'the raw material indispensable for crop genetic improvement', and thus were important for global food security. In 2010, the Nagoya Protocol on Access to Genetic Resources, and the fair and equitable sharing of benefits arising from their utilization, was developed as a bilateral mechanism for ABS under CBD. It called upon nations to develop effective legislative, administrative and policy measures to provide, bilaterally, those genetic resources that were within their jurisdiction and which were accessed in accordance with prior informed consent and on mutually agreed terms between two parties (FAO, 2011; Hodgkin *et al.*, 2016).

India's Response to Changing Paradigms

India has been one of the first countries to develop and enact laws relating to biodiversity, in response to new regimes in international law concerning access, conservation and property rights on genetic resources. These processes have not been easy. A formidable task was to maintain a balance between new and traditional rights. Accordingly, three Acts were passed by the Indian Parliament at the beginning of the current century in an attempt to protect the nation's biological diversity, IPR and the interests of researchers, be they plant breeders or farmers/farming communities (Batur and Dedeurwaerdere, 2014). The three Acts were: (i) the Protection of Plant Varieties and Farmers' Rights Act (PPV&FRA) 2001; (ii) the Biological Diversity Act (BDA) 2002; and (iii) the Geographical Indication of Goods (Registration and Protection) Act 2000. These legislative measures, in addition to providing enhanced intellectual property protection, emphasized the importance given to rights of farmers, the traditional knowledge and the biological resources of the country. The PPV&FRA is a unique Act, being the first in the world to provide rights to farmers to produce, sell and use their own seeds, equivalent to those of breeders and researchers over the valuable genetic resources conserved by them. Hence, the law aims to protect plant varieties developed through public and private sector research as well as those developed and conserved by farmers and farming communities. Accordingly, under the provisions of this Act, a PPV&FRA authority has been established that not only registers new varieties developed by breeders and farmers but also ensures fair and equitable benefit sharing through the provision of a national gene fund (Paroda and Agrawal, 2017).

The primary objective of the Biological Diversity Act 2002 is to protect India's rich biodiversity and associated traditional knowledge against their use by others without sharing the benefits arising out of such use. It provides for the establishment of a National Biological Authority (NBA), state biodiversity boards and biodiversity management committees with extensive powers to promote conservation, sustainable use and documentation of biological resources. Foreign organizations require NBA approval to access biological resources. Provisions have also been made to set up biodiversity funds and management committees at the national, state and local level.

The Geographical Indication of Goods (Registration and Protection) Act 2000 aims to provide a comprehensive framework to facilitate registration, conservation and protection of goods with a unique geographical identity. The Act provides for the establishment of a Geographical Indication Registry and an Appellate Board to take any necessary action against infringement.

Germplasm Flow under New Regimes

In the pre-CBD era, all biodiversity was considered, managed and used as a global public good, with easy access, and was exchanged freely. In the present context, it would have been difficult for N.I. Vavilov to carry out his historical collection expeditions. India imported 60–70,000 accessions p.a. prior to CBD, which reduced significantly. India also exported around 20–25,000 accessions p.a., which also declined. Imagine what would have been the food options for us had these regulations been in place prior to CBD when seeds of food crops like corn, potato, tomato, pepper, soybean etc. were shared and became our major food crops.

As a consequence of enacting legislative measures, invariably the process of germplasm exchange declined globally. In retrospect, the whole process of germplasm exchange and use has slowed down. Fortunately, many countries did share the germplasm with CG centres, which are a major resource for multilateral exchange of crops listed under Annex I of the treaty (ITPGRFA). During the treaty negotiations, it was decided that a call would be taken later to include more crops, but more than 20 years have passed and not a single species has been added to the annex list of 64 crops. Under the CBD, germplasm beyond the multilateral system (under the FAO umbrella of the CG system) can be exchanged under bilateral agreements and collaborative research projects. For bilateral exchange, the Nagoya Protocol has been developed, which now needs to be understood and followed by all the parties to the CBD for access to, and benefit sharing derived from, germplasm. But not much progress has been made. Exchange of genetic resources, which was earlier decided by scientists, after the CBD, the ITPGRFA and the Nagoya Protocol, has been taken care of by bureaucrats and lawyers. Obviously, this is one of the major paradigm shifts that has occurred in GR management.

It has halted exchange of germplasm, affecting agriculture a great deal (Batur and Dedeurwaerdere, 2014; Paroda and Agrawal, 2017).

In India, there is still debate concerning exchange of germplasm, both for public and private seed sectors engaged in plant breeding. Even SMTA has not yet been put into practice for want of procedural clearances and lack of proper understanding. During the mid-1980s, the ICAR, as a policy, allowed free access to parental lines of hybrids bred by the public system, understanding well that seeds of these hybrids would otherwise not reach end-users, i.e. the smallholder farmers. This very policy decision not only accelerated coverage with hybrid seeds, resulting in increased crop productivity, but also strengthened the existing private seed sector in India. Nevertheless, there is an obvious hesitation to share germplasm, out of fear of either biopiracy or loss of ownership. Hence there is a need for much-needed trust building and partnership. This demands an enabling policy environment and a clear understanding for sharing germplasm as well as information between public and private parties engaged in plant breeding.

Think Globally, Act Locally

In the present scenario, it is necessary that we think globally but take concrete measures to act locally. Action at the national/regional level is extremely critical for research, documentation and conservation of the available germplasm before it is lost for ever. Despite 2016 being the International Year of Pulses, there is greater realization that research on pulse crops has been inadequate. India is a gene-rich centre. As one of the eight mega-gene centres of the world, it also has a strong NARS with adequate human resources. With respect to genetic resources, there are five bureaux dealing separately with plants, animals, fish, insects and microbes. Scientific and economic value of genetic materials is difficult to assess, as future problems and needs cannot be precisely anticipated. Moreover, feeding the ever-increasing population would require either intensification of existing agricultural systems or expansion into new areas. This means that optimal management of agricultural ecosystems and diversity of genetic resources would be an essential part of any overall strategy for achieving this goal. In the past, NARS had strong national breeding programmes for developing improved varieties and hybrids. However, subsequently, there was greater dependence on pre-breeding materials provided by the CG centres (Khush, 2016; Paroda and Agrawal, 2017). Unfortunately, over the years, efforts on pre-breeding materials have also declined at these centres due to resource constraints. A paradigm shift from household food security to that of household nutritional security demands much higher investment in intensified scientific understanding of agriculturally important species (be they crops, animals, insects, aquatic species or microbes) as future genetic resources of great potential.

Conservation through Continuum

During the second half of the 20th century, *ex situ* methods of germplasm conservation, especially seed gene banks, were considered a panacea in the management of genetic resources. Everybody thought that because there was a danger of extinction of diversity of plant genetic resources, their seeds should be collected and conserved in gene banks, irrespective of whether they were useful. In most cases, once collected, seeds were retained in these banks for long periods with not much effort given to their evaluation for useful traits or documentation for use by researchers. These gene banks were often considered as 'black boxes'; but unless you know the useful traits of the germplasm collections, how can these be utilized for crop improvement? Also, less emphasis was given to protect vegetatively propagated plants or those that were considered recalcitrant. As a consequence, in many cases, useful variability was lost for want of alternative scientific storage systems such as tissue culture banks or cryobanks.

There is a need to establish a clonal bank repository at the national level along the lines of a national seed gene bank, where, in one place, or its designated regional centres in the country's various agro-ecological zones, most of the vegetatively propagated plants can be maintained, researched and conserved for present and future use. In retrospect, there is now a need for conservation measures that are low-cost and

more sustainable at various ecosystem levels, involving communities known to be 'gene saviours'. Also, there is an urgency to develop a 'conservation continuum', encompassing *in situ*, on-farm, *ex situ*, permafrost and other conservation methods with adequate funding support (Paroda and Agrawal, 2017).

Further, it is of prime importance that farmers, livestock keepers, aquaculture practitioners and foresters engaged in conserving useful varieties, breeds and species derive direct (financial) or indirect (livelihood security) benefits in order to remain occupied in such conservation activities. There must be a compensation mechanism for farming communities employing their unique conservation practices to serve society continuously. Hence, national leaders/policy makers have a responsibility to ensure that the process of natural evolution remains well supported in the best interests of future generations.

The first and second reports on the State of the World's Plant Genetic Resources, brought out by the FAO, provided an authentic assessment of various conservation methods and the state of germplasm collections of plant genetic resources. It documented more than 1750 individual gene banks worldwide, of which about 130 hold more than 10,000 accessions each. Currently, about 7.4 million accessions are maintained in gene banks globally. Analyses suggest that 25–30% of the total holdings (1.9–2.2 million accessions) are unique, the remainder are duplicates held either in the same or, more often, at a different collection. Crop wild relatives (CWR) comprise 10% of these collections, but not many of them have been used so far. Around the globe, genetic resources are maintained in the gene banks at local and national level by governments, universities, botanical gardens, NGOs, companies, farmers and others in the private and public sectors. They house a wide range of different types of collections: national collections maintained for the long term; working collections maintained for the medium or short term; collections of genetic stocks; and others. When we look at the national gene banks around the world, the N.I. Vavilov Genebank in Russia (VIR) was the largest. Lately, the gene bank in the USA is the biggest, followed by those in India, China, Russia, Brazil, Japan and South Korea. In some countries of central Asia and the Caucacus, such as Armenia, Georgia, Kazakhstan,

Turkmenistan and Kyrgyzstan, not even two to three scientists were deployed to work on their valuable genetic resources, and there was practically no infrastructure for the gene banks. Such national systems need support, both in terms of infrastructure and capacity building. It is satisfying to note that in the past decade or so, each of these countries has established functional gene banks (Padulosi *et al.*, 2002).

Programmes on *in situ* and on-farm conservation have recently gained tremendous impetus as these protect germplasm in the natural habitat and take into account social and cultural factors such as farmers' perceptions and knowledge. On-farm conservation entails active participation of local communities in the documentation and description of local species and varieties in a catalogue or register, establishment of nurseries for multiplication and distribution of unique plant or seed material, promotion of nutritional values and traditional recipes, development of enterprises and market linkages for sale of products or services based on the local unique crop diversity, and safeguarding of unique species and varieties found on farms. Thus, *in situ* and on-farm conservation efforts remain ineffective without the participation of the local community. Traditionally, local farmers are known to maintain several indigenous crops on their farms, especially fruit species or varieties. Such farmers have been designated 'custodian farmers', identified for actively maintaining and promoting agrobiodiversity and related indigenous technical knowledge at the farm and community level. Linking such farmers to research institutions and gene banks for characterization and evaluation of elite genotypes, and providing technology for rapid multiplication and distribution of plants is the need of the hour. Documentation of traditional knowledge is another activity that ensures its protection against theft and ensures financial benefits to knowledge holders when commercial sectors exploit that knowledge. Scientific validation of such traditional knowledge is also essential for improved understanding of the ecological functions of agrobiodiversity, especially in the context of the physical environment and socio-economic factors. There is an urgent need to promote the use of more nutritious species such as millets, indigenous fruits, vegetables, roots and tubers, compared with the past when major

emphasis was given to only a select few staple varieties. We now need to ensure upscaling and outscaling of innovations to achieve dietary diversity and improved nutrition at household level. Information systems are still weak, and capacity-building is urgently required (Paroda and Agrawal, 2017).

It is indeed satisfying that permafrost conservation for plant genetic resources has now been put in place. The Svalbard Global Seed Vault (SGSV), established in 2008 inside a mountain on a remote island in the Svalbard archipelago, half-way between mainland Norway and the North Pole, provides a duplicate storage facility for all seeded PGRFA. It is a state-of-the-art seed storage facility built to withstand natural and man-made disasters. The seed vault is managed by the government of Norway. The seed samples are stored in a reinforced concrete tunnel drilled 70 m into a mountain, stored in foil packets at $-18°C$, and are expected to remain viable for thousands of years. Unlike the hundreds of existing seed banks, the vault does not rely solely on artificial refrigeration systems; even if the power fails, the temperature is expected to never rise above freezing. The SGSV has been built to store a massive 4.5 million varieties of crops, with each variety containing around 500 seeds. The Global Crop Diversity Trust works in conjunction with the government of Norway to manage seeds in the vault. The vault currently holds 880,837 seed samples of 5403 species belonging to 71 institutes. These seeds were donated by almost every country in the world, so there is a massive variety of represented seeds. All germplasm from CGIAR gene banks has been safely duplicated here. If a crop is lost through a natural disaster or a war, and a seed bank is destroyed, the government can request replacement seeds from the vault. A recent example of this was when the International Center for Agricultural Research in the Dry Areas (ICARDA) retrieved part of its seed collection from the SGSV to fulfil requests for germplasm use. ICARDA's original gene bank in Aleppo, Syria, was forced to be shifted after the war in the area. ICARDA had replicated over 80% of its collection in the SGSV prior to the conflict. The seeds held in ICARDA are globally sought due to unique landraces and wild relatives of cereals, legumes and forages, collected from the fertile crescent of western Asia. A total of 38,073 seed samples were sent to ICARDA's new sites for gene bank facilities in the cropping seasons 2016 and 2017 in Lebanon and Morocco. Of these, 15,000 accessions (including bread and durum wheat, lentil, faba bean, chickpea and grasspea), multiplied in 2016, were sent back for safe duplication to SGSV on 22 February 2017. This proved to be a classic demonstration of collective wisdom of policy makers, scientists and farmers.

Genetic Diversity – Use It or Lose It!

It is a well-established fact that there is less use of genetic diversity today than in the past, which led to ushering in the Green Revolution. The FAO has, therefore, initiated, with the support of the Bill and Melinda Gates Foundation (BMGF), a project to strengthen plant breeding capacity and research on a global scale, so that use of genetic resources is enhanced globally. This project, known as the Global Partnership Initiative for Plant Breeding (GIPB), is a multi-partner platform with an aim to improve institutional capacity for effective crop variety development and their distribution through seed systems. More details are available at: http://www.fao.org/in-action/plant-breeding/en/ (Paroda and Agrawal, 2017).

It is well documented that the use of PGR has declined globally. Many countries are not laying enough emphasis on pre-breeding and generation of genetic variability for crop improvement. They are largely dependent on import of pre-breeding materials, mainly from CG centres. In view of this, plant breeding must be brought to the forefront. Many stalwarts like Drs Norman Borlaug, G.S. Khush and S.K. Vasal achieved great strides in varietal improvement and adaptation, mainly due to extensive use of genes from landraces and wild relatives. No doubt, working with wild relatives and species is more difficult and requires good infrastructural facilities, yet they are very important in the current context of climate change.

Of course, there are several other reasons for the decline in the use of germplasm. As already mentioned, access to useful germplasm is becoming more difficult due to existing new regulatory regimes. In addition, research on traits of interest and partnership in sharing germplasm is badly lacking. Overall, efforts on pre-breeding are declining due to lack of funding to the National Agricultural Research Systems and CG centres. On the other hand, the requests for germplasm by the breeders have also declined due to lack of digitization, proper evaluation for useful traits, germplasm characterization and existing regulatory systems.

Advances in New Science for Agrobiodiversity Management

We are currently in an exciting scientific era where genome decoding of organisms is becoming almost a routine activity and the possibility of precisely tailoring structure and function of an organism is becoming a reality with new tools of biotechnology, especially gene editing using Crisper-Cas technology, advances in omics, space technology and bioinformatics. New technologies pervading agriculture in terms of smartphones, satellite imaging, phenotyping using drones, IPM, automated farm practices and decision support systems for nitrogen use efficiency (NUE) are helping farmers to grow more food on their land while reducing cost of water, fertilizer, pesticides, etc. (Paroda and Agrawal, 2017).

However, the availability of appropriate seed and planting material/breeds remains the most critical factor for enhancing productivity, adaptability and resilience of agro-ecosystems. Developments in science and technology in genetic engineering, genomics, biotechnology, nanotechnology, bioinformatics and synthetic biology have increased the speed, scale and efficiency in research outputs. These technologies are the game-changers that will dictate how genetic resources are researched in future and used effectively. Nonetheless, existing agrobiodiversity would remain the 'hardware and software codes of nature', requiring systematic deciphering for designing agricultural crops and breeds for their use through new science. Before the emergence of the modern era of use of 'gene guns' by biotechnologists or plant breeders to transfer desirable new genes into designer crops, farming households could assess in their fields and courtyards the semi-wild and semi-cultivated plants for their existing strengths and weaknesses, and select desirable traits while minimizing undesirable ones.

Nevertheless, the products of biotechnology will also have to be field-tested besides undergoing biosafety tests before their identification and release as superior varieties for commercial cultivation. An important aspect with the application of new technologies for agricultural production would be to generate awareness and dispel fears in the minds of the general public about the use of new products (e.g. golden rice) that are the outcome of cutting-edge technologies as international public goods. With new advances in gene editing, the opportunities to accelerate crop breeding and use of germplasm will increase significantly.

References

Batur, F. and Dedeurwaerdere, T. (2014) The use of agrobiodiversity for plant improvement and the intellectual property paradigm: institutional fit and legal tools for mass selection, conventional and molecular plant breeding. *Life Sciences, Society and Policy* 10, 14.

FAO (1997) Paraguay – financial and economic implications of no-tillage and crop rotations compared to conventional cropping systems. TCI Occasional Paper Series No. 9. Investment Centre Division, FAO, Rome.

FAO (1999) Agricultural biodiversity. Background paper for conference *The Multifunctional Character of Agriculture and Land*. Maastricht, The Netherlands, 12–17 September.

FAO (2011) *Nagoya Protocol on Access to Genetic Resources and the Fair and Equitable Sharing of Benefits Arising from Their Utilization to the Convention on Biological Diversity.* Secretariat of the Convention on Biological Diversity, United Nations Environmental Programme, Montreal, Canada.

Frison, E.A., Cherfas, J. and Hodgkin, T. (2011) Agricultural biodiversity is essential for a sustainable improvement in food and nutrition security. *Sustainability* 3(1), 238–253.

Hodgkin, T., Baily, A., Bemhart, A., Nicol, R.D. and Rao, V.R. (2016) Exploring the benefits of system based approach to plant genetic resources conservation. *Indian Journal of Plant Genetic Resources* 29(3), 253–257.

Khush, G.S. (2016) Biodiversity use for food security. *Indian Journal of Plant Genetic Resources* 29(3), 251–252.

Kotschi, J. and Lossau, A. von (2011) Agrobiodiversity – the key to food security and adaptation to climate change. *Diversity*, 22.

Mohapatra, T. (2016) Indian agrobiodiversity system. *Indian Journal of Plant Genetic Resources* 29(3), 230–233.

Padulosi, S., Hodgkin, T., Williams, J.T. and Haq, N. (2002) Underutilized crops: trends, challenges and opportunities in the 21st century. In: Engels, J.M.M., Ramanatha Rao, V. and Brown, A.H.D. (eds) *Managing Plant Genetic Resources*. CAB International, Wallingford, UK and IPGRI, Rome, pp. 323–338.

Paroda, R.S. and Agrawal, A. (2017) Managing agrobiodiversity through use: changing paradigms. *Indian Journal of Plant Genetic Resources* 30(1), 5–12.

Philippe, B. (1972) *United Nations Conference on the Human Environment*, Stockholm, 5–16 June.

Swaminathan, M.S. (2011) *In Search of Biohappiness: Biodiversity and Food, Health and Livelihood Security*. World Scientific Publishing, Singapore.

Swaminathan, M.S. (2016) Agrobiodiversity and achieving zero hunger challenge. *Indian Journal of Plant Genetic Resources* 29(3), 249–250.

Upadhyay, H.D. (2016) Germplasm management for enhanced genetic gain. *Indian Journal of Plant Genetic Resources* 29(3), 261–264.

18

The Growth of the Indian Seed Sector: Challenges and Opportunities

Seed is the basic and most critical input for sustainable agriculture. The response of all other inputs depends on quality of seeds to a large extent. For agriculture to prosper, farmers must have a reliable supply of high-quality seeds and seedlings of superior varieties, at an affordable price. Fortunately, recent advances in the technology of seed and seedling production are helping to improve both the quality and range of planting materials. It is estimated that the direct contribution of quality seed alone to the total production is about 15–20% depending upon the crop, and it can be further raised to 45% with efficient management of other inputs (Poonia, 2013). Seeds of varieties with appropriate characteristics are required to meet the demand of diverse agroclimatic conditions and intensive cropping systems. Sustained increase in agriculture production and productivity is dependent, to a large extent, on development of new and improved varieties of crops and an efficient system for timely supply of quality seeds and planting materials to farmers. The seed sector has made impressive progress over the past five decades (Hanchinal, 2017). In traditional agriculture, farmers save the seed from their own crops to use the following year. Now that most of the farmers are producing seed for commercial purposes, they are buying seed of improved varieties, which have high market value. New technological developments are helping production of better seed at a lower cost. The ICAR-SAU system continues to make available breeder seed of all notified varieties and parental lines of hybrids through the Department of Agriculture and Cooperation (DoAC), Ministry of Agriculture and Farmers' Welfare (MoA & FW), and Government of India (Anonymous, n.d.).

Historical Perspective

The Green Revolution, ushered in during the late 1960s, became a turning point for Indian agriculture. The introduction of high-yielding, semi-dwarf and fertilizer-responsive varieties of wheat and rice was a turning point, changing the status of the country from 'food scarce' to 'food secure' (Paroda, 2013). It is well known that the success of Indian agriculture was on account of an effective combination of the excellent support of policy makers, capable agricultural scientists and administrators, and hard-working farmers to bring in needed change. In order to meet the growing demands for our increasing population, likely to be 1.6 billion by 2050 (Anonymous, 2009), the country needs to double farmer's income by 2022 (as stated by Shri Narendra Modi, the Prime Minister). This can only be possible by bridging the existing yield gaps through improved productivity and by integrated natural resource management. Hence, the Evergreen Revolution would demand much faster growth of the seed sector, especially to meet the demand of hybrid seeds and to replace old varieties with new high-yielding ones.

The post-independence era of the 1950s and 1960s is invariably perceived as the one bubbling with national plans for industrial development and growth. National leaders placed the highest priority on agriculture, considering food self-sufficiency as the major goal. As a result, concerted research and development efforts were initiated. Whilst, on the research side, NARS was reorganized and strengthened, for development-related activities considerable emphasis was given to building national systems for seed development, irrigation infrastructure and establishment of fertilizer cooperatives. While the All India Coordinated Research Project (AICRP) on Maize was the first coordinated research project to begin, in 1957, the first public sector seed company, National Seeds Corporation (NSC), came into existence in 1963, with a mandate to provide good-quality seeds of improved varieties and hybrids to farmers at a reasonable price. Under the National Seed Project, the NSC played a lead role in the establishment of the state seed corporations (SSCs) during the mid-1970s (Hanchinal, 2017). Subsequently, the State Farms Corporation of India (SFCI) came into existence and played a critical role in achieving the country's food security. Contrary to the interest of the private sector in hybrid technology, these public sector companies mainly dealt with open-pollinated varieties (OPVs) and played a very significant role in increasing crop productivity. It is the lead national organization today in the country and has played a significant role towards achieving the Green Revolution in India. Its basket is full of seeds with nearly 600 varieties of 60 crops involving more than 8500 registered seed growers and 2800 dealers throughout the country. It has the distinction of being the largest producer of certified seeds of field crops with an annual turnover of Rs 6.33 billion (Anonymous, 2016).

Recognizing the fact that the adoption of the new high-yielding varieties (HYVs), introduced in the Green Revolution era, depended mainly on adequate availability of quality seeds at affordable prices, steps were taken by the government to establish the public sector seed industry. Subsequent developments in the seed industry in India, particularly in the past 60–65 years, are very significant. A major restructuring of the seed industry was carried out by the government through the National Seed Project (NSP) Phase-I (1977–78), Phase-II (1978–79) and Phase-III (1990–1991), which strengthened the seed infrastructure that was most needed and relevant around those times (Anonymous, n.d.). This could be termed the first turning point in the shaping of an organized seed industry. Further, NSP set up seed processing plants in 17 states. These huge processing plants were supposed to provide 'certified' seeds of food crops, mainly self-pollinating crops, to farmers. The plants operated mostly below capacity and, for all practical purposes, turned into 'white elephants'.

During this period, there were relatively few private companies involved with agricultural seeds (mainly small enterprises confined to the production of some vegetable and ornamental flower seeds), and government policies focused on the public sector with limited private sector participation.

The new policy on seed development in 1988 heralded a new era of private enterprise in the seed sector in India. It stimulated the growth of a number of private seed companies. This coincided with the fourth loan from the World Bank to India's seed sector to make it more 'market responsive'. The loan of US$150 million aimed to encourage the private seed industry and open India to multinational seed corporations (GRAIN and Sharma, 2005). The most significant impact of the new seed policy was an increase in collaborative agreements between domestic and foreign companies, aimed at importing technology and parental material. Under the 1988 policy, vegetable seeds could be imported freely while seeds of oilseeds, pulses and coarse grains, like maize, sorghum and millet, could be imported for two years by companies that had technical and financial collaboration agreements for production of seed with companies abroad (Hanchinal, 2017). Importing was allowed, subject to the provision that the foreign supplier agreed to supply parent-line seeds or breeder seeds to the Indian company within two years of the date of the first commercial consignment.

The Seeds Act 1966, the Seeds Control Order, promulgated thereunder, and the new policy on seed development, 1988, formed the basis for the promotion and regulation of the seed industry and ensured availability of good-quality seeds to farmers. Around the same time, considering the recommendations of the Agricultural Research Review Team (1964), the ICAR was reorganized (Borthakur and Singh, 2012). Starting with the first SAU, at Pantnagar in 1960, several SAUs also came into existence in quick succession in different parts of the country. In 1969, the Tarai Development Corporation (TDC) was established at Pantnagar campus with funding support from the World Bank. In the seed sector, these two institutions, NSC and TDC, along with the Rockefeller Foundation at the Indian Agricultural Research Institute (IARI) in New Delhi, played a very important role in developing the required trained human resource to build the seed sector in the country. The Indian seed industry underwent structural changes with the entry of private seed companies, mostly family owned, during the 1980s, and this trend continued in the 1990s. As expected, private seed companies focused mainly on hybrid seeds and a few large companies diversified into R&D to increase their share of the seed market. The new seed policy of 1988 and the economy-wide reforms of 1991 attracted multinational companies to India in a major way. Most of them entered through partnership with the national companies, and a few established their independent seed businesses.

Even in self-pollinated crops like paddy, the public sector share was nominal at that time, and the private sector supplied 60–80% of commercial seed in the states of Haryana and Andhra Pradesh. Low marginal cost and risk in producing paddy seed and a potentially lucrative market for hybrid rice could explain the greater private sector participation. In the case of inaccessible hilly areas, also, the private sector supplied a significant proportion of commercial maize and vegetable seed. Only in the case of high-volume seed crops like potato and groundnut was there less participation by the private sector, as borne out in the aggregate statistics. All potato seed in Uttar Pradesh, Himachal Pradesh and elsewhere is supplied by the government, although there are some private seed producers, mostly in Punjab and western Uttar Pradesh, but their share is negligible (Singh et al., 2008). In the case of groundnut, besides state seed corporations and the government, there are some producers' organizations and oil-trading public agencies who also supply seed to farmers.

Today, the Indian seed sector is poised to emerge as a matured sector with three major changes. First, private seed companies consider R&D as an important mechanism to differentiate their product and enhance their market share. This trend is likely to intensify further. The second major change had arisen due to the process of globalization and liberalization. The resource-rich multinational companies with well-established R&D programmes overseas are expanding their activities through mergers and a few acquisitions, and compete without the back-up of national research institutions. Third, the industry is being governed now by multiple regulations, with the protection of IPR emerging as the single most important factor to shape its growth and performance. Fortunately, India has put in place by now all the necessary legislation and institutions to strengthen the IPR regime to comply with the WTO. Concomitantly, other regulations, like those dealing with development and commercialization of genetically engineered crop varieties, protection of varieties, and access to and use of genetic resources were enacted to ensure faster growth of the seed industry. Therefore, the IP landscape for the seed industry should be smoother with reference to issues relating to the National Biodiversity Authority (NBA), IPR, Protection of Plant Varieties and Farmers' Rights Authority (PPV&FRA), especially in the context of export/import and patents.

Policy Guidelines: A Progressive Step

During the 1970s and 1980s, in spite of the release of a large number of crop varieties and hybrids (including vegetables), the growth of the seed sector somehow reached a plateau with an annual turnover of Rs 5–6 billion by 1987. To overcome this, as stated earlier, the new policy for seed development was initiated as a liberal approach for importing seed and planting materials for the benefit of Indian farmers. At the same time, a progressive decision was taken in the late 1980s to make available the breeder seeds of parental lines of publicly bred hybrids, even to the private sector, for accelerating production of hybrid seeds and the pace of the seed sector's development. This resulted in much faster growth of the seed sector, especially of private seed companies, resulting in four to five times increase in overall turnover. Today, the country's total seed business is about 35 million quintals, amounting to a turnover of nearly Rs 200 billion, of which the private sector share is more than 60%.

Post-WTO Scenario

The aim of the WTO, established in January 1995, is to provide a regulatory and institutional framework for the world trade system. In the post-WTO era, the Indian seed industry went through a process of rapid change as the government enacted the PVP&FR Act 2001 to ensure faster growth of the seed sector (Venkatesh *et al.*, 2015). These initiatives triggered major investments in the seed sector in India. As a result, use of hybrid seeds increased significantly in different crops (in some cases up to 95%), mainly through the contribution of private seed companies, including the multinational companies. Simultaneously, the ICAR strengthened its hybrid research programme and accelerated breeder seed production of field crops by implementing a Mega Seed Project in 2004–05. As a result, breeder seed production surpassed 100,000 quintals by 2009–10.

The most dramatic change in the seed sector has been experienced since 2002, with the introduction of Bt cotton in India (Choudhary and Gaur, 2010). With over 95% of cotton area currently under Bt cotton hybrids (11 million ha), productivity has increased by 139% and demand for cotton hybrid (Bt) seed has increased by 220% in just one decade. This entire boom has mainly been possible through the proactive role of the private seed sector, which accessed the technology from the MNCs and other sources, making huge investments and profits (Plewis, 2014). On the contrary, the public seed industry, which mainly dealt with varieties and some hybrids released by the public research system, could not harness the benefits of such developments. Industry has become a capital-intensive industry. Indian companies have established relationships/joint ventures with foreign companies. Competition in the seed market has become stronger, with many foreign/multinational companies now operating in India and competing with local seed organizations. Availability of varieties/hybrids/transgenics from various sources and their testing and introduction in India in agricultural and horticultural crops has encouraged many Indian seed companies to start plant-breeding research. Deriving strength from the post-WTO scenario, PPV&FR legislation and modern technological developments in evaluation of improved varieties, advances in seed production technology, advantages of agroclimatic conditions (for tropical, sub-tropical and temperate crops seed production), experienced seed growers, skilled manpower and easy availability of labour (male/female) on affordable wages, India has significant prospects to increase its seed export share from 1% to 10% by 2025. Public sector investment can play a critical role in promoting innovation in specific areas where the private sector cannot justify investment. The policy environment and the governance system must be in place to achieve the goal of maximizing benefits of agricultural R&D. Hence, there is a need for ample and sustained government funding, along with robust and agile institutional innovations that foster public and private investment in agriculture.

The Seeds Bill

The Seeds Bill seeks to regulate the production, distribution and sale of seeds. It requires every seller of seeds (including farmers) to meet certain minimum standards. Constituted in 1998, a Seed Policy Review Group in India has recommended a shake-up and reform of the Indian seed laws. A new Seed Bill needs to be passed at the earliest opportunity, which would replace the current 1966 Seeds Act. The Seeds Bill 2004 has been pending since December 2004 for approval by Parliament. The government is planning to revive the Seeds Bill, which was first introduced in the Indian Parliament in 2004 and last discussed in 2014, but has never seen the light of day. In between, the government proposed new amendments to the Bill in April 2010 and November 2010, accepting most of the recommendations given by the Standing Committee. After the new Seeds Act is in force, it will regulate the sale, import and export of seeds and planting materials of agricultural crops including fodder, green manure and horticulture, and supply quality seeds and planting materials to farmers throughout the country. Also, it will ensure that farmers selling or exchanging seeds from other farmers are exempt from this requirement. A National Seeds Board (NSB) needs to be established in place of the existing Central Seed Committee and Central Seed Certification Board, which will have responsibility for executing and implementing the provisions of the Seeds Bill (Act) and advising the government on all matters relating to seed planning and development. All varieties (both domestic and imported) that are placed on the market for sale, and distribution of seeds and planting materials, will be registered under the proposed Seeds Act. However, for vegetable and ornamental crops, a simple system of varietal registration based on a 'breeder's declaration' could be adopted. The Board will undertake registration of kinds/varieties of seeds that are to be offered for sale in the market, on the basis of identified parameters for establishing value for cultivation and use (VCU) through testing/trialling. Registration of varieties will be granted for a fixed period on the basis of multi-location trials to determine VCU over a minimum period of two seasons, or as otherwise prescribed, as in the case of long-duration crops and horticultural crops. Samples of the material for registration will be sent to the National Bureau of Plant Genetic Resources (NBPGR) for retention in the national gene bank. Varieties that are in the market at the time the revised Seeds Act comes into effect will have to be registered within a fixed time period and subjected to such testing as will be notified. The NSB will accredit the ICAR, SAUs and public and private organizations to conduct VCU trials of all varieties for the purpose of registration, as per the prescribed standards. Under the new Seeds Act, the NSB will maintain the National Seeds Register containing details of varieties that are registered. This will help the Board to coordinate and assist activities of the states in their efforts to provide quality seeds to farmers. The NSB will prescribe minimum standards (of germination, genetic characteristics, physical purity, seed health etc.) as well as suitable guidelines for registration of seed and planting materials. Provisional registration would be granted on the basis of information filed by the applicant relating to trials over one season to tide over the stipulation of testing over three seasons before the grant of registration. The government will have the right to exclude certain kinds or varieties from registration to protect public order, or human, animal and plant life and health, or to avoid serious prejudice to the environment.

All GE crops/varieties will be tested for environment and biosafety before their commercial release, as per the regulations and guidelines of the Environment Protection Act (EPA) 1986, including the recommendations of the Genetic Engineering Approval Committee (GEAC). The EPA 1986, read with the Rules 1989 (and subsequent amendments), would adequately address the safety aspects of transgenic seeds/planting materials. A list will be generated from the Indian experience of transgenic cultivars that could be rated as environmentally safe. Seeds of transgenic plant varieties for research purposes will be imported only through the NBPGR as per the EPA 1986. Transgenic crops/varieties will be tested to determine their agronomic value for at least two seasons under the All India Coordinated Research Project Trials of ICAR, in coordination with the tests for environment and biosafety clearance, as per the EPA, before any variety is commercially released in the market. After the transgenic plant variety is commercially released, its seed will be registered and marketed in the country as per the provisions of the Seeds Act. After commercial release of a transgenic plant variety, its performance in the field will be monitored for at least three to five years by the Ministry of Agriculture and state departments of agriculture. Transgenic varieties can be protected under the PPV&FR legislation in the same manner as non-transgenic varieties after their release for commercial cultivation. All seeds imported into the country will be required to be accompanied by a certificate from the competent authority of the exporting country regarding their transgenic character or otherwise. If the seed or planting material is a product of transgenic manipulation, it will be allowed to be imported only with the approval of the Genetic Engineering Appraisal Committee (GEAC), set up under the EPA 1986. Packages containing transgenic seeds/planting materials, if and when placed on sale, will carry a label indicating their transgenic nature. The specific characteristics including the agronomic/yield benefits, names of the transgenes and any relevant information shall also be indicated on the label. Emphasis will be placed on the development of infrastructure for the testing, identification and evaluation of transgenic planting materials in the country.

The effective implementation of the PPV&FR Act is expected to promote private plant breeding in the country in the long run. The immediate effect could be in terms of increased access to seeds developed by transnational seed companies. These companies may sell seed on their own or link up with the national companies for multiplication and marketing of their material. It is also likely that transnational seed companies would establish joint research ventures with the national companies, such as that between Monsanto and Mahyco, or market-notified varieties of food-security crops released by the public system as non-exclusive licensees. Whatever may be the path, Indian farmers are expected to gain from having multiple choice and access to improved seed, which can have a positive effect on crop productivity and raise farm incomes. At the same time, this could create some degree of concentration in the seed market because of substantial investments made by some of the transnational seed companies. To safeguard from monopolistic seed trade, necessary provisions are made in the Government of India legislations and policies. The provision of compulsory licensing and presence of a strong public breeding programme for developing varieties, which can be delivered by public and private seed agencies, are effective mechanisms to control monopolistic tendencies. The provisions for mandatory registration of plant varieties, farmers' rights to sell unbranded seed of any variety, and disclosure of information on parents of hybrids are being discussed by the private seed industry platforms, and these may significantly influence relations among seed entities. Nevertheless, issues like protection of genetic resources, a provision under the Convention on Biological Diversity (CBD), and to encourage free access to seed among farmers, are quite important from the system's perspective. It is even more important for the crops where traditional seed systems are dominant, such as minor millets, underutilized crops and varieties with specific limited use. Public ownership of genetic resources could also be used to

bargain for access to proprietary technology to promote a competitive seed industry.

Public–Private Partnership

The past ten years have witnessed a lot of collaborations and direct partnerships between the public and private sectors. Since the mid-1990s, the ICAR had been liberal in the supply of breeder seed of all notified open-pollinated (OP) varieties and parental lines of the hybrids of field crops to both public and private sector partners against a centralized or individual indent, at the price fixed by the respective breeder seed committees. However, the scenario changed somewhat after the implementation of the PPV&FR Act 2001. Now, a large number of seed companies from the private sector are becoming licensed partners with the public research organizations, to multiply and market seeds of their varieties by paying predetermined and mutually agreed terms of benefit sharing through a licence fee and royalty. Some research institutions, e.g. IARI; New Delhi Indian Institute of Horticultural Research (IIHR), Bangalore; and University of Agricultural Sciences, Dharwad (UAS-D), have worked out an effective mechanism to foster strong public–private partnerships. The Institute of Technology Management Units (ITMU), along with the institutional Business Promotion & Development Units (BPDs), played a key role. There are still a number of challenges to be addressed – development of research capacity both on conventional breeding and molecular technologies, biosafety and management of genetically engineered crops, and public dialogue on controversial issues. Establishment of biotechnology capacity is relatively capital- and human resources-intensive. Both the public and private sectors will have to play an important role, and there is much potential for forging public–private links to enhance overall impact. These links could be useful as advances in biotechnology have blurred the differences between pure science and agricultural science, requiring close links with general science and technology providers. It is more the case when a major responsibility for promotion of biotechnology in India rests with the Department of Biotechnology in the Ministry of Science and

Technology. These public–private links can be fostered by setting appropriate mechanisms for the sharing of cost and benefits, establishing joint ventures, and management and ownership of intellectual property. Given the current debate on biotechnology in India and elsewhere, effective biosafety regulations must be put in place that are credible, cost-effective and properly coordinated. There is no easy solution to these issues.

Biotechnological research in India is governed by a number of acts, namely the Seeds Act, Environment Protection Act (EPA), PVP&FR Act and Biological Diversity Act (BDA). It is important that there is coherence between these, otherwise some of the positive aspects of these provisions could be neutralized, hampering the growth of public/private sector research. Finally, there is inadequate flow of information about new technologies to farmers. Since much of this information is a public good, public institutions and government will have to take the major responsibility for disseminating information and educating farmer consumers.

In the case of hybrid rice, the ICAR, the SAUs, the IRRI and national private seed companies collaborated for development of male sterile lines, development of hybrids and refinement of seed-multiplication technologies. The partnership upscaled the hybrid rice technology and intensified plant-breeding and seed-multiplication activities in the private sector. The technology has been commercialized and is being adopted mainly in marginal areas of eastern India because of significant yield advantage. Currently, hybrid rice area covered is around 2.5 million ha, whereas the potential is much greater.

A group of private seed companies, both national and international, have formed a consortium to fund the plant-breeding programme of ICRISAT for pearl millet and sorghum. The member companies pay an annual fee and have access to advanced breeding material. The material is available to the public plant-breeding programmes but not to non-member seed companies. Private seed companies benefit from advanced breeding material and minimize their research cost, while ICRISAT is able to generate resources to fund its breeding programmes for the crops. The Council of Scientific and Industrial Research (CSIR)-National Botanical Research Institute (NBRI), Lucknow, has taken the lead in

this direction for Bt cotton, but their initiative is constrained by the lack of freedom to operate.

Technology-led Growth

The most dramatic change in the seed scenario was experienced in the first decade of the current millennium. Again, this could be attributed to a combination of two important policy decisions. First, the introduction of the PPV&FR Act 2001, and second, the release of Bt cotton in India in 2002. The enactment of the PPV&FR Act has instilled much-needed confidence in the seed industry both in terms of intellectual property and higher investment in R&D. The rapid expansion of the Bt cotton production area (reaching ~95%) has enhanced the demand for Bt cotton hybrid seed by 220% (Dravid, 2011). The adoption of Bt cotton technology increased production by 139%. India could turn into a net exporter of cotton from being an importer just a decade ago. In this case, the private sector took the lead in accessing the technology from the multinational companies. The public seed sector, solely depending on public research technology/varieties and hybrids, and not having the necessary financial strength and flexibility, could not reap the benefits. Thus, time and again, Indian farmers have shown their receptivity and inclination to adopt any new technology that promised higher production and profitability. It is also evident that if the technology is promising, the farmers are willing to invest. All these factors led to higher growth of the Indian seed industry (around US$2 billion), with a potential to grow by 60% in the next five years.

At present, a decelerating productivity growth rate, increasing prices and demand for food grains, fragmented land holdings, shrinking natural resources and the challenges of climate change have emerged as the major concerns for policy makers and scientists alike. For raising the agricultural productivity, seed is recognized to be the cheapest, yet most critical single input. Use of good-quality seed can result in as much as 15–20% yield increase. Therefore, any attempt to turn around our agricultural productivity will depend largely on a higher replacement rate of quality seeds of high-yielding varieties/hybrids. Unfortunately, in spite of several efforts to ensure availability of these, the replacement rates in most of the field crops are still much below the optimum levels. Hybrid seeds in cross-pollinated crops give higher yield; hence, to improve crop productivity, greater emphasis should be laid on the development of hybrid seeds. The aggressive promotion measures undertaken for the use of hybrid seeds resulted in the increased demand and production of hybrid seeds in the country. The crop-wise requirement and availability of hybrid seeds during each of the last five years is shown in Table 18.1, which is sufficient to meet our requirements. It is encouraging that both private and public sectors are contributing towards seed production (Table 18.2).

However, often the seeds of new improved varieties are not available to farmers. Thus, the seed is replaced, but not the variety. There is also a need to undertake an authentic assessment of the state-wise seed requirement of different crops, actual availability of quality seed of new improved varieties and the desirable seed replacement rates (SRRs). In certain cases, the subsidy linked to the certified seed of field crops of large volume and low value has proved counter-productive to the improvement of seed/variety replacement rates (SRR/VRR). Instances of purchase of seed-grade groundnut or pulses for consumption as food commodities are also common, as the prices of the certified seed are at times lower than the commercial grain, which are generally in short supply. Extending the scope of government subsidy even to truthfully labelled seed of promising hybrids, produced by private companies, following the model of the government of Bihar, would be yet another bold policy decision, as already recommended by the Hooda Committee Report to the Government of India (2010). The production of breeder, foundation and certified seed during 2006–07 to 2015–16 is given in Table 18.3.

Seed Rolling Plan

Production of the right kind of certified seed requires at least three years' advance planning for undertaking activities in a systematic manner, beginning with the production of breeder seeds and followed by that of the foundation and certified seeds as described in Box 18.1.

Seed production needs systematic planning and implementation of a series of activities. It has been estimated that to achieve food production targets, there is a need to replace the existing seeds (seed replacement ratio) at the rate of 33% for self-pollinated crops, 50% for cross-pollinated crops and 100% for hybrids. It is well known that many farmers do not have access to certified seeds and depend on farm-saved seeds for boosting agricultural production. Further, the country is often affected by natural calamities of different magnitude

Table 18.1. Year-wise requirement and availability of certified/quality seeds of hybrids ('000 t). (From: *State of Indian Agriculture, 2015–16*, Ministry of Agriculture and Farmers' Welfare, Government of India)

Year	Requirement	Availability
2011–12	179.1	210.1
2012–13	198.7	212.3
2013–14	192.5	201.8
2014–15	173.2	203.4
2015–16	211.2	259.1

Table 18.2. Total seed production by the public and private sectors. (From: Singh and Chand, 2011, and Seeds Division, Department of Agriculture & Cooperation, Ministry of Agriculture, GoI)

Year	Total seed production (million t)	Share of private sector (%)	Quantity of seed produced by private sector (million t)	Quantity of seed produced by public sector (m t)
2003–2004	1.3227	47.48	0.6280	0.6947
2004–2005	1.4051	45.02	0.6326	0.7725
2005–2006	1.4818	46.80	0.6935	0.7883
2006–2007	1.9431	41.00	0.7967	1.1464
2007–2008	1.9423	42.59	0.8272	1.1151
2008–2009	2.5040	39.78	0.9961	1.5079
2009–2010	2.8000	38.93	1.0900	1.7100

Table 18.3. Production of different categories of seed (t). (From: *State of Indian Agriculture, 2015–16*, Ministry of Agriculture and Farmers' Welfare, Government of India)

Year	Breeder seed	Foundation seed	Certified/ quality seed
2006–07	6823	74,800	1,405,000
2006–07	7382	79,654	1,481,800
2007–08	9196	85,254	1,943,100
2008–09	9441	96,274	2,503,500
2009–10	10,683	114,638	2,797,200
2010–11	11,921	180,640	3,213,592
2011–12	12,338	222,681	3,536,200
2012–13	11,020	161,700	3,285,800
2013–14	8229	174,307	3,473,130
2014–15	9849	157,616	3,517,664
2015–16	8621	149,542	3,435,248

and type, e.g. flood, drought, cyclones etc., during the normal cultivation season (kharif/rabi), when agricultural programmes are stranded and a contingency crop plan is employed for salvaging the situation. State governments are giving weight to the creation of seed banks, wherein seed for emergency situations can be stored for use during unforeseen and unknown situations. To meet these objectives, states have been advised to prepare a long-term Seed Rolling Plan that envisages the identification of the right varieties of seed for the seed chain and of the agencies responsible for the production of seeds at every level. The plan should take into account the nature of the crop cultivated, existing and desired SRR and requirements for contingencies like flood, drought, cyclones etc.

Box 18.1. Certified seed chain.

The process of production of certified seeds begins with the production of nucleus seed of a notified variety. In India, starting with breeder seed, a three- to four-generation seed multiplication chain is adopted.
Nucleus seed: The nucleus seed is a genetically pure seed without any off-type produced directly by the concerned breeder. They are obtained from a handful of healthy and true-to-type plants selected carefully (following plant-to-row or ear/panicle-to-row progenies) and then grown strictly in isolation. All morphological characteristics, such as plant size, growth habits, colour and shape of various plant parts at various stages, days taken in maturity etc. are taken into account and recorded. This stage is the most important phase in seed production, because any erroneous selection of the nucleus seed plants at this stage would adversely affect all successive generations. Once these plants are selected; their seeds are obtained and threshed separately. If it is a crop with important seed trait, which determines the agronomic value of the variety, such as oil quality or quantity, grain cooking quality, high Fe or Zn content etc., the quality of seeds, yield etc. are recorded and those not meeting the expected values are removed. These seeds are properly packed and regrown to get the breeder seed.
Breeder seed: Breeder seed is the direct progeny of nucleus seed, produced under the supervision of a qualified breeder, maintaining the highest genetic purity achievable in a given species. The true-to-type plants with desired quality are harvested and threshed separately to obtain the breeder seed. The breeder seed crop is monitored at all critical growth stages by a monitoring team constituted of breeders, seed technologists, pathologists, agronomists, entomologists and representatives from the NSC/SSC, private companies, DoAC and ICAR. Breeder seed is further multiplied as foundation and certified seed. In India, breeder seed is produced by the ICAR, NSC, SFC and SAUs, and such other designated centres in India.
Foundation seed: Progeny of the breeder seed, which can be clearly traced to breeder seed, is called foundation seed. The foundation seed is used to produce certified seed. The production of the foundation seed must be undertaken as per the guidelines laid down by the concerned certification agency, which also organizes periodic inspections and accepts or rejects the seed crop. A person or company who grows and distributes the certified seeds in accordance with the procedure and specifications of the certification agency is called the certified seed producer. The NSC, SFCI and other designated agencies in the public and private sectors have the responsibility to produce foundation seed to meet the demand of national varieties. The SSCs mostly produce the foundation seed to suit local demands.
Certified seed: This is the last stage, which actually reaches the farmer. Certified seed is the progeny of foundation seed produced under regular inspection and monitoring of the seed crop and testing of the resultant seed by the certification agencies. It must meet the minimum standards of seed quality parameters prescribed in the Indian Minimum Seeds Certification Standards, 1988/2013.

Seeds and Planting Materials of Horticultural Crops

To meet the nation's food security needs, it is important to make available to Indian farmers a wide range of seeds and planting materials of superior quality, in adequate quantity and on a timely basis. India has to put strenuous efforts into developing a seed industry. The demand by farmers for quality seeds of new high-yielding varieties in the changing seed scenario is expecting miracles from plant breeders and the seed trade. They want more and more new improved varieties to be released, multiplied and made available to them. The assignment facing this industry is staggering. The fate of improved varieties evolved by scientists using modern advanced technology/tools – hybrid vigour, tissue culture, GM plants, molecular breeding, organic seeds etc. – depends on the quality of seeds supplied to farmers. The concept of seed quality under these new technologies is also under phenomenal change (Bhaskaran and Raja, 2017). A totally new approach is the immediate need to assess the quality and thereby upgrade the same. Advances have been made in seed health testing, verification of species/cultivars, quick viability, X-ray analysis, coated seed testing, seed vigour, GM seed testing, seed cleaning and size grading.

The success of any orchard mainly depends upon availability of the right type of planting material. Any shortfall in quality of planting materials causes lasting damage to productivity and orchardists' income. The horticultural wealth of India needs special mention. India is the largest producer of mango, banana, coconut, arecanut and cashews. A number of improved varieties have been developed for the domestic market, export and dessert use. Under the provisions of the Seed Act, for horticultural crops, separate sub-committees, for vegetable crops, flowers, ornamental plants, medicinal plants, spices, fruits, plantation crops and potato/tobacco, have been constituted. In this endeavour, public sector seed institutions will be encouraged to enhance production of seed towards meeting the objective of food and nutritional security. The Indian seed programme adheres to the limited three-generation system of seed multiplication, namely, breeder, foundation and certified seed. Breeder seed is the progeny of nucleus seed. Nucleus seed is the seed produced by the breeder to develop a particular variety and is used directly for multiplication as breeder seed. Breeder seed is the seed material directly controlled by the originating or sponsoring breeder or institution for the initial and recurring production of foundation seed. Foundation seed is the progeny of breeder seed. Foundation seed may also be produced from foundation seed. Production of foundation seed, stage-I and stage-II, may thus be permitted if supervised and approved by the Certification Agency and if the production process is so handled as to maintain specific genetic purity and identity. Certified seed is the progeny of foundation seed or the progeny of certified seed. If the certified seed is the progeny of certified seed, then this reproduction will not exceed three generations beyond foundation stage-I and it will be ascertained by the Certification Agency that genetic identity and genetic purity has not been significantly altered (Tunwar and Singh, 1988).

Certification standards for fruit crops, namely, banana, lychee, grape, pineapple, jackfruit, passion fruit, mango, sweet orange, lemon, guava, apple, pear, plum, peach and, for tissue culture, raised potato micro-tubers, apricot, cherry, walnut, almond, aonla, custard apple, ber, bael, papaya and pomegranate have been formulated. Progress has also been made in the formulation of certification standards for important spices. These standards are formulated by respective sub-committees and are adopted after the approval of the Central Seed Committee (CSC). The standards are already approved in some cases, whilst for the remainder it is underway. Certification standards for important flowers and ornamental plants, namely marigold, sunflower, chrysanthemum, carnation, gerbera and aster, were formulated by the sub-committee on flowers and ornamental plants and approved for adoption by the CSC.

Accelerating Growth of the Seed Sector

At the present time, a decelerating productivity growth rate, increasing prices, demand for foodgrains, shrinking natural resources and the emerging challenge of climate change have all become major concerns for policy makers and scientists alike. To attain a national GDP growth rate of 8%, it is necessary that agricultural growth is increased from the current 2% to a minimum of 4%. As also recommended by the Hooda Committee (2010), the best way to achieve this will be through bridging the existing yield gap for most of the crops through higher coverage under high-yielding varieties (HYVs) and hybrids combined with efficient crop-management practices. Moreover, seed is recognized as a cheap, most critical single input. Use of good-quality seed can result in as much as 15–20% yield increase. Therefore, any attempt to turn around agricultural productivity will largely depend on higher replacement rate of quality seeds of high-yielding varieties/hybrids. Unfortunately, the replacement rates in most of the crops, both for varieties and hybrids, are below the optimal levels defined. This is more so in the case of pulses, groundnut and forage crops. Also, for hybrids, coverage has yet to be accelerated (almost doubled) in the case of single-cross maize hybrids, rice hybrids and some oilseeds like sunflower, castor and rapeseed-mustard.

The Way Forward

The Indian seed sector is backed by a strong crop-improvement programme in both the public and private sectors. At the moment, the industry is highly vibrant and energetic and is well-recognized in the international seed arena. Several developing and neighbouring countries have benefitted from quality seed imports from India. India's seed programme has a strong seed production base in terms of diverse and ideal agroclimatic conditions spread throughout the country for producing high-quality seeds of several tropical, temperate and subtropical plant varieties in enough quantities at competitive prices. Over the years, several seed crop zones have evolved with extreme levels of specialization. Similarly, for post-harvest handling, the Indian seed processing/conditioning industry has perfected the techniques of quality upgrading and maintenance to ensure high standards of physical condition and quality of seed. By virtue of the diverse agroclimates, several geographical zones in the country have emerged as ideal seed storage locations under ambient conditions. In terms of seed marketing

and distribution, more than 20,000 seed dealers and distributors are active in the industry.

Over the years, seed-quality specifications comparable to international standards have been evolved and have been adopted by the Indian seed programme in both the public and private sectors. The country has a sound mechanism for seed quality control through voluntary seed certificates and compulsory labelling, monitored by provincial-level seed law enforcement agencies (SLEAs). For seed technology research, India has a national-level institute under the ICAR as well as research units set up in the SAUs. In seed education, four to five prominent SAUs offer post-graduation in seed technology, leading to an MSc/PhD. The seed industry has well-reputed national-level associations apart from several provincial-level groups to take care of the interests of the industry. The NSC, which is the largest single seed organization in the country, with such a wide product range, pioneered the growth and development of a sound industry in India. The NSC, SFCI, SSCs and other seed-producing agencies are continuously and gradually expanding all their activities, especially in terms of product range, volume and value of seed handled and level of seed distribution to the un-reached areas, etc. Over the past six decades, these seed-producing agencies have built up a hard core of competent and experienced seed producers and seed dealers in various parts of the country having adequate levels of competence in handling and managing various segments of seed improvement on scientifically sound and commercially viable terms. India has one of the oldest associations of seed professionals, the Indian Society of Seed Technology (ISST), established in 1971, with more than 900 members. It plays an important role in identifying issues needing scientific deliberations and generating awareness of issues relating to seed technology.

The Indian seed industry has come a long way, occupying sixth position in global trade with about Rs 200 billion crores turnover and an annual growth rate of 12–15%. The public sector, which dominated the seed industry till the 1970s, has reflected a declining trend in the current millennium, whereas the private seed sector has gained momentum over the past two decades. On the contrary, the pace of growth of public sector seed organizations, including the

NSC, has remained slow. Hence, the process of revitalizing the entire seed sector, especially the NSC, needs urgent attention to harness the fruits of available innovations for increased productivity. To accelerate productivity growth, the entire seed sector needs revamping at the national level. Also, there is an urgent need to overcome complacency, create a new enabling environment, think 'outside the box' and develop a clear road map to achieve new heights in the area of seed research and development in India.

Over the years, the seed industry has become a capital-intensive industry. Indian companies have established relationships/joint ventures with foreign companies. Competition in the seed trade is quite strong and many foreign and multinational companies are now operating in India and competing with local seed organizations. Availability of varieties/hybrid/transgenics from various sources, and their testing and introduction in India for food and horticultural crops, has promoted Indian seed organizations to start plant-breeding research. The Indian seed sector is now poised for continued growth in years to come. We now need the policy makers, scientists, seed sector personnel and farmers to join hands with a missionary zeal to revitalize Indian agriculture through the implementation of the following ten-point action plan:

- For accelerating the progress of the seed industry (both public and private), it is critical to have the Seed Bill 2004 approved by Parliament. Considerable time has elapsed since the Seed Act was first passed in 1966. Unfortunately, many of the reforms expected to ensure faster growth of the seed sector are held up for want of a proposed new Seed Bill.

- A National Mission on Seed has been launched by the government, as proposed in the Hooda Committee Report (2010) for increasing agricultural productivity. This would provide an enabling environment for the growth of the seed sector. In this Mission, adequate support for hybrid and quality seed production must be ensured through strong public–private partnership and also through the active involvement of progressive farmers, following the Dharwad model or IARI model. The mission should aim for faster growth of the seed sector in a holistic manner.

- According to the National Seed Plan (NSP), the projected seed requirement of 2.54 million t by 2009/10 has already been achieved. However, as stated earlier, there is a considerable mismatch and the availability of seeds of new HYVs and hybrids is still not sufficient to accelerate the pace of increasing productivity. Therefore, advanced planning through a five-year rolling seed plan should be developed jointly by the ICAR and the DoAC, in consultation with state agriculture departments and SAUs.

- There is a need to achieve a second Green Revolution, especially through greater emphasis on hybrid seeds. Hybrids also need to be promoted aggressively to improve crop productivity, especially in crops like maize, rice, sorghum (rabi), pearl millet, pigeon pea, rapeseed-mustard, sunflower, cotton, castor etc. Farmers will benefit if sufficient quality hybrid seed is produced jointly by private and public sector institutions. Seeds produced need to be treated on a par for subsidy since greater coverage of area under hybrids will be in the best national interest. Involvement of progressive farmers in hybrid seed production will be a step towards healthy competition.

- Complementarity of public sector policy and infrastructure and private sector dynamism can be capitalized upon through appropriate PPP. Successful models of PPP, as experienced in the recent past, could be replicated. Again, long-term contracts can be entered into between state seed corporations, ICAR research institutes/SAUs, the private seed sector, cooperatives of farmers and self-help groups to undertake production and supply of quality seeds.

- Our system for seed-quality assurance requires considerable investment in terms of modern infrastructure, equipment and competent human resources. Seed-certification agencies have to be equipped and provided with trained human resources for an efficient certification process. The six seed-testing laboratories in the private sector that are accredited by the International Seed Testing Association (ISTA) could also be notified by the government. Efforts need to be intensified to get ISTA accreditation of all seed-testing laboratories in the public sector, including IARI and the IISS (Indian Institute of Seed Science).

- India has the potential to become a major player in the seed sector, globally. Its present share in the global seed market is less than 2%. There is good export market potential, particularly in the African, South Asian Association for Regional Cooperation (SAARC) countries and south-east Asian countries. Therefore, India must aim to achieve a target of 10% by 2020, as envisaged in the National Seeds Policy (2002). Some of the Indian seed companies are already doing business in other countries, but the potential is still much greater. Adoption of OECD seed certification schemes has increased the export potential of seeds of Indian varieties, many of which could perform well in other countries. For this, there is need for a forward-looking, long-term seed export policy. This golden opportunity during an era of globalization should not be missed. To achieve this, an enabling environment must be created through a single-window system for the processing and clearance of all seed-related export proposals.

- Germplasm conservation through use can help in achieving both sustainable agricultural growth and development. Hence, it is emphasized that the national germplasm collection available at the National Bureau of Plant Genetic Resources (NBPGR), being national public goods, should be made available more freely to Indian scientists/institutions/seed companies engaged in R&D programmes. For this, the Standard Material Transfer Agreement (SMTA), as adopted by the FAO International Treaty on Plant Genetic Resources for Food and Agriculture (ITPGRFA) for multilateral access, should be adopted also for bilateral exchanges urgently, with

necessary safeguards. Further, all data on available germplasm in the public domain must be documented/catalogued and posted on the NBPGR website for utilization. Also, in the national interest, all released and notified verities of different crops should be registered with PPV&FRA at the earliest opportunity.

- Biotechnology offers great potential in increasing production at reduced cost of inputs. Bt cotton has clearly revealed that such technologies are beneficial to smallholder farmers. Similar benefits can be reaped in other crops like soybean, rapeseed-mustard, maize, rice and some vegetables. There is a need to have a clear policy and road map defined on GM food crops as a matter of national priority or else our farmers will be deprived of greater benefits from this new science. Public perception and policy decisions need to be based on scientific data and not otherwise. Current mistrust in promoting biotechnology, especially GM food crops, is detrimental to the further growth and development of Indian agriculture. Further, the Biotechnology Regulatory Authority of India Bill should be cleared at an early date for the benefit of the seed sector.

- In this context, the leadership role of the NSC should be recognized by other state seed corporations. It should assume an oversight function and play an important role for project monitoring and evaluation (PME) in the national context. With the advent of new cutting-edge science and technology, there is a need to develop good R&D support through well-trained scientific manpower. Restructuring of the NSC is also required towards product diversification, core competence and improved governance.

References

Anonymous (2009) How to feed the world in 2050. Available at: http://www.fao.org/fileadmin/templates/wsfs/docs/expert_paper/How_to_Feed_the_World_in_2050.pdf (accessed 6 June 2018).

Anonymous (2016) Annual report 2015–16 of Department of Agriculture Cooperation & Farmers' Welfare, Government of India. Available at: http://agricoop.nic.in/sites/default/files/Final%20Annual%20Report%20English.pdf (accessed 6 June 2018).

Anonymous (n.d.) Indian seed sector. Available at: https://seednet.gov.in/Material/IndianSeedSector.htm (accessed 6 June 2018).

Bhaskaran, M. and Raja, K. (2017) Novel approaches in seed testing and seed quality assurance. *XIV National Seed Seminar 2017*, 28–30 January, New Delhi.

Borthakur, A. and Singh, P. (2012) Agricultural research in India: an exploratory study. *International Journal of Social Science & Interdisciplinary Research* 1(9), 59–74.

Choudhary, B. and Gaur, K. (2010) Bt cotton in India: a country profile. *ISAAA Series of Biotech Crop Profiles*. ISAAA, Ithaca, New York.

Dravid, P.S. (2011) Future growth diverse for Indian seed industry. *Indian Seed and Planting Materials* 4(4), 21–29.

GRAIN and Sharma, D. (2005) *India's New Seed Bill*. Available at: https://www.grain.org/es/article/entries/457-india-s-new-seed-bill (accessed 6 June 2018).

Hanchinal, R.R. (2017) The Indian seed industry: achievements and way forward. *XIV National Seed Seminar 2017*, 28–30 January, New Delhi.

Hooda Committee Report (2010) Report of Working Group on Agriculture Production under the chairmanship of Bhupinder Singh Hooda to recommend strategies and an action plan for increasing agricultural production and productivity, including long-term policies to ensure sustained agricultural growth. Available at: http://commodityindia.com (accessed 27 May 2018).

Paroda, R.S. (2013) Indian seed industry: the way forward. Lecture at Indian Seed Congress, 8 February, organized by NSAI at Hotel Leela Kempinsky, Gurgaon, India. Available at: http://www.taas.in/documents/pub31.pdf (accessed 6 June 2018).

Plewis, I. (2014) Indian farmer suicides: Is GM cotton to blame? *Significance* 11, 14–18. DOI: 10.1111/j.1740-9713.2014.00719.x.

Poonia, T.C. (2013) History of seed production and its key issues. *International Journal of Food, Agricultural and Veterinary Sciences* 3(1), 148–154.

Singh, H. and Chand, R. (2011) The Seed Bill, 2011: some reflections. *Economic and Political Weekly* 46, 22–25.

Singh, H., Mathur, P. and Pal, S. (2008) Indian seed system development: policy and institutional options. *Agricultural Economics Research Review* 21(1), 20–29.

Tunwar, N.S. and Singh, S.V. (1988) Indian Minimum Seed Certification Standards. Central Seed Certification Board, Department of Agriculture & Cooperation, Ministry of Agriculture, Government of India.

Venkatesh, P , Sangeetha, V. and Pal, S. (2015) India's Experience of Plant Variety Protection: Trends, Determinants and Impact. 2015 AAEA & WAEA Joint Annual Meeting, 26–28 July, San Francisco, California.

19

Revitalizing the Indian Seed Sector

Introduction

Seeds play a major role in the growth of the agricultural sector. Indian Agriculture has grown impressively and the National Agricultural Research System (NARS), in collaboration with the Indian seed industry, has played a critical role in its growth. Seeds represent the basic and most critical input for sustainable farming. The response of all other inputs, to a large extent, depends on the quality of seeds. It is estimated that the direct contribution of quality seeds alone to total production is about 15–20% depending upon the crop; and it can be raised further to 45% with efficient management of other inputs (Poonia, 2013). Developments in the seed industry in India, particularly in the last 30–35 years, have been very significant. A major restructuring of the seed industry by the government through the National Seed Project (NSP) Phase-I (1977–78), Phase-II (1978–79) and Phase-III (1990–1991) was carried out for strengthening seed infrastructure, which was most needed and relevant around those times. During the past five years, the seed market in India has grown considerably, nearly twice as fast as the global seed market. At present, in terms of market volume, rice, wheat and maize account for most of the market; cotton represents the biggest segment in terms of market value. Traditionally, farmers in India used seeds that were saved from the previous year's harvest. Although, such seeds lose their yield and strength with time, they still account for nearly two thirds of the total seed market in India. The commercial market for seeds, though comparatively small, is lately experiencing a healthy growth rate. The seed sector consists of both organized and unorganized manufacturers. The organized sector consists of public and private companies. Public companies currently have a stronger focus on producing high-volume, low-cost seeds. On the contrary, private manufacturers produce low-volume, high-cost seeds (Anonymous, n.d.a). Adequate quantities of quality planting material at the appropriate time and at an affordable price are to be made available to every farmer for bringing about radical changes in the agricultural scenario of the country. Importantly, this remains one of the most important challenges before the seed sector today.

Production and Supply Scenario

During the past 50 years, India recorded an unprecedented growth in agricultural production. The first phase of noticeable agricultural growth was during the Green Revolution in the late sixties and seventies, with the introduction of semi-dwarf, high-yielding varieties of wheat and rice. As a result, substantial increases in foodgrain production, from 50.3 million t in 1952 to 88.1 million t in 1971, was realized (Abrol, 2000).

During this period, a significant role was played by public sector seed organizations – NSC, State Farms Corporation of India (SFCI), state farm corporations (SFCs), ICAR institutes, SAUs – in achieving quality seed production of HYVs. The second quantum jump in production was realized in the late eighties and early nineties, mainly through a Special Foodgrain Production Programme and a new policy for seed development by the government, as well as an ambitious programme on hybrid research by the ICAR. In addition, a significant policy decision of the ICAR to share freely the parental seeds of hybrids with the private sector sparked increases in productivity as well as cropping intensity (from 118.6% in the early seventies to 133.8% in the nineties) (Paroda *et al.*, 2015). Subsequently, in 2001 the government enacted the Protection of Plant Varieties and Farmers' Rights Act (PPV&FRA) to ensure faster growth of the seed sector. All these initiatives helped the private seed sector in India to play a much bigger role, especially for the promotion of hybrid seed technology (Brahmi *et al.*, 2004). As a result, use of hybrid seeds and the share of the private sector increased significantly in different crops (in some, like maize, cotton and vegetables, up to 80%). Simultaneously, the ICAR accelerated breeder seed production of field crops, which surpassed the total requirement of nearly 80,000 quintals in 2009/10. Vegetables are the fastest-growing sector in agriculture. Being high value per unit weight, the vegetable seed segment has a significant share in the overall seed market (World Vegetable Center, 2017). Though the actual contributions made by the public and private seed companies in the vegetable seed market are difficult to assess, according to industry estimates, the hybrid vegetable seed market in India is around Rs 15 billion (Damodaran, 2009). The vegetable seed market segment grew at a rate of 10–15% p.a. There was an increase of 194% in the vegetable hybrid seed market during 1998–2008, and it is continuing (Koundinya and Kumar, 2014). Most public research institutes initiated vegetable variety improvement and seed production programmes in the country. However, the private sector is very active in R&D in vegetable crops. Out of about 151 vegetable hybrids released by the All India Coordinated Research Project on Vegetables (Singh *et al.*, 2014), nearly 55% are developed by the private sector. With

increased availability of quality seeds of improved varieties and hybrids produced by several reputable companies, the seed replacement rate (SRR) in most of the vegetables, which was ~20% in the early eighties, rose to 60–90% by early 2000. Further, with the rapid pace with which biotech innovations are being tested, market share of vegetable hybrids is expected to rise.

The seed supply system in India is comprised of 'informal' and 'formal' supply systems at a ratio of 70:30. The informal supply system is also referred to as the 'farmer-driven seed system', where the farmers obtain, produce, conserve, improve and distribute seed. The informal sector provides a dynamic and flexible supply of seed required by smallholder farmers. The key characteristics of this system are: (i) it operates at local level; (ii) it deals with small seed quantities; (iii) it has a wide range of exchange mechanisms; (iv) it has no regulatory control; and (v) it addresses farmers' immediate needs and operates mainly on 'social certification' based on mutual trust. The formal seed supply system is represented by the public and private seed sector. In India, presently, the public sector comprises one national-level corporation the NSC – 15 state seed corporations, 22 seed certification agencies, two central seed-testing and 122 state seed-testing laboratories (3 ISTAs are accredited and 20 have ISTA membership), which are providing requisite strength in serving the seed industry and farmers. The R&D in the public sector is dependent upon public research under the aegis of the ICAR institutes and the SAUs. The SAUs and the ICAR institutes are engaged in breeder seed production and in the production of foundation and certified/truthfully labelled seed of varieties bred by them. Besides, seed is also produced by farmers under the Farmers' Participatory Programme of several institutes and under the Seed Village Programme of the government. Thirty-four SAUs and 22 ICAR institutes across the country are engaged in seed-production activities. SAUs are taking up breeder seed production involving its KVKs to bring about a seed revolution in the country. The private sector has more than 600 players (including domestic and multinational companies) and the top ten seed producers account for more than two thirds of the domestic market. Over the past two decades, private sector seed companies have collected germplasm and also built their R&D capabilities. Some of these have realized the importance of R&D and now spend about 5–10% of their revenue on it. These players have developed many hybrids based on the local needs of the farmers and have been able to gain significant market share.

The impact of the seed industry on Indian agriculture is clearly evident from the steep rise in production and availability of all classes of seed in the country. With availability of quality seeds, India had been able to achieve a very good seed replacement rate in the major crops, i.e. wheat (32.55%), paddy (40.21%), maize (56.58%), sorghum (23.85%), pearl millet (60.4%), chickpea (9.35%), urdbean (34.41%), mungbean (29%), pigeonpea (22.16%), groundnut (22.5%), rapeseed-mustard (78.88%), soybean (52.75%), sunflower (32.47%) and jute (42.11%) during 2012–13 (Anonymous, 2017). But, it is not the ultimate goal, and the proportion of seed supply by the formal and informal seed sectors needs to be reversed for sustainable yield enhancement for national food security.

Raising Agricultural Productivity

As stated earlier, a decelerating productivity rate, increasing prices, demand for foodgrains, shrinking natural resources and the challenge of climate change are major concerns for policy makers and scientists alike. To obtain a national GDP growth rate of 8%, it is necessary that we raise agricultural growth from 2% to a minimum of 4%. The best way to achieve this is to bridge existing yield gaps for most of the crops through greater coverage under HYVs and hybrids. For raising agricultural productivity, seed is recognized as the cheapest source, yet it is the most critical single input. Use of good-quality seeds can result in as much as 15–20% higher yields. Therefore any attempt to enhance agricultural productivity would depend largely on the higher replacement rate of quality seeds of HYVs/hybrids. Except for a few crops, seed replacement rate in India is very low. In a few states it is less than 5% across the crops, which is a big challenge to the seed industry and to Indian agriculture. To overcome this situation, there should be a rolling plan of the government

where targets should be fixed to cover the area under quality seed. Unfortunately, replacement rates in most of the crops, both for varieties and hybrids, are much lower than the optimum. The government has initiated several programmes/schemes for ensuring availability of good-quality seed of high-yielding and improved varieties (to sustain biotic and abiotic stresses) to farmers at affordable cost. Besides providing subsidy support for certified seeds, the government has proposed a new Seed Bill 2004, which would likely address several emerging issues and technological innovations on the one hand and the interests of farmers on the other. Despite the Indian seed industry being vibrant and robust, one thing is clear, that currently we are at a 'crossroads' and a process of revitalization is needed urgently to accelerate the pace of seed development to harness the fruits of available innovations for increased productivity.

Strategies for Revitalizing the Seed Sector

In order to increase agricultural productivity, the following strategies need urgent attention:

• The National Seed Plan (NSP) projection for seed requirement was 2.54 million quintals for 2009–10, against which 2.8 million quintals could be achieved. However, there is considerable inconsistency; seeds of new HYVs and hybrids are still not sufficient to increase area coverage under them. Therefore, seed replacement has to be linked with new variety replacements. Moreover, almost 70% of seed used by farmers continues to be farm-saved (Anonymous, n.d.b). For many crops, the SRR is much below the desired level. Therefore, a rolling seed plan for each five-year cycle would help with advance planning. The Federation of Seed Industries of India (FSII) and the National Seed Association of India (NSAI), in collaboration with the Department of Agriculture and Cooperation of the Ministry of Agriculture and Farmers' Welfare (DAC & FW) and the ICAR should aim to develop a seed production strategy for the next five to ten years. Accordingly, a policy paper on seed development needs to be attempted as a matter of priority.

- For achieving desirable levels of SRR, there is an urgent need for adequate seed production. In each state, a seed production programme should be organized under a comprehensive and integrated state seed plan appropriate to the region's requirements. States should ensure production, multiplication and replacement of seed to increase the seed multiplication ratio (SMR) and SRR progressively, particularly with respect to regionally important crops/varieties. In this context, a proactive role of concerned seed corporations in each state would be desirable. Production of hybrid seeds needs to be promoted aggressively to improve crop productivity, especially of crops like rice, maize, sorghum (rabi), pigeonpea, rapeseed-mustard and castor. In this context, efforts of the private sector should also be covered by the government for incentives on a par with public sector seed-production agencies, especially by identifying promising hybrids of different crops that would enhance productivity. Acceleration of hybrid seed production of these crops by the private seed companies is the need of the hour, and hence an aggressive approach by FSII, NSAI, DAC&FW and ICAR would go a long way in meeting national targets.

- A huge public sector infrastructure including NSC, SSCs, SCAs, STLs, SAUs and ICAR institutes, which are engaged in seed production, maintenance and quality evaluation, is a great resource of the Indian seed industry. As mentioned above, there are more than 600 seed companies including multinational companies that are engaged in the seed business, and this is a big asset to the Indian seed industry. Complementarity of public sector policy, and infrastructure and dynamism of the private sector, can be maximized through appropriate PPPs. Successful models of PPP, as experienced in the recent past, could be replicated. Again, long-term contracts can be entered into between state seed corporations, ICAR research institutes/SAUs and the private seed sector, cooperatives of farmers or self-help groups to undertake production and supply of quality seeds.

- The success of partnership lies in trust, openness and transparency. This can be built by regular interactions and dialogue, and an appropriate policy framework to strengthen PPP. Therefore, a standing working group in the Ministry of Agriculture and Farmers' Welfare (DAC&FW/DARE-ICAR) or a National Seeds Board (NSB), as envisaged in the National Seeds Policy (2002), needs to be constituted. This could include representations from DAC&FW, ICAR and NSAI and may act as a 'think tank' playing an overseeing as well as an honest broker role in promoting PPP. The proposed working group would also review existing guidelines for incentives and rewards and suggest ways to build new partnerships while taking care of the access and benefit-sharing (ABS) mechanisms, as well as the interests of smallholder farmers. Eventually, this body would draw up a clear road map for accelerated growth of the Indian seed sector.

- Good models and success stories on PPP existing in the NARS and the CGIAR systems, such as that of the Indian Agricultural Research Institute (IARI), the National Research Centre on Plant Biotechnology (NRCPB), the Indian Institute of Horticultural Research (IIHR), the International Crop Research Institute for the Semi-Arid Tropics (ICRISAT) and the International Rice Research Institute (IRRI), can be replicated or further fine-tuned, as desired by other institutions/universities. However, some of these institutions have expressed concerns about the break in the continuous requisition of breeders' seed by the contracted parties, which has to be addressed to build up the needed confidence.

- For access to new hybrids/varieties/genetic materials, the private sector would be willing to pay a royalty between 3% and 7% of sale proceeds, depending on exclusive/non-exclusive rights. The public sector seed corporations may also join hands in a similar way for the promotion and popularization of improved varieties, especially hybrids. In view of the poor conversion of breeder seed to foundation and certified seed, the public sector is required to opt for PPP mode to convert breeder seed into maximum certified seed. The public sector should pay a reasonable royalty on breeder seed on mutually agreed terms, as is applicable for private seed companies.

- There is an urgent need to build crop-based/institution-based technology parks/'incubators' so that scientists from both public research institutions and the private seed sector can work together from the start of the partnership in evaluating germplasm and breeding lines, developing hybrids/varieties, and evaluating and producing quality seed for food and nutritional security through PPP by commercializing varieties with desirable traits. This may encompass development of transgenics, exhaustive biosafety assessment, field evaluation, public dialogues and release of final products, keeping in mind the national interest. The private seed sector must come forward to accelerate this process and take advantage of the congenial environment that exists at present. A lead role by the NSAI in facilitating and catalysing the process would go a long way towards strengthening PPP.

- Seed quality assurance requires considerable investments in terms of proper infrastructure, equipment and competent human resources. Seed certification agencies have to be adequately equipped and need to be more efficient for certification of quality seeds. The seed-testing laboratories in the private sector, which are accredited by the International Seed Testing Association (ISTA), also need to be notified by the government for seed testing and certification purposes. In this context, devolution of the current centralized certification system by involving the private sector, as was proposed in the new Seed Bill 2004, needs to be revisited.

- Adequate infrastructure for seed processing needs to be created by all states at the SSCs, SAUs and private seed agencies. For this, the National Agriculture Development Plan – *Rashtriya Krishi Vikas Yojna* (RKVY) funds can be used to provide assistance up to 50% of the cost of seed-processing facilities and construction of seed godowns.

- The Indian Seed Bill 2004, first introduced in 2004, and the last in 2014, are still awaiting Parliamentary approval. It had taken more than five decades to revise the Seeds

Act of 1966. Hence, there is an urgency to have it enacted soon. Provisions of compulsory variety registration would go a long way in meeting the demand of good-quality seeds of varieties, and provision of farmers' exemption shall take care of their interest in reusing their own saved seeds as well as sale of their produce. Also, the interest of both farmers and the private sector for seed development is being taken care of under the Protection of Plant Varieties and Farmers' Rights Act 2001 (PPV&FRA).

- Restructuring and revamping public sector seed-producing undertakings is also required for product diversification/upgrade and for improving their governance, core competence and competitiveness. Most of the 15 state seed corporations (SSCs) are currently almost non-functional. The SSCs should either be reformed/reorganized to make them vibrant organizations or should be closed to allow alternative mechanisms such as the private seed sector to operate. At all costs, complacency in the seed sector must be avoided. State seed farms, having substantial capital assets and being an important mechanism for efficient multiplication of seeds, require urgent attention as several of them are in a state of neglect. An approach for making optimal use of these farms is to involve progressive farmers in collaboration with the SAUs/ICAR institutes to produce certified seeds under contractual arrangements with assured incentives. Also, strengthening of the Seed Village Scheme would help in accelerating the pace of good-quality seed production. This approach needs to be adopted as a high priority.

- India has the potential to become a leading player in the seed business sector in the developing world but its present share in the global seed market is less than 2%. There is good export market potential, particularly in the African and South Asian Association for Regional Cooperation (SAARC) region as well as south-east Asian countries, where India can aim to achieve the target of 10% by 2020 as envisaged in the National Seeds Policy (2002). Some Indian/multinational companies are already doing seed business in these countries, but the potential is much bigger. Specific interventions through active involvement of the FSII and NSAI, to boost seed exports, need urgent consideration. For this, an enabling environment through a single-window system is essential to accelerate agricultural productivity growth rates.

- Seed being the prime catalyst of increasing productivity, it is recommended that the National Mission on Seed launched recently by the government provides an enabling environment for a faster and efficient quality seed production programme at the national level. In this mission, adequate support for hybrid and quality seed should be ensured for higher seed replacement rates. A strong PPP through active involvement of farmers and for improved seed systems and seed increase is a need of the current time. Subsidy of hybrid/ quality seed production has to be extended to the private sector. The farmers' participatory role for seed increase would ensure availability of quality seed at a faster pace.

- Germplasm conservation through use can help in achieving both sustainable agricultural growth and development. Hence, it is emphasized that the national germplasm collection available at the NBPGR, being national public goods, be made available more freely upon request to Indian scientists/ institutions/seed companies engaged in crop improvement (R&D) programmes. For this, the Standard Material Transfer Agreement (SMTA), as adopted recently by the FAO International Treaty on Genetic Resources for Food and Agriculture (ITGRFA) for multilateral access, can be adopted for immediate implementation with necessary safeguards. All the data on available germplasm needs to be documented/catalogued and placed on the NBPGR website.

- There are the seven north-eastern states of India where hardly any seed corporations and seed certification agencies exist. The vast areas of rice-fallow lying vacant can be efficiently utilized under various rabi oilseeds and pulses once the good-quality seed of such varieties is taken to these states. Special efforts are required with the seed production of different varieties of various crops of the region to make the dream of the second Green Revolution come true. For

this, the various voluntary organizations, self-help groups and small farmers should be associated, which will help in improving the conditions of small entrepreneurs in these states.

- Some of the policy-related issues for consideration by the central government are as follows:
  ○ The Ministry of Agriculture must harmonize seed-related regulations both at the central and state levels. The Seed Bill, once passed by Parliament, would provide an enabling environment for faster seed sector growth in the country.
  ○ All quality-assured seeds must qualify for seed subsidy. Subsidies need to be linked to promote area coverage under new HYVs and hybrids for increasing productivity, irrespective of whether they are produced by the public sector or the private sector. This would benefit both farmers and the nation and would avoid existing discrimination, which is counter-productive.
  ○ For accelerating hybrid seed production, the present system of receiving indents of the parental lines of notified hybrids by the public/private sector (through NSAI to DAC) and fixing one uniform price, irrespective of their commercial value, needs urgent revision in consultation with the ICAR and the NSAI.

- R&D investment in the public sector has lately increased and many seed organizations have developed excellent research infrastructure and human resource. The share of private sector investment in plant breeding and seed development has increased in recent years but is still far below (15–20%) that in the developing countries. This has to be accelerated, especially in view of the declining trend in plant-breeding research in the public system.

- Untimely rains and frequent fluctuations in temperature have made the conventional seed production sites unfit or less remunerative in terms of production and quality of seed. Abrupt rises and falls in temperature during the crop season in the conventional seed-production areas of various crops is again a serious issue that warrants searching for alternative safe sites for climate-smart, quality seed production. For crisis management, there is a need to establish regional seed banks as a contingency measure. Also, there is an urgent need for the creation of modern seed processing and storage facilities in different regions to be accessed by both the public and the private sector on a fixed processing fee basis.

Conclusion

The Indian seed programme is now occupying a pivotal place in Indian agriculture, and is well poised for continued growth in years to come. The National Seeds Corporation, which is the largest single seed organization in the country, with such a wide product range, has pioneered the growth and development of a sound seed industry in India. The NSC, SFCI, SSCs and other seed-producing agencies are continuously and gradually expanding all their activities especially in terms of product range, volume and value of seeds handled, and level of seed distribution to inaccessible areas. Over the past four decades, these seed-producing agencies have built up a hard core of competent and experienced seed producers and seed dealers in various parts of the country and have adequate levels of specialization and competence in handling and managing various segments of seed improvement on scientifically sound and commercially viable terms. The competitive advantages that Indian agriculture possesses are: (i) favourable agroclimatic zones; (ii) large irrigated lands; (iii) a gap between present productivity and potential productivity; and (iv) availability of skilled, educated, technical and scientific manpower. To leverage global competitive advantage, Indian agriculture in general, and the seed supply chain in particular, need intervention in the areas of policy, technology and market access. Finally, our policy makers, scientists, seed sector workers and farmers need to join hands in a true spirit of partnership to revitalize Indian agriculture. This requires a 'missionary zeal' and strengthening of the Seed Technology Mission to accelerate the pace of Indian agriculture.

References

Abrol, I.P. (2000) Agriculture in India. Centre for Advancement of Sustainable Agriculture. Available at: http://www.planningcommission.nic.in/reports/sereport/ser/vision2025/agricul.doc (accessed 6 June 2018).

Anonymous (2017) Accelerating seed delivery systems. Available at: http://www.naasindia.org/documents/Base_paper_ASDS.docx (accessed 6 June 2018).

Anonymous (n.d.a) Indian seed sector. Available at: http://seednet.gov.in/Material/IndianSeedSector.htm (accessed 6 June 2018).

Anonymous (n.d.b) State of Indian farmers: a report. Available at: http://www.lokniti.org/pdf/Farmers_Report_Final.pdf (accessed 6 June 2018).

Brahmi, P., Saxena, S. and Dhillon, B.S. (2004) The Protection of Plant Varieties and Farmers' Rights Act of India. *Current Science* 86(3), 392–398.

Damodaran, H. (2009) Business Line. *The Hindu*. Available at: www.thehindubusinessline.com (accessed 6 June 2018).

Koundinya, A.V.V. and Kumar, P.P. (2014) Indian vegetable seeds industry: status and challenges. *International Journal of Plant, Animal and Environmental Sciences* 4(4), 62–69.

Paroda, R., Singh, N.N., Gupta, N. and Kochhar, S. (2015) *Building Trust: The Journey of TAAS (2001–2015)*. Trust for Advancement of Sciences, IARI, Pusa, New Delhi. Available at: http://www.taas.in/documents/pub44.pdf (accessed 6 June 2018).

Poonia, T.C. (2013) History of seed production and its key issues. *International Journal of Food, Agricultural and Veterinary Sciences* 3(1), 148–154.

Singh, B., Chaubey, T., Singh, P.M., Srivastava, R.K. and Bhushan, C. (2014) *Four Decades... Accomplishment of AICRP (Vegetable Crops)*. ICAR-Indian Institute of Vegetable Research, Varanasi, India. Available at: http://www.iasri.res.in/aicrpvc/Decade/cover%20page%20and%20content.pdf (accessed 6 June 2018).

World Vegetable Center (2017) *Annual Report 2016*. Publication 17–814. World Vegetable Center, Tainan, Taiwan.

20

Managing and Improving Soil Health

_____

Introduction

Soil is the basic building-block of life on earth, and is the most wonderful gift of nature to humankind. It is a dynamic and multi-functional system, which exists as a relatively thin layer on the earth's crust. No healthy life is possible without healthy soil. We all are dependent on the soil for our basic needs of food, feed, fibre, medicine and fuel. Hence, maintaining healthy soil would help in achieving quite a few important SDGs. Historically, old civilization flourished only when soils were fertile and appropriate water resources were available for irrigation. This is true for northern India where civilization prospered in the Indo-Gangetic plains. In fact, healthy soils of the Indo-Gangetic plains led to the Green Revolution in the mid-1960s, transforming India from a food-deficient to a food-sufficient country. However, the challenges before agricultural scientists, farmers and policy makers to meet future food and nutritional security needs are quite different and complex as compared to the pre-Green Revolution era. After the Green Revolution era, soil health declined considerably. Around 94% of agriculturally suitable land in the country is already under cultivation, limiting the scope for horizontal expansion. Soil organic carbon (SOC) content of the soils of the Indo-Gangetic plains is less than 5 g/kg compared to 15–20 g/kg of uncultivated virgin soils (Bhattacharyya et al., 2000), mainly attributable to intensive tillage,

removal/burning of crop residues, mining of soil fertility and intensive cereal-based cropping systems.

Indian soils are broadly classified into five main groups – alluvium-derived soils, red soils, black soils, soils of arid region and soils of Himalayan and Shiwalik regions. These soils differ in their productivity and need different management practices depending on their physical and chemical properties, biological conditions, rainfall/availability of water for irrigation and the existing crops and cropping systems. Large acreage of cultivated lands is presently showing fertility fatigue and multiple nutrient deficiencies. This is a challenge for making farming profitable, sustainable and resilient, and for ensuring future food security.

Soil properties and associated environmental conditions govern many ecosystem functions such as decomposition and transformation of organic wastes, mediating nutrient cycles and influencing population of soil organisms. While it is difficult to describe how well any soil performs to its interrelated ecosystem functions, earlier definitions included 'fertility', and later ones implied that 'quality' and 'soil health' are more inclusive performance indicators. Soil health is critical for ecosystem functioning, since it plays a key role in the carbon cycle, storing and filtering water, and adapting to, and mitigating, climate change. As we know, in the past century, food production increased dramatically due to enhanced crop yields as the result of widespread adoption of technologies – new high-yielding and disease-resistant crop varieties, irrigation, mechanization and especially the use of mineral fertilizers. Unfortunately, the impact of modern agriculture on soil health has become a major global concern because of inappropriate as well as inefficient use of inputs and unscientific farming practices. Notable among the consequences of the Green Revolution are: (i) decline in quantity and quality of soil-organic matter; (ii) deterioration of soil physical health; (iii) nutrient withdrawal in excess of replenishment; (iv) increased atmospheric concentration of carbon dioxide (CO_2) from 316 ppm (parts per million) to 400 ppm; and global temperature by 0.12°C per decade; (v) problems of soil degradation by erosion, salinization, depletion of soil organic matter (SOM) and nutrient imbalance; (vi) depletion, pollution and eutrophication of natural waters; and (vii) decrease in microbial population and its diversity. The key factor in soil quality is SOM fraction, which, although relatively small, has a strong impact on the overall well-being of the soil and its beneficial functions. Soil organic matter controls soil microbial population and its many functions such as decomposition and nutrient cycling. Fertilizer use can have a positive or a negative effect on soil health. Depending on the tillage system used, regular additions of N fertilizers can enhance SOM levels. Organic matter can help increase soil aggregate stability and would contribute to resistance to erosion and soil degradation. While crop yields were the primary focus in the past, awareness of the requirement to produce more food with limited resources and to bring additional land into production led to the concept of cropping sustainability or sustainable intensification, i.e. consistently achieving high crop yields without damaging the soil's capacity to produce high yields. To meet the demand and prevent hunger and malnutrition, production of agri-food has to be increased by 70% by 2030 using the same or even fewer land and water resources, while decreasing the environmental footprint of agriculture. Future challenges for agriculture would not only be to meet the food requirements of an expanding population but also to undertake it in a manner that is sustainable for present and future generations, particularly facing the projected climate change scenario. Good soil health would be of paramount importance to feed the projected Indian population of 1300 and 1700 million by 2020 and 2050, respectively, under the threat of extreme weather events induced by climate change. Future food production targets would thus be met only when potential of improved varieties and good agronomic practices are combined with improved soil management.

Soil Health Degradation

Understanding the continuing capacity of soil to function as a vital living system, by recognizing that it contains biological elements that are the key to ecosystem functioning within land-use boundaries, is important. These functions are able to sustain biological productivity of soil, maintain or enhance water and air quality, and promote plant, animal and human health. To define this, the terms 'soil quality' and 'soil health' are used interchangeably; although it is important to distinguish that soil quality is related to soil function, whereas soil health presents soil as a finite, non-renewable and dynamic living resource. Soil health is an integrated function of biological, chemical and physical properties. Healthy soils maintain a diverse community of soil organisms that would help control plant disease, insects and weed pests; form beneficial symbiotic associations with plant roots (e.g. nitrogen-fixing bacteria and mycorrhizal fungi); recycle essential plant nutrients; improve soil structure with positive repercussions for soil water- and nutrient-holding capacities and, ultimately, improve crop production.

Because soil organic matter is concentrated on the soil's surface, accelerated soil erosion can lead to progressive depletion of soil carbon. SOC loss owing to soil erosion from cultivated lands is estimated to be 16.4 t/ha p.a., resulting in an annual total soil *loss* of 5.3 billion t in the country (Bhattacharyya *et al.*, 2015). SOC and its dynamics are key determinants of soil health and for providing essential ecosystem services. In India, crop production relies largely on chemical fertilizers because of limited availability of animal manures and crop residues. Crop residues are invariably removed by animal grazing, and in some cases residues are even burned for growing the next crop. A significant proportion of animal excreta is used as household fuel. Low or unbalanced fertilization leads to depletion of soil nutrients and soil degradation owing to low organic carbon in such soils. There are reports of stagnation or decline in yields and crop response to applied fertilizers. Stagnation in yield can be due to degrading soil health as a result of reduced supply of soil nutrients, causing macro- and micro-nutrient deficiencies. In general, Indian soils are poor in nitrogen (N) and phosphorus (P) and relatively better in potash (K). The deficiency of several micro-nutrient elements such as zinc (Zn), iron (Fe) and manganese (Mn) appears quite extensive in the soils. About 41%, 49%, 33%, 13%, 12% and 5% of the soils are deficient in sulphur (S), Zn, boron (B), molybdenum (Mo), Fe and Mn, respectively. The current status of nutrient-use efficiency is quite low, especially for P (15–20%), N (30–50%), S (8–12%), Zn (2–5%), Fe (1–2%) and copper (Cu) (1–2%) owing to deterioration in chemical, physical and biological soil health. Most soils, especially in the Indo-Gangetic plains, are quite low in SOM. Continuous cropping without appropriate crop management reduces soil organic carbon levels by 50–70% in most intensively cultivated areas to equilibrium levels dictated by climate and precipitation. There is evidence that SOM decreases due to intensive tillage, burning and/or removal of crop residues, and limited use of animal manure. The main reasons for soil-health deterioration are: wide gaps between nutrient demand and supply, high nutrient turnover in the soil-plant system coupled with low and imbalanced fertilizer use, emerging deficiencies of secondary micro-nutrients in soils, soil acidity, nutrient leaching in sandy soils, impeded drainage in swell-shrink soils, soil salinization, sodification, intensive tillage and lack of recycling of organics in soils.

Soil Health Assessment

A healthy agricultural soil is capable of supporting production of commodities to a specified level and of a quality sufficient to meet human requirements and sustain functions essential for maintaining quality of human life and biodiversity conservation (Kibblewhite *et al.*, 2008). Soil health is an integrative property reflecting capacity of the soil to respond to agricultural interventions. Intrinsic to this concept is the maintenance of soil quality and avoiding processes of erosion and nutrient mining, which degrade soil. A healthy soil has a stable system with a high level of biological diversity and activity, internal nutrient cycling and resilience to disturbance. Management factors that can modify soil quality include tillage, residue management, crop rotations, nutrient management etc. Therefore, soil health assessment has been suggested as a tool for evaluating sustainability of soil and crop management practices. Soil health is evaluated in terms of indicators, measuring specific physical, chemical and biological properties, which are affected in turn by agricultural practices. Soil health attributes are expressed by productivity, nutrient- and water-use efficiency and quality of produce. Despite being a relatively small component of soil in terms of volume, the single most important soil property relating to soil health is SOM as it exerts a profound influence on chemical, physical and biological properties of the soil. Micro-organisms appear excellent indicators of soil health because they respond quickly to changes in the soil ecosystem. In some instances, changes in microbial population or activity can be preceded by detectable changes in soil physical and chemical properties, thereby providing an early sign of soil improvement or degradation. Microbial indicators of soil health cover a diverse set of microbial measurements due to multi-functional properties of microbial communities in the soil ecosystem. Soil micro-organisms thus play a key role in maintenance, functioning and sustainability of agro-ecosystems, mainly regulating C and N cycling, with direct implications for soil fertility and plant nutrition. We must visualize the future by asking what needs to be known about soil biology that is currently not recognized or fully understood, and how these needs could be addressed using emerging research tools.

Soils and Greenhouse Gases

The Agriculture, Forestry and Other Land Use (AFOLU) sector emits just under a quarter (approx.

10–12 Gt CO_2 p.a.) of all anthropogenic GHG; the largest emissions are from deforestation, followed by agricultural emissions from livestock, soil and nutrient management. Annual GHG emissions from agricultural production in 2000–2010 accounted for about 11% of the total anthropogenic GHG emissions, globally, and between 5 and 5.8 Gt CO_2 p.a. Organic matter in the world's soils represents a major stock of organic C, storing about 1500 Gt C (equivalent to 5500 Gt CO_2) at a depth of 1 m, and a further 900 Gt C in the next 1 m. Land-use changes, especially clearing natural vegetation to expand areas for crop production, have significantly depleted global soil C stocks and contributed enormously to increased CO_2 emissions. Agriculture, and the changes in land use associated with it, is one of the principal contributors to climate change, accounting for one third of the global GHG emissions.

India is the world's fourth-largest economy and fifth-largest GHG emitter (5% of global emissions); and there is a likelihood of further increases in the future. According to a 2008 estimate, agriculture was the second-largest source of GHG emissions, accounting for 18% of gross national emissions. It is, therefore, desirable to find out ways to slow or reverse this trend through land management practices. Soils are capable of generating (source) or storing (sink) greenhouse gases, depending on how they are managed and how healthy they are. The role of the soil as a carbon sink and as a carbon store can be strategically optimized through proven farming techniques and methods that reduce emissions simultaneously. The carbon that is removed from the atmosphere and captured in the soil and plant biomass is the same as that which makes agricultural soils more fertile and leads to higher margins of profit for producers. The Intergovernmental Panel on Climate Change (IPCC) indicated that carbon sequestration would account for about 90% of global agricultural mitigation potential by 2030. Promoting good agronomic practices and sustainable land management practices would thus help trap and store more emissions in farmland and natural spaces. Improving the health of the world's soils would help store extra carbon, equivalent to 8 billion t of greenhouse gases, thus slowing the impact of climate change (Hill, 2016). After the International Year of Soils in 2015, the National Agricultural Research Systems (NARS) are now focusing on the climate change agenda, which needs greater attention.

Soil Health Management

It is a known fact that enhancement and maintenance of soil health is essential for the sustainability of Indian agriculture for meeting the basic needs of an ever-increasing human as well as livestock population. Increasingly, there is awareness of a direct link between soil and human health in terms of the elements that improve soil health and mitigate climate change. The major strategies and technologies for maintaining healthy soils are: (i) developing scientific land-use plans; (ii) minimizing tillage (reduced/no-till); (iii) recycling organic/crop residues; (iv) integrating legumes in cropping systems; (v) precision in nutrient and water management; and (vi) adopting a farming systems approach. SOM is the main indicator of soil heath for agricultural (fertility, yields) and environmental functions. Sequestration of SOC is an important strategy to improve soil quality and to mitigate climate change. Maintenance and/or improvement in soil health in terms of SOM content and supply of various macro- and micronutrients can be mediated through application of organic nutrient sources such as manures and crop residues, available on the farm, and supplementing them with mineral fertilizers in the balanced forms. Tillage tends to accelerate oxidative breakdown of organic matter with release of increased volumes of CO_2 into the atmosphere, beyond the limits from normal soil respiration processes. This results in reduced organic matter and also explains why it is difficult to build up organic matter content with tillage operations. Combining retention of crop residues (rather than removal or burning) with direct seeding of crops without 'normal' tillage leads to increased organic carbon, as crop residues are precursors of SOC pool.

Soil health can be improved by adopting conservation agriculture (CA) (based on three key principles: minimum or no tillage; covering the ground using organics; and diversifying cropping systems); managing efficiently pests/diseases and nutrients; and preventing soil compaction, crusting and salinization. Improving soil fertility by INM for healthy crop growth and biochemical transformation of biomass C into SOM or humus in association with CA is the key strategy (Lal, 2014). In CA, sustainable land management delivers carbon benefits as carbon sequestration, in which growth of agricultural and natural biomass actively removes carbon from the atmosphere and stores it in soil and biomass. Maintaining soil microbial biomass (SMB) and microflora activity and diversity are fundamental to sustainable agricultural production systems. There are many examples from around the world where properly implemented CA systems have improved soil health including biological quality and agronomic yields (Choudhary *et al.*, 2018). Changes in tillage, residue and rotation practices induce major shifts in the number and composition of soil fauna and flora, including pests and predators (Verhulst *et al.*, 2010). Increase in SOC concentration in the surface layer, reduction in soil erosion and improvement in biological properties due to CA also enhance soil fertility and chemical attributes, ameliorate sodic soils and improve agronomic productivity. Carbon sequestration under CA has been estimated at 224 kg/ha p.a. for Asia (World Bank, 2012). Penetration of CA-based management systems in both irrigated and rainfed areas is presently showing acceleration in India. Since soil health hinges on SOC content, sparing adequate quantities of farmyard manure remains a major challenge. Hence, it was recommended to have interventions on maximizing returns of various organic sources by: (i) evolving community/village biogas units and increasing supply of liquefied petroleum gas (LPG) for home cooking to replace dung as fuel; (ii) enacting legislative measures that obligate a total ban on burning vegetative materials of all kinds; (iii) adoption of short-duration, multi-purpose varieties of legumes as catch crops in cereal-cereal rotations; (iv) popularization of CA practices; (v) promoting integrated soil- and nutrient-management practices; and (vi) rewarding and incentivizing those who adopt these interventions. The improvement in SOC is less likely to be achieved without proper policies being in place.

Nutrient Management and Soil Health

Fertilizer use can have positive or negative effects on soil health. The key for positive effects on soil health lies in good science-based nutrient-management practices, especially efficient use of nutrients. Recent development of the 4R Nutrient Stewardship Principles (IPNI, 2012), applying the right source of the nutrient at the right rate, at the right time and by the right method, is the foundation of the improved nutrient management practices connected to economic, social and environmental benefits to society. Adoption of such practices ensures economic crop production compatible with minimizing environmental effects. Imbalanced and inappropriate fertilizer application adversely affects soil fertility, a major component of soil health, and limits the capacity of the soil to produce optimally at spatial and temporal scales. Depending on the tillage system used, regular additions of N fertilizer can enhance SOM levels. Negative effect of fertilizers on the soil microbial population depends on the N source and method of application, but it is localized and short-lived. For sustained agricultural production and attaining higher fertilizer use efficiency, soil fertility evaluation is vital. A national soil policy is urgently needed to maintain a healthy soil resource for the country. Fertilizer subsidy policy needs a review as preferential subsidy for one plant nutrient over others has led to nutrient imbalance in farmers' fields. Besides reducing crop productivity and farm profitability, this has led to severe depletion of nutrients in the soil. Notwithstanding the fact that there are several other factors besides nutrient management that affect soil health, like organic recycling, the use of legumes, biofertilizers as well as application methods, which all have a significant impact on it. A multi-pronged approach is therefore required to address all soil health-related issues in the realm of nutrient management by smallholder producers. Fertilizer use efficiency continues to be deplorably low. Poor nutrient use efficiency causes enormous economic loss and results in degradation of soil health. This practice continues, despite the availability of good knowledge on plant nutrient management. Support of innovative approaches and tools (Nutrient Expert decision support system, Green Seeker sensors, remote sensing, GIS), UAVs (drones) and spectroscopic techniques (dry chemistry) for rapid analysis of soil fertility are critical to empower the existing extension system. In this context, an achievable target of 10% increase in nitrogen-use efficiency (increasing from an average of 40–50%), can save 1.7 million t of fertilizer – N costing around Rs 20 billion, annually.

Soil Health and Crop Production

An essential component of sustainable agriculture, as embedded in the definition of soil health, is to balance ecosystem functions in such a way that the agricultural production target is achieved without compromising other ecosystem functions with respect to present and future needs. Soil health and degradation affect agricultural productivity, but quantifying these relationships has been rather difficult (Wiebe, 2003). However, it is clear that necessary increase in food production will have to come from the increased productivity of the existing land rather than from horizontal expansion. In this context, restoration of degraded soil and improvements in soil health would be extremely important, recognizing that soil health is a strong determinant of agronomic yield. Thus, good agronomic practices, especially a CA system, with proactive ways and means of residue recycling to enhance soil health and sustain productivity, will have to be promoted.

In view of the above, it is clear that we must have an action plan with the focus to improve soil health for sustainability of production systems. For this, the following strategies are proposed for priority implementation:

Research-oriented efforts

- The assessment of soil health requires quantification of critical soil attributes – physical, chemical and biological. Presently, there is no operationally defined integrated index of soil health. A framework for soil health evaluation is critical for the development of a useful monitoring programme covering different functions and land uses, and it must identify priorities and indicators relating to policy-relevant end-points.
- Long-term research sites across different agro-ecologies should be set up for progressive indexing of soil health, providing space

for need-based adjustments in soil- and nutrient-management prescriptions, suiting conventional and sustainable management interventions. Long-term studies would provide a useful database for simulation modelling as many changes have taken place in soil health over the years.

- Although studies of soil health hinge on organic matter content, nearly all studies deal with total SOM rather than more reactive labile or biomass C fractions that purportedly are sensitive indicators of changes in SOM. Present knowledge is grossly insufficient, especially to describe the impact of CA/sustainable soil management on the soil C pool and dynamics at the local and regional level. These aspects need to be studied more systematically.
- There is a need to establish appropriate fertilizer management strategies in such systems so that soil health is maintained or further improved. Nutrient management research should be launched to increase nutrient-use efficiency in the context of climate change-induced variations in soil processes and to explore the adaptation and mitigation potential of balanced and site-specific nutrient management.
- Management of farm wastes and their utilization through INM is important for improving soil health. Short-duration, dual-purpose mungbean for grain and green manuring has potential for fertilizer-saving as well as improving soil health in intensive cereal-based systems on the Indo-Gangetic plains.
- Efficient soil- and nutrient-management strategies need to be developed to improve the health of acid soils as well as salt-affected soils.
- As the soils act as source and sink for GHGs, their precise measurement and monitoring, more so under CA, is required for potential mitigation and quantification purposes.
- The relative value of *in situ* composting of organic residues and green manure vis-à-vis their natural turnover from the point of acceptability to farmers and their effectiveness in terms of improvement in soil health should be evaluated.
- There is a great need to develop an 'organic fertilizer calculator', which could be employed

as a tool for comparing cost, nutrient value and nutrient availability of organic materials with improved understanding of the real-time nutrient-release characteristics. Such findings are useful for reckoning the effective nutrient value of diverse organic materials, and their role in improving management of fertilizers and reducing environmental footprints.

Policy initiatives

- There is a need to establish a National Mission on Sustainable Soil Health Management. One goal would be to evaluate constraints impeding sustainable soil health. Another aim would be to suggest a time-bound action plan and long-term strategies to overcome limitations.
- In the absence of public support for farmers, poor agricultural land management would accelerate land degradation, increase farmers' vulnerability to climate change and lead to emission of additional GHGs into the atmosphere.
- There is a need to play a proactive role towards farmers for adopting effective use of fertilizers, soil conservation and carbon build-up measures as ecosystem services. The ongoing soil-health sub-mission could consider accommodating such initiatives.
- All states must inculcate professionalism in soil-testing services and enhance the quality of outputs. Hence, the soil-testing laboratories must be strengthened in terms of infrastructure and trained human resources.
- At a broad societal level, there needs to be greater public awareness regarding the importance of soil for primary crop production, quality of food and health. The public should be made aware of the impact of poor soil on water quality and the environment and how it mediates GHGs and climate change. Soil health affects human, animal and environmental health, food and nutrition security, and climate change, and is linked with SDGs. Accordingly, societal sensitivity and awareness has to be created.
- At the national level, there is an urgent need to target at least 10% replacement of

chemical fertilizers by biofertilizers and organic cycling in the next five years. In pursuance of this goal, it is necessary to strengthen quality standards, efficient production methods, shelf-life-enhancing storage, followed by proper distribution and marketing.

- Location-(biophysical attributes) and situation-(socioeconomic state of farmers and markets) specific land-use alternatives, which are more competitive and less exploitive of natural resources than those currently in practice, should be developed.
- A new fertilizer use policy needs to be devised in the current context based on available scientific knowledge. It is high time to comprehensively review the existing differential allocation of subsidy to urea as compared to P and K. Moreover, the change in fertilizer subsidy pattern must not be done in one go; it must be gradual and consistent with the rise in the MSP announced by the government from time to time.
- The demand for de-control of urea should not be considered in the same way as for phosphatic and potassic fertilizers. There is a need to take care of the interests of both the farmers and industry. Whilst ensuring reasonable returns on investments by industry, the provision of a variable subsidy and stable maximum retail price (MRP) would certainly help farmers.
- There is full justification to create a new mechanism to incentivize efficient fertilizer users within the ambit of the direct fertilizer subsidy system. The implications of direct transfer of subsidy to farmers in place of subsidy to manufacturers needs to be analysed. This shift has real implications on timely availability of finances to smallholder, resource-poor farmers, since they would have to pay up front when buying fertilizer, whereas reimbursement may take time and require the farmer to run from 'pillar to post'. Hence, a comprehensive dialogue/understanding is needed while implementing a direct subsidy scheme, especially to ensure provision of advance payments by banks for fertilizer purchase.

References

Bhattacharyya, R., Ghosh, B.N., Mishra, P.K., Mandal, B., Srinivasa Rao, C. *et al.* (2015) Soil degradation in India: challenges and potential solutions. *Sustainability* 7, 3528–3570.

Bhattacharyya, T., Pal, D.K., Mandal, C. and Velayutham, M. (2000) Organic carbon stock in Indian soils and their geographical distribution. *Current Science* 79, 655–660.

Choudhary, M., Datta, A., Jat, H.S., Yadav, A.K., Gathala, M.K. *et al.* (2018) Changes in soil biology under conservation agriculture based sustainable intensification of cereal systems in the Indo-Gangetic Plains. *Geoderma* 313, 193–204.

Hill, O. (2016) Improved soil health could cut greenhouse gas emissions. *Farmers Weekly*, 10 April.

IPNI (2012) *4R Plant Nutrition: A Manual for Improving the Management of Plant Nutrition* (ed. by Bruulsema, T.W., Fixen, P.E. and Sulewski, G.D.) International Plant Nutrition Institute, Norcross, Georgia.

Kibblewhite, M.G., Ritz, K. and Swift, M.J. (2008) Soil health in agricultural systems. *Philosopical Transactions of the Royal Society* B 363, 685–701.

Lal, R. (2014) Societal value of soil carbon. *Journal of Soil and Water Conservation* 69(6), 186A–192A.

Verhulst, N., Govaerts, B., Verachtert, E., Castellanos-Navarrete, A., Mezzalama, M., Wall, P., Deckers, J. and Sayre, K.D. (2010) Conservation agriculture, improving soil quality for sustainable production systems? In: Lal, R. and Stewart, B.A. (eds) *Advances in Soil Science: Food Security and Soil Quality*. CRC Press, Boca Raton, Florida, pp. 137–208.

Wiebe, K. (2003) Linking Land Quality, Agricultural Productivity, and Food Security. Agricultural Economic Report 823. Resource and Economics Division, Economic Research Service, United States Department of Agriculture, Washington, DC.

World Bank (2012) Carbon Sequestration in Agricultural Soils. World Bank Report No. 67395-GLB. Agriculture and Rural Development (ARD), The World Bank, Washington, DC.

21

Increasing Water-use Efficiency

_____

Introduction

The population in most of the tropical developing countries is increasing by leaps and bounds. According to the FAO, over 800 million people currently lack adequate food, and by 2025 the food requirements of an additional 3 billion people will need to be met. Hidden hunger, such as protein and micro-nutrient deficiencies, is expected to become increasingly serious, particularly for women and children. Therefore, food and nutritional security continue to be a priority for the nations across the globe. India has made rapid strides in agriculture, achieving self-sufficiency in food requirement by recording a five-fold increase in production from the base line of 1950–51 through the Green Revolution. Efforts have also resulted in achieving an increase of 11 times in horticulture production, six times in milk, 25 times in egg and nine times in production of fish, from the base line of 1950–51. The cultivated area has remained static at around 142 million ha for the last 40 years, but production has increased many-fold, not only of cereals but also of all other agricultural commodities. This has been possible due to proper policy support and scientific advancements in developing high-yielding cultivars and farmer-friendly production technologies as well as through the efforts of farmers. It is also a fact that India, with

2.4% of the global geographical area and only 4.5% of the water resources, currently supports about 17% of the human and 11% of the livestock populations of the world. The success story of Indian agriculture is remarkable, but farmers' distress continues, and to address the concerns, the country has called for the doubling of farmers' incomes (Singh and Singh, 2017a).

Per capita availability of land for producing agricultural commodities has declined from 0.48 ha in 1951 to about 0.2 ha in 1981 and 0.15 ha in 2000 and it is expected to decline further to about 0.09 ha by 2050. In India, water is the key issue to sustain the required growth of agriculture in future (Naidu and Singh, 2004). The availability of fresh water for agriculture is expected to decline from the current level of 80% to about 70% by 2050 due to increasing demand for water for industrialization and urbanization. Global warming will further reduce the availability of fresh water for agriculture. Looking into population growth, declining land and water quality coupled with the challenges of climate change, has created much concern over feeding the ever-growing population. Thus, the challenge before the country is much greater than before, and has to be addressed with a strategic approach utilizing innovations in science and technology (Singh and Singh, 2016).

Emerging Challenges

Declining water resources

Water is a scarce and valuable resource. Globally, only 0.26% fresh water is available for irrigation and drinking purposes; the rest is either saline (7.5%) or not available in a useful form. It is estimated that the total world demand for fresh water in the year 2050 would be 1447 billion m^3 as compared to 634 billion m^3 in 2000. Availability of utilizable water is going to decrease from 1020 m^3/person in 2004 to 770 m^3/person in 2050. In India, currently about 80% of available water is consumed for agriculture production, whereas this share will be reduced to about 70% by 2025 due to increasing demand for water for industry and drinking purposes. As competition for water for different purposes increases, the need for additional storage as a proportion to total water consumed will increase in future. To meet future food needs, the FAO projected a 14% increase of water withdrawals in 93 developing countries and 45 million ha of net expansion of global irrigated area from 2000 to 2030. It is well accepted that even with the fullest economic utilization of water resources, the success of agriculture will continue to be governed by the vagaries of rainfall. Hence, it is imperative to enhance water productivity of both rainfed and irrigated ecosystems for sustainable progress (Naidu and Singh, 2004).

In India, surface water potential is about 180 million ha and groundwater resource is about 44 million ha. With annual precipitation of about 400 million ha, the average annual natural flow is about 188 million ha. The annual requirement of fresh water is estimated at 105 million ha by 2025, which is nearly equal to the ultimate water resource level of the country. Out of this, 77 million ha has been considered for irrigation purposes. In terms of area, the ultimate irrigation potential of the country has been assessed at 155 million ha (58 million ha from major/medium projects, 17 million ha from surface water minor irrigation projects and 80 million ha from groundwater projects). Although India has the largest irrigation system in the world, its water-use efficiency has not been more than 40% (Rajput and Patel, 2017). If the situation continues like this, the water crisis would result in reduced production and productivity, which would affect our food and nutritional security. This calls for more productive use of water and more crop yield per drop of water (Singh and Singh, 2016).

Currently, only 38% of cultivated area is irrigated and more efforts are needed to cover additional area to enhance productivity. Policies in the past have invariably been for the creation of water potential, whereas utilization of created potential and enhancement of irrigation efficiency received little attention. Thus, with current levels of efficiency, even after exploitation of all the available resources, more than 50% area may still remain rainfed (Naidu and Singh, 2004). With increasing population, economic prosperity, industrialization and climate change, the pressure on water resources available for agriculture may be much greater. This scenario will demand increasing water-use efficiency and water productivity in agriculture, both under irrigated and rainfed systems. Studies have shown that improving water productivity by 40% on rainfed and irrigated land could reduce the need for additional withdrawals over the next 25 years to zero (Narayanamoorthy, 1996, 2001; Dhawan, 2002; Kumar, 2003; Rank and Gontia, 2017).

Climate Change: A Challenge

Climate change, a cause of concern globally, is likely to have an impact on agricultural crops, due to erratic rainfall, more demand for water and enhanced biotic and abiotic stresses (Agarwal, 2009). The global climate model predicted that the Indian sub-continent will be warmer by about 1.5°C during the middle of the current century, and the second half of the winter will be warmer than the first half. It is also predicted that the Indian sub-continent would receive about 6% more rain, which could be irregular and rather more intense. There will be some reduction in the incident radiation and increase in the concentration of CO_2 and other GHG. According to the emissions inventories that different governments submitted to the UN Framework Convention on Climate Change (UNFCCC), agriculture accounts for around 15% of global GHG emissions. The increase in deforestation in developing countries for agricultural purposes could raise its contribution to between 26% and 35%. However, all these changes will only be harmful, as enhanced CO_2 concentration may enhance photosynthesis in C3 crop species, but increased temperature may increase water use and hasten the process of maturity. Innovations and strategic approaches may convert weaknesses into opportunities. Increased temperatures will have an effect on reproductive biology and reduced water may affect productivity, but adaptive mechanisms like time adjustment and productive use of water will reduce these negative impacts. These challenges could be addressed through identification of the genes tolerant to high temperature, flooding and drought, development of nutrient-efficient cultivars and production systems for efficient use of nutrients and water (Singh, 2015). The adoption of conservation agricultural technologies and good agricultural practices directed at improving the resource base and vegetative cover, the commissioning of cattle dung-based biogas plants, proper management of cattle breeds and feeds to reduce emission of GHG, and the development of stress-tolerant varieties of field, fruit and tree crops could be important climate change-related mitigation and adaptation technological interventions. This would need a reorientation of the research agenda to address emerging challenges due to climate change (Agarwal, 2009). There is a need to enhance knowledge at all levels to address all challenges and convert them into opportunities. Concerted effort will be essential to meet the ever-increasing demand for food, fibre, fuel and energy for the increasing population in developing countries.

Crop Management Options

Water productivity and water-use efficiency

Water productivity denotes the output of goods and services derived from the unit volume of water. It demonstrates how efficiency of water use can be enhanced to maximize yield. With increasing water scarcity, the water productivity will have to be measured in terms of energy, protein and carbohydrate production to maximize the production in terms of these parameters.

There is no single definition of water productivity that suits all situations. However, in all situations, water productivity could be enhanced by saving water, by preventing water loss or by increasing productivity per unit (crop transpiration in agriculture) and by allocation of water to higher-value uses. Reallocation of water from low-value to high-value uses would generally not result in any direct water saving, but can directly increase the economic productivity of water. Suitable water application methods, varieties and management practices will have to be evolved to achieve this goal. Use of liquid forms of fertilizers through fertigation appears to be promising for deeper application with sizeable input saving. For example, in Arka Anmol mango, fertigation with 75% of the recommended dose of fertilizer resulted in yield on a par with that of a 100% dose (Singh *et al.*, 2000; Singh and Singh, 2016).

The term 'efficiency' is usually dimensionless, but in the case of water-use efficiency it is not. This will vary with scale as well as the purpose for which it is being quantified. In the case of the steel industry, it will be the amount of water required to produce one tonne of steel. In the case of agriculture, water-use efficiency at the field level will amount to crop output in physical terms, i.e. crop yield in kg divided by the amount of water consumed, or in monetary terms, i.e. crop yield multiplied by its price divided by the amount of water used; in other words, 'crop per drop'. Water productivity will, therefore, be a function of price that the economic product commands in the market. Thus, while banana will have a higher productivity in terms of kg/ha-mm, grapes will have a higher productivity in terms of Rs/ha-mm. Hence, improving water productivity is equivalent to obtaining more value from each drop of water (Vishwanathan *et al.*, 2016).

Water productivity through crop improvement

The productivity of water, irrespective of environment, will be governed by those factors that minimize water losses from the soil system and improve transpirational water use by crops. The water may be lost from the soil system by evaporation, deep percolation below the root zone, run-off and utilization by weeds. Thus, management practices that are directed to conserve water by reducing these losses would increase the proportion of water used by the crops in transpiration. Crop productivity will depend on the development of leaf area to intercept radiant energy and the role of photosynthesis to convert it into dry matter. However, distribution of assimilates within the plant will determine the proportion of the total dry matter that is harvested as economic yield. Thus to improve productivity of water in a water-deficit environment, one must increase the water passing through the crops in transpiration (T), increase water-use efficiency (W) and/or increase the proportion of total dry matter going to grains, i.e. harvest index (H).

Water productivity with respect to evapotranspiration (WPET) varies considerably for different crops. The WPET ranged from 0.6–1.9 kg/m^3 for wheat; 1.2–2.3 kg/m^3 for maize; 0.5–1.1 kg/m^3 for rice; 7–8 kg/m^3 for forage sorghum; and 6.2–11.6 kg/m^3 for potato under experimental conditions. However, the values of large-scale field-level water productivity are lower than those at experimental level. There are also large variations in productivity of water under different environmental conditions. Modern rice varieties have about a three-fold increase in water productivity as compared to tall traditional varieties due to their improved harvest index. The same is true for modern dwarf wheat and other drought-tolerant genotypes developed through improved plant breeding and biotechnology, wherein the partitioning efficiency of total biomass to economic yield has been improved remarkably. Biotechnical approaches have potential to enhance water productivity (Singh, 2015).

Potential production rates of C3 plants are around 200 kg dry matter/ha/day and those of C4 plants between 200–400 kg dry matter/ha/day, while crassulacean acid metabolism (CAM) plants are known for their very high water-use efficiency. Pineapple, aloe vera and fodder cactus are good examples of cultivated CAM plants. In general, stomata in water-stressed plants remain open during early morning hours when energy load on the crop canopy is low, and close as the solar radiation and vapour pressure deficit of air increases during the mid-day hours. This interaction is manifested to an extreme in the evolution of CAM in some succulents when stomata remain open during the night and close during

the day. Such adaptations are important from a carbon- and water-economy point of view, but their real benefit in field and fruit crops is yet to be more understood and exploited.

To mitigate the impact of drought and heat tolerance in a climate-change scenario, the putative traits that could be beneficial over a long timescale should include phenology, osmotic adjustment, rooting characteristics and assimilate transfer from vegetative parts to grains. The importance of these and other traits and their simplified manifestations in other simple, measurable plant characteristics (leaf-relative water content, canopy temperature, transpirational cooling (canopy minus air temperature difference), use of potassium iodide solution as foliage desiccant for assimilate transfer) have been standardized and utilized in screening crop germplasm for drought tolerance in field crops. Molecular markers and quantitative trait loci (QTLs) identified for osmotic adjustment and rooting characteristics in rice could open the way for easy screening of genotypes for these traits in other crops. The induction of mRNAs suggests that there is a molecular control that might be manipulated genetically, thus altering the development of desiccation tolerance of young seedlings and embryos. Use of dreb gene for drought tolerance in cereal crops is offering good potential through a genetic engineering approach (Singh, 2015).

In India most of the water diverted to agriculture is used for growing staple food crops and only about 10% is used for horticultural crops. The improvement of water productivity of horticultural crops has the potential to reduce the water requirement in terms of providing unit energy, unit protein and other nutrients as well as economic returns. As the majority of the horticultural crops are perennial in nature, they invariably have deep and extensive root systems capable of extracting water from deeper layers and a large canopy to harvest optimum natural resources. Hence they have better productivity than field crops.

Farm Management Options

On individual farms, higher water productivity requires selection of appropriate crops and cultivars, proper soil- and water-management technology and improved planting methods. All cultural and agronomic practices that reduce soil evaporation, run-off, deep percolation, transpiration by weeds, application of mulches and micro-irrigation could improve water productivity. Efficiency of irrigation can be increased by converting intensive irrigation into extensive irrigation. Deficit irrigation, in which less water is applied than that required to meet the full crop water demand, results in a small reduction in photosynthesis and yield as compared to the relatively more concomitant reduction in transpiration and crop water use, leading in turn to high water-use efficiency.

In a flood system of irrigation, selection of proper crop geometry and switch-over from a border to a furrow system of irrigation results in 30–60% saving of water and significant increases in productivity, especially in the case of widespread crops such as mustard and cotton. Similar benefits for improving the productivity of water could be realized by improved planting methods (sunken and raised-bed, raised-bed furrow-irrigated (FIRB), broad-bed furrows (BBF), terracing), water harvesting and recycling, synchronization of water applications with the most sensitive growth period of crops, a holistic approach of watershed and improved drainage for water table control.

Pressurized irrigation systems along with fertilizer application (fertigation) resulted in remarkably high water-use efficiency and yield and thus high productivity of water. Sprinkler and drip systems of irrigation, in the case of some field and horticultural crops, especially under water scarcity and poor groundwater quality, have helped in increasing yield and saving water.

The micro-irrigation scheme implemented by the government was evaluated, covering about 3900 farmers spread over 26 districts in the states of Andhra Pradesh, Haryana, Karnataka, Maharashtra, Orissa and Tamil Nadu. The study revealed that farmers invariably introduced high-value horticultural crops like grape, banana, strawberry, citrus, mango, cashew nut and coconut after installing the drip system. There was a yield increment for crops like banana, grape, citrus and pomegranate, which ranged from 41% (grapes) to as high as 141% (pomegranate). Economic analysis of 695 beneficiary farmers and 76 non-beneficiary farmers indicated that the cost was recovered in a period

of less than three crop seasons. The study also revealed the benefit: cost ratio being more than 2.5:1 in most cases. Keeping in view these benefits, the Ministry of Agriculture launched a national mission on micro-irrigation in 2010, which aims at increasing water-use efficiency, crop productivity and, above all, farmer's income (Viswanathan *et al.*, 2016).

Rainfed cultivation is still practised over 60% of cropped area throughout the country. Selection of appropriate horticultural crops based on land, soil and climatic conditions is the first step for horticulture development in the rainfed areas (Jose and Singh, 1998). Based on drought tolerance, the horticultural crops can be classified as: hardy (bael, ber, boradi, aonla, custard apple, apricot), less hardy (avocado, citrus, coconut), moderate (fig, apricot, breadfruit, cashew nut, chestnut, chironji) and susceptible (apple, arecanut, banana, cherry, strawberry). Horticultural developmental activities through perennial fruit orcharding have already paid high dividends, bringing stability in fragile ecosystems (e.g. apple in Himachal Pradesh, mango and cashew nut in the western Ghats in Maharashtra and large cardamom in Sikkim). The niche potential of marginal mountain lands, if properly nurtured with scientific horticultural practices, can bring fortunes and can convert non-viable subsistence farming to economically viable farming. The success story of the Konkan region in the western Ghats in commercialization of mango, cashew, black pepper etc. demonstrates the possibility of converting once-barren, hilly tracts into economically viable regions. Success stories of seed spices and medicinal and aromatic plants in the arid zones of Rajasthan and Gujarat are pointers in the right direction. A holistic approach of watershed management, water harvesting for groundwater recharge in grey areas and life-saving irrigation or supplemental irrigation can enhance productivity to the tune of 20–30% and could be helpful in increasing cropping intensity and livelihood security of farmers in the rainfed ecosystem.

Micro-irrigation, which includes both drip and sprinkler, has given hope for improving productivity and profitability of agricultural farms. As recommended by the Task Force on Micro-Irrigation, to cover 69 million ha (Naidu and Singh, 2004), the government is committed to improve the efficiency of water use. Micro-irrigation was implemented under Prime Minister Krishi Sinchayi Yojna, with the focus on 'more crop per drop' (Ministry of Agriculture and Farmers' Welfare). With a focus by the government, area is expanding. Until December 2016, 8.6 million ha have been covered (Table 21.1), which clearly indicates that drier zones have adopted micro-irrigation more; however, in these areas, especially in Rajasthan and Haryana, a much larger area is under sprinkler, largely for field crops.

In horticultural crops, a drip system is preferred because it provides opportunity for fertigation to save water. Trials conducted across the country have categorically shown that yield enhancement ranging from 40–80% can be achieved plus water saving and improvement in quality (Rank and Gontia, 2017). Trials conducted at GB Pant University of Agriculture and Technology (GBPUA&T), Pantnagar, have clearly demonstrated that yield enhancement in rice and wheat of 30–50% could be achieved by using micro-irrigation. The trials clearly indicated that only 1200–1500 l of water will be required against 3000–3500 l of water needed for the production of one kg of rice (Bhardwaj et al., 2017; Singh, 2017). To have maximum benefit from micro-irrigation, an integrated approach is needed, through an institutional support system linked to public and private enterprise, involving all stakeholders and keeping technology at the forefront and farmers at the centre in order to achieve faster and inclusive growth (Singh and Singh, 2017b). Micro-irrigation has to be infrastructural (Jain, 1998; Naidu and Singh, 2004)

to facilitate growth, save water and enhance productivity. The government has declared, micro-irrigation under infrastructure, considering it is needed in the development of agriculture (personal communication, 2017).

Augmenting poor-quality waters

In future, reclamation and proper use of brackish- and sewage water could be an additional option for increasing water productivity and resource-use efficiency both in field and fruit crops. Groundwater constitutes the most important source of supplemental irrigation in arid and semi-arid regions in India. Unfortunately, the water in 32–84% of the aquifers surveyed in different states of the country has been observed to be of poor quality. In India, Punjab, Rajasthan, Haryana, Uttar Pradesh, Madhya Pradesh, Gujarat and Karnataka, groundwaters have been found to be highly concentrated with salts. Considerable research has been carried out to utilize poor-quality water in alternate and mixed mode with fresh canal water in field crops. More work is still needed to harness brackishwater in horticultural crops by using pressurized systems of irrigation. In general, the chloride-rich saline waters are more harmful than sulphate-dominated waters. However, there exists great scope for using brackishwater by drip irrigation in horticultural crops. Brackishwater, in future, could also be needed for inland aquaculture through diversified agriculture.

Table 21.1. Area coverage under micro-irrigation in India (state-wise). (From: DOAC & FW, 2016)

	State	Coverage under drip (%)	Coverage under sprinkler (%)	Total coverage in micro-irrigation (m ha)	State-wise coverage (%)
1	Rajasthan	12	88	1.75	20.3
2	Andhra Pradesh	72	28	1.32	15.3
3	Maharashtra	71	29	1.31	15.2
4	Gujarat	50	50	1.07	12.4
5	Karnataka	51	49	0.95	11.1
6	Haryana	4	96	0.58	6.7
7	Madhya Pradesh	52	48	0.43	0.5
8	Tamil Nadu	90	10	0.36	4.2
9	Chhattisgarh	7	93	0.27	3.1
10	Bihar	9	91	0.11	1.3
11	Others	45	55	0.46	0.5
All of India		**45**	**55**	**8.61**	**100.0**

A large number of river stretches are severely polluted as a result of discharge of domestic sewage. Treatment of domestic sewage and subsequent utilization of treated sewage for irrigation can prevent pollution of water bodies, reduce the demand for fresh water in the irrigation sector and result in huge savings in terms of nutritional value of sewage in irrigation. A lot of sewage treatment plants have been established by the government but they are not in proper operational condition for reclamation of sewage water for safe use in irrigation.

Indiscriminate use of wastewater loaded with toxic elements and harmful pathogens pollutes our natural resources and impairs human health. The problem is more serious in vegetables and fodders grown in peri-urban areas by using sewage water, and this needs urgent attention by scientists and policy makers. Both water quantity and nutrients contained in urban and peri-urban wastewater make them attractive alternative water sources for agriculture and aquaculture. Treated wastewater from off-site treatment plants can be re-used for irrigation of parks and gardens, agriculture and horticulture, tree plantation and aquaculture, if these exist or can be established not far from the wastewater treatment plants. To prevent potential negative impacts on human health and the environment, the importance of wastewater re-use in urban and peri-urban agriculture has to be recognized and clear policy guidelines for its proper treatment and re-use need to be established.

Promoting greenhouse technology and plasticulture

The greenhouse technology and use of plasticulture using a drip system of irrigation along with fertigation is one of the most modern technologies at present to grow high-value crops with remarkable saving of water. However, the design of greenhouses has to be location-specific. Due to controlled environmental conditions, high-value crops and off-season fruits and vegetables in nurseries can be grown throughout the year under protected conditions with water economy of 40–50%. Greenhouse technology has been used successfully in the hilly states of the north and north-east as well as in water-scarce states like Maharashtra and Karnataka by optimizing the energy, water and fertilizer application in high-value floriculture and vegetable crops. Rainwater harvesting and its re-use in drip systems of irrigation and artificial recharging of groundwater has been used quite successfully in Karnataka.

The utilization of plastic mulch along with a drip line underneath has been very successful in controlling soil evaporation and water use by weeds. High-value crops like strawberry followed by another crop of chilli under the same plastic cover using soil mulch has been found very remunerative in water-scarcity areas of Haryana, with almost total control of soil evaporation, which is about 60–65% of total evapotranspiration in this region. A similar high degree of water economy has been achieved by fitting drip lines under plastic mulch for growing several horticulture and vegetable crops in other parts of India. Studies have reported a 30–40% increase in yields of tomato by using straw and polythene mulch. Similar encouraging results have been reported in the case of groundnut, cotton etc.

Diversification and intensification

Options for improving productivity and economic efficiency of water further lie in the production of timber, energy plantation, agro-horticulture systems, silvipasture and growing of medicinal plants as intercrops or sole crops. The shifting from field crops to low-water fruit crops and medicinal plants has tremendous scope to improve productivity of water as well as more remuneration to farmers. The low-water fruit crops like ber, aonla, custard apple, pineapple, pomegranate and tamarind are drought-tolerant, hardy fruit crops. If they are supplemented with a drip system of irrigation covered with plastic mulch, very little watering at critical phases of growth could give good yield of fruit with very high water-use efficiency. The drip system of irrigation followed in pomegranate crop in Maharashtra during the water-deficit period (March to mid-June), with just 20–30% wetted area, gave 4.5–5.7 t/ha fruit yield by consuming just 3560–5322 l of water. Similarly, *in situ* harvesting of rainfall through run-off collection from the micro-catchments with 0.5%, 5% and 10% slopes ensured 300–1000 mm run-off supplement in different fruit crops with average annual rainfall of 360 mm on sandy catchment areas around Jodhpur (Rajasthan), India.

The design criteria indicating the percentage slope and size of catchment for different fruit crops have been devised for the sandy and rocky catchments of the Jodhpur region of Rajasthan. The catchment size should be such that the runoff from it, at one recharge, does not ordinarily exceed the moisture storage capacity of the soil profile under the tree canopy. Similarly, pitcher irrigation has been found very useful to establish young fruit plants, especially melons, in areas having little rainfall or no irrigation facilities. In the state of Madhya Pradesh, different medicinal plants grown in agroforestry/agrihorticulture systems gave a very high cost:benefit ratio.

Integrated farming systems

Farming systems as a concept takes into account the components of soil, water, crops, trees, livestock and other resources, keeping farm families at the centre. Integrated farming systems are, therefore, more productive, profitable and sustainable. For centuries, Indian farmers have raised crops and livestock for better livelihood.

The water conservation-based integrated farming system model integrating field and horticulture crops plus livestock plus biogas plants is quite prevalent in several parts of India. In Chitrakoot, Madhya Pradesh, farmers grow crops and medicinal plants and raise livestock. These practices help in better resource utilization, employment generation, livelihood security and welfare of small land holders for holistic rural development. Such models need to be replicated in other parts of the country for conservation of soil and water resources and increasing productive and economic efficiency of water and other costly inputs for the welfare of the rural community. Also, attention needs to be given to value addition, processing and marketing of produce to reduce losses and attract remunerative prices for farmers' produce, thus increasing their income.

Future Strategy

Out of several available on-farm water-management strategies, providing irrigation at the most sensitive stages of growth; use of well-designed surface, sub-surface and pressure systems of irrigation; mulching; fertigation; plasticulture; conjunctive use of sewage and brackishwater; and a holistic approach of watershed development would help in increasing water productivity. An integrated approach to farming systems, value addition, processing and marketing are immensely important as they help increase productivity, profitability, employment, habitat conservation and livelihood security for small and marginal farmers.

Addressing the issue of efficient water use would need a well-articulated future road map to move forward. The following are some specific suggestions:

- Water must be treated as a national asset. National water policy should, therefore, be governed by central government or else state-level disputes will always hamper future agricultural growth. Like other inputs, water should be priced in order to ensure judicious use of this valuable natural resource.
- The current practice of flood irrigation needs to be discouraged as a matter of national priority. Instead, practices for efficient water use need to be promoted, with greater focus on micro-irrigation, especially sprinkler and drip irrigation, to enhance water productivity.
- The policy of 'khet ka pani khet main' needs to be adopted by encouraging bunding of fields. On-farm water conservation must be an aim of the Mahatma Gandhi National Rural Employment Guarantee Act (MGNAREGA), and other schemes need to consider farm bunding a priority goal under this ambitious national scheme.
- Agricultural diversification, especially involving horticulture, agroforestry and silvipastoral approaches, should be a major strategy and will lead to considerable saving of valuable water resources.
- A strategy for conjunctive use of brackishwater (up to 20%) in canal-command central and north-western states that receive less than 500–700 mm precipitation should be put into practice through enabling policy interventions and needed incentives to farmers adopting these water-saving practices.

- A massive public awareness campaign to promote water-use economy is urgently warranted. Good watershed management practices need to be promoted through community involvement and by forming water-user associations. Also, pricing of water would help to realize the value of this precious resource for increasing water productivity.

- Concerted efforts need to be made urgently for outscaling of innovations that save water such as conservation agriculture, plastic mulching, direct-seeded rice, alternate furrow irrigation, micro-irrigation, fertigation etc., through much-needed policy-, research- and development-related initiatives.

References

Agarwal, P.K. (2009) *Global Climate Change and Indian Agriculture – A Case Study from ICAR Network Project*. ICAR, New Delhi.

Bhardwaj, A.K., Bhomik, T. and Prajapati, B. (2017) Effect of drip irrigation and fertigation on growth and productivity of wheat in mollisols (abstract). *Book of Abstracts* 9, 84. ASM Foundation, New Delhi.

Dhawan, B.D. (2002) Technological Change in Indian Irrigated Agriculture: A Study of Water-saving Methods. Commonwealth Publishers, New Delhi.

DOAC & FW (2016) Department of Agriculture, Cooperation and Farmers' Welfare, Ministry of Agriculture and Farmers' Welfare, New Delhi.

Jain, B.H. (1998) *Why Drip Irrigation to be Infrastructure Industry? An Approach Paper*. Jain Irrigation System Ltd, Jalgaon, India.

Jose, C.S. and Singh, H.P. (1998) *Current Trend on Micro-irrigation Development in Indian Horticulture*. Workshop on Micro-irrigation and sprinkler irrigation system. Central Board of Irrigation and Power, New Delhi.

Kumar, M.D. (2003) Food security and sustainable agriculture in India: water management challenge. Working paper 60. International Water Management Institute, Colombo, Sri Lanka.

Naidu, C. and Singh, H.P. (2004) *Report of the Task Force on Micro Irrigation*. Ministry of Agriculture, Government of India.

Narayanamoorthy, A. (1996) *Evaluation of Drip Irrigation System in Maharashtra*. Gokhale Institute of Micrograph series no. 42. Agro-Economic Research Centre, Gokhale Institute of Politics and Economics, Pune, Maharashtra, India.

Narayanamoorthy, A. (2001) *Impact of Drip Irrigation on Sugarcane Cultivation in Maharashatra*. Research Report. Agro-Economic Research Centre, Gokhale Institute of Politics and Economics, Pune, Maharashtra, India.

Rajput, T.B.S. and Patel, N. (2017) Strategies for better water utilization – a step forward doubling farmers' income. *Sodh Chintan* 9, 173–181.

Rank, H.D. and Gontia, N.K. (2017) Enhancing farmers' income through micro-irrigation adoption. *Sodh Chintan* 9, 182–190.

Singh, H.P. (2015) Dynamics of biotechnological approaches for food and nutritional security. *Sodh Chintan* 7, 150–175.

Singh, H.P. (2017) Rice water management under micro-irrigation – an efficient option for high water productivity (abstract). Book of Abstracts 9, 84.

Singh, H.P. and Singh, B. (2016) Enhancing water productivity – micro-irrigation an option. *Sodh Chintan* 8, 51–64.

Singh, H.P. and Singh, B. (2017a) Technological changes, innovations and policy reforms for enhancing farmers' income. *Sodh Chintan* 9, 1–17.

Singh, H.P. and Singh, B. (2017b) Micro-irrigation an option for improving productivity of water and farm profitability (abstract). Book of Abstracts 9, 72.

Singh, H.P., Kumar, A. and Jose C.S. (2000) Micro-irrigation for horticultural crops. *Indian Horticulture*, 21–29.

Viswanathan, P.K., Kumar, M.D. and Narayanamoorthy, A. (eds) (2016) *Micro Irrigation Systems in India: Emergence, Status and Impacts*. Springer Singapore.

22

Agroforestry in India: The Way Forward

Introduction

Agroforestry is an effective land-use system contributing to food, nutritional and environmental security. Besides its multifarious use as food, fuel, fodder, fibre and timber, it enables optimization of land use by smallholder farmers. Agroforestry has noteworthy potential to provide employment and additional income to farmers. Many countries, through agroforestry, could increase their forest/tree cover to meet specific national targets, which would otherwise be quite difficult to achieve.

In the context of climate change, agroforestry mitigates GHG emissions through micro-climatic modifications and carbon sequestration (Albrecht and Kandji, 2003; Dhyani *et al.*, 2016). In landscape management, it plays an important role in reducing GHG emissions and acts as an effective means of checking environmental pollution. In fact, agroforestry can help in achieving resilience in agriculture while addressing effectively the threat of climate change (Rao *et al.*, 2007). As land-holding size is shrinking, tree farming combined with agriculture may perhaps be the only way forward to optimize farm productivity and thus enhance livelihood opportunities of smallholder farmers and landless labourers, especially women.

At present, there is growing concern over natural resource degradation (soil erosion, salinity, sodicity, water-logging, environmental pollution,

desertification etc.) owing to indiscriminate use of agricultural chemicals/other inputs and/or inappropriate land-use systems. As a result, substantial land area has gone out of production. Agroforestry, as an alternate land-use option, holds promise in such situations (Dhyani *et al.*, 2005). A closer integration of agricultural crops and forest trees would check additional adverse effects of climate change and land degradation and would also ensure timber and firewood availability in rural areas, thus allowing farmyard manure use for agricultural production and organic recycling for improving soil health.

Despite its obvious benefits, agroforestry continues to face challenges of an unfavourable policy environment, lack of scientific knowledge and public awareness, legal constraints and poor coordination, as well as convergence among multiple organizations/ministries involved (Agriculture and Farmers' Welfare; Environment, Forest and Climate Change; Rural Development and Trade) (NAP, 2014). Inadequate investment, lack of required extension strategies and weak market linkages are the real concerns in improving the livelihood of smallholder farmers. Moreover, developments in agroforestry are impeded by legal, policy and institutional arrangements. Its environmental benefits mostly go unnoticed, and investments in it are often linked with long gestation periods. As a result, the potential of agroforestry has not been clearly understood by farmers and farming communities.

In the recent past, fortunately, a paradigm shift towards environmental protection and sustainable land use has taken place, and tree-based production systems are being promoted in India. The call for planting trees on each bund by Prime Minister Narendra Modi is a needed policy directive to promote agroforestry. In 2014, the World Congress on Agroforestry was organized jointly by the World Agroforestry Center-International Centre for Research in Agroforestry (ICRAF) and the ICAR in New Delhi with the central theme 'Trees for Life: Accelerating the Impacts of Agroforestry'. During the Congress, India announced its national agroforestry policy. As a result, agroforestry, evergreen agriculture and smallholder production systems have attracted considerable attention. The World Congress made many useful recommendations, which, when implemented, will go a long way in stimulating large-scale adoption of agroforestry by farmers and would provide much-needed raw material to wood-based industries on the one hand, and provide energy and environmental security on the other. It was also emphasized that agroforestry can be the only alternative in India to meet the target of increasing forest cover to 33% from the present less than 25%; thus also restoring the lands considered degraded presently. Therefore, a major role for agroforestry would be to provide environmental services such as increased coverage under vegetation and higher carbon sequestration.

Role of Agroforestry

Agroforestry plays a vital role in the Indian economy. It rehabilitates degraded lands and increases farm productivity. It also meets almost half of the demand for fuel wood, 60–70% of small timber, 70–80% of plywood, 60% of raw material for paper pulp and 9–11% of green fodder for livestock, besides meeting subsistence needs of households for food, fruit, fibre and medicine. This, however, is not the main contribution of agroforestry. To measure this, the first step is to find out the actual area under agroforestry. Estimates vary from 11.15 million ha (ISFR, 2013) (Table 22.1) to 25.32 million ha (Dhyani and Handa, 2013). The estimates are not based on the revenue records or actual measurements. Therefore, the first priority is to initiate an assessment through geospatial technologies as manual (traditional) methods of mapping are expensive and would take a long time. The most important agroforestry species for promotion in different ecoregions of the country are *Populus* spp., *Eucalyptus* spp., *Tectona grandis*, *Prosopis* spp., different bamboo spp., *Acacia* spp., *Gmelina* spp., *Grewia* spp., *Melia* spp., *Ailanthus* spp., *Dalbergia sissoo, Casuarina* spp., *Leucaena leucocephala, Azadirachta indica, Anthocephalus cadamba, Albizia* spp., *Terminalia* spp., *Salix tetrasperma* and *Hardwickia binata*.

Table 22.1. Physiographic zone-wise tree-green cover under agroforestry (ISFR, 2013).

Physiographic zone	Geographical area (sq. km)	Tree green cover (sq. km)	Tree green cover (% to GA)
Western Himalayas	329,255	7131	2.17
Eastern Himalayas	74,618	1818	2.44
North-east Zone	133,990	7513	5.61
Northern plains	295,780	8740	2.96
Eastern plains	223,339	9872	4.42
Western plains	319,098	7450	2.33
Central highlands	373,675	9168	2.45
North Deccan	355,988	6949	1.95
East Deccan	336,289	12,450	3.70
South Deccan	292,416	7771	2.66
Western Ghats	72,381	7465	10.31
Eastern Ghats	191,698	5102	2.66
West coast	121,242	13,523	11.15
East coast	167,494	6602	3.94
Total	**3,287,263**	**111,554**	**3.39 (av. % of total GA)**

Agroforestry Research

Organized research on agroforestry started worldwide about 35 years ago. Many south-Asian countries, including India, are at the forefront of this research. The ICAR initiated the All India Coordinated Research Project (AICRP) on agroforestry in 1983. At present, there are 37 centres of AICRP on agroforestry, representing all agroclimatic conditions of the country. The research initiatives gained momentum with the commencement of forestry education programmes in SAUs during 1985/86 and with the establishment of the National Research Centre for Agroforestry (NRCAF) at Jhansi, Uttar Pradesh, in 1988, later upgraded in 2014 to the Central Agroforestry Research Institute (CAFRI). In addition, the Indian Council of Forestry Research & Education (ICFRE) focuses on agroforestry research through its institutes and advanced research centres. The research indicates considerable variability in nature, its components and the ecological and socioeconomic conditions of agroforestry practices; identified 184 tree species suitable for agroforestry; and provided improved accessions of poplars, eucalypts, *Dalbergia*, neem, *Acacia*, *Leucaena*, *Ailanthus*, *Pongamia*, *Casuarina* and *Mangium* hybrids, which were supplied to farmers. The significant achievements of the research resulted in the development of location-specific agroforestry practices for different agroclimates. Agroforestry systems for rehabilitation of degraded wastelands and problem soils have also been developed. The other areas of research included economic analyses of agroforestry systems, extent of the agroforestry area and the carbon sequestration potential of agroforestry species. Lately, research work is being done on environmental protection, post-harvest technology, fishery, apiculture and lac in relation to agroforestry systems (Dhyani *et al.*, 2015).

In general, agroforestry will be able to make available diversified foodstuffs, averting malnutrition and providing organic food materials, for which there is an emerging market even in the developing countries. More recently, however, such systems are considered important from the perspective of augmenting economic returns to the growers. Agroforestry practices including tree-based smallholder production systems offer great potential to create new jobs in the rural areas. The diversity of products from agroforests provides opportunities for development of small-scale rural industries and for creating off-farm employment and marketing opportunities. Spread of agroforestry, such as poplar-based in the northwestern states, *Casuarina*- and *Eucalyptus*-based in the southern states and *Ailanthus*-based in the western states, and associated industrial development in these areas, indicates the trend. In addition, when strategically applied on a large scale, with appropriate mix of species, agroforestry enables agricultural land to sustain extreme weather events such as floods and droughts as well as climate change.

Agroforestry Policy

Agroforestry policy is a path-breaker in making agroforestry an instrument for transforming the life of the rural farming population, protecting ecosystems and ensuring food security through sustainable means. The major highlights of the policy are: establishment of institutional set-up at national level to promote agroforestry under the mandate of the Ministry of Agriculture and Farmers' Welfare; simplifying regulations related to harvesting, felling and transportation of trees grown on farmlands; ensuring security of land tenure and creating a sound base of land records and data for developing a market information system (MIS) for agroforestry; investing in research, extension and capacity-building and related services; access for quality planting material; institutional credit and insurance cover for agroforestry practitioners; increased participation of industries dealing with agroforestry produce; and strengthening MIS for tree products. The policy also suggests massive extension programmes to broadcast the outcomes of intensive R&D activities in the field of agroforestry (NAP, 2014). The major hurdle in the implementation of the national agroforestry policy and the recommendations of the World Congress is the lack of appropriate tree-harvesting and transportation rules between and within a state. Furthermore, there is lack of marketing, credit and insurance infrastructure in the agroforestry sector. Research and technology gaps also need to be addressed. The Indian National

Agroforestry Policy (INAP), 2014, has catalysed actions in many countries particularly affected by land degradation and drought. Neighbouring countries like Nepal and Bangladesh, and six other SAARC countries, swiftly began developing policies, and also Rwanda, Comoros, Seychelles and Vietnam. The policy is a model for countries to sustainably intensify agriculture and address water crises and Intended Nationally Determined Contributions (INDCs) – the basis of post-2020 global emissions reduction commitments included in the climate agreement within the Paris Agreement (2015).

Agroforestry Status after the Implementation of the Policy

As per the latest Forest Survey of India (FSI) report (ISFR, 2015), there is an increase of 110.34 million m^3 in total growing stock of the country as compared to the last assessment (ISFR, 2013). The noteworthy feature of this is the healthy contribution of 88.66 million m^3 from trees outside forests, which primarily indicates the agroforestry contribution. The potential of agroforestry for sustainable development has been recognized in many international policy declarations. For example, the UN Framework Convention on Climate Change (UNFCCC) and the Intergovernmental Panel on Climate Change (IPCC) acknowledged it as a component of climate-smart agriculture, and frequently mentioned it as having strong potential for climate change adaptation and mitigation. The UN Convention to Combat Desertification (UNCCD) acknowledges agroforestry's potential to control desertification and rehabilitation. It is also seen as an important element in the ecosystem approach promoted by the Convention on Biological Diversity (CBD) for agrobiodiversity conservation.

Since the launch of the policy in 2014, considerable progress has been made in terms of putting it into practice. To implement recommendations, an inter-ministerial committee has been set up. The DAC&FW, under the MOA & FW, is now the nodal ministry for implementing agroforestry programmes. It is playing a significant role in the promotion of agroforestry. It has taken a policy decision to include trees in all its programmes,

and this would significantly increase tree-planting on farms. Efforts are on to issue guidelines on the production and supply of high-quality planting material and accreditation of nurseries producing agroforestry planting material. Until recently, felling, transit and processing of trees grown on farms required approvals and permits from government agencies, and this was a strong impediment in the establishment of agroforestry systems. To promote agroforestry, 20 multi-purpose tree species (MPTS) commonly grown by farmers were prioritized by the ICAR to be exempted from the regulatory regime. Now there is strong political support for agroforestry. The prime minister frequently uses the phrase 'Har Med par Ped' ('trees on every field bund/boundary'). A Sub-Mission on Agroforestry (SMAF) has been implemented with an outlay of Rs 9.9 billion. Relaxation of transit regulations is a prerequisite for assistance under the Mission; 15 states including Haryana, Punjab, Himachal Pradesh, Tamil Nadu, Gujarat and Madhya Pradesh have denotified a number of tree species from felling and transit regulations, thus relaxing regulatory regime; and all states are being motivated in this direction. This would make it much easier for landowners and farmers to practise agroforestry. The project is expected to assist all the states to scale up agroforestry in a targeted manner. In another significant move, the corporate social responsibility (CSR) laws of India were modified and notified in 2014. Accordingly, agroforestry became a legitimate part of the recognized CSR activities (http://finance.bih.nic.in/Documents/CSR-Policy.pdf). However, awareness among CSR investors about the potential and need for agroforestry in the country needs to be created to get the benefit of the funds. Recently, under CSR, the Oil and Natural Gas Commission (ONGC) has funded an agroforestry project, which is being implemented jointly by ICAR, ICRAF and ONGC in partnership with the local farming community in Rajasthan. More such efforts are needed to attract CSR funds.

The agricultural sector has the facilities of insurance and credit from the financial institutions and also has an organized marketing structure. However, the farmers practising tree-based farming are devoid of any such facilities and this is a major hurdle in boosting agroforestry among resource-poor farmers. There have been some tree insurance initiatives taken in Tamil Nadu, Kerala and a few other states by the insurance companies. However, such efforts are needed at country level to encourage involvement of farmers in expanding the area under agroforestry. For large-scale adoption, it is essential to provide credit and marketing facilities to farmers adopting tree-based farming. This needs urgent attention at the union as well as the state level to take additional initiatives.

Road Map for Agroforestry

Agroforestry is the best option to achieve the target of forest cover, to mitigate GHG emissions and to provide additional options for income to smallholder farmers in India. However, to achieve these objectives, a clear road map is needed. The following actions are needed both at the national and regional level:

Actions at national level

- The agroforestry policy adopted by the government in 2014 is indeed a desired step in the national interest. In this context, a National Agroforestry Mission on similar lines to the National Horticulture Mission needs to be established to ensure an aggressive approach to promoting agroforestry.
- As envisaged in the policy document, an agroforestry board has to be established similar to the rubber board, coffee board, tea board etc., mainly to facilitate pricing, processing, value addition, procurement, credit, insurance, marketing, and to provide incentives to agroforestry farmers and other stakeholders for their environmental services.
- Agroforestry practices can, potentially, contribute towards increasing the present 24% tree cover of the country to meet the national target of 33%. To achieve this, sufficient investment would be required to cover about 12–14 million ha of land, i.e. 8–10% of the total cultivated area under agroforestry practices. Also, to achieve the set national targets, other niche areas, such as degraded

lands, arid areas etc., would have to be covered under agroforestry practices.

- The most important agroforestry tree species for each agro-ecoregion in the country need to be identified as a priority, based on data generated under the All India Coordinated Agroforestry Project, and all-out efforts are needed to promote their planting in larger areas. Also, action should be initiated to 'denotify' them as forest species.
- To accelerate efforts on agroforestry at the national level, it would be desirable to have a subject matter specialist in each Krishi Vigyan Kendra (KVK) in each district earmarked for agroforestry.
- An expert group needs to be constituted to suggest agro-ecological, region-wise, scientific land-use planning, so as to promote only the most remunerative tree species, based on research results already available.
- Special efforts are needed to produce high-quality seed and planting materials of elite stocks identified by research institutions. All such planting materials need to be made available through much-needed certification and accreditation systems.
- Most of the national sustainable development strategies must embrace agroforestry for poverty alleviation, rural livelihood security, skill development, natural resources management, agricultural productivity enhancement and restoration of degraded landscapes to contribute more effectively towards India's Intended Nationally Determined Contributions (INDCs) to the UN Framework Convention on Climate Change (UNFCCC).
- With India's INDCs pointing towards climate justice, agroforestry becomes a key instrument for resilience-building for vulnerable and resource-poor communities. In fact, agroforestry for adaptation to climate change needs to be mainstreamed and highlighted to generate public awareness.
- Investments in agroforestry projects and programmes by the public and private sectors (including the corporate and small cooperatives) for research, extension, enterprise and education need to be encouraged and incentivized. Innovative financial mechanisms, including climate finance, for agroforestry

would benefit small agri-business enterprise and smallholders and would encourage greater participation of people.

Actions at regional level

- The nodal ministry/focal point for dealing with matters relating to agroforestry needs to be clearly defined at the national and sub-national levels.
- Development of country-specific national policies on agroforestry and enabling mechanisms for their implementation need to be given high priority. The expertise of the International Centre for Research in Agroforestry (ICRAF), the facilitating role of the Asia-Pacific Association of Agricultural Research Institutions (APAARI) and assistance from international agencies would be useful in furthering this initiative.
- A regional consortium-cum-network on agroforestry, with a facilitation role of the ICRAF, in partnership with APAARI, needs to be initiated to ensure policy advocacy, public awareness, research collaboration, sharing of knowledge and germplasm, capacity development and other collective actions.
- The proposed regional network should place a high priority on the development of a sound regional agroforestry database, information systems and ecoregion-based

decision-support systems. Documentation and sharing of success stories in the regions need to be encouraged through open access to information.

- An independent scientific study would be useful for identifying and assessing suitable determinants to scale up innovations in agroforestry, including market mechanisms, import and export policies and support-price mechanisms.
- Investment, being critical to promoting agroforestry research, education, training and extension, must be at least doubled at the national and regional level.
- Medium- to long-term collaborative studies to quantify the contribution of agroforestry to ecosystem services, carbon sequestration and climate change mitigation need to be institutionalized by the IARCs and national institutions.
- Development of agroforestry value chains are critical for scaling up promising innovations and for creating win–win situations for the agroforestry sub-sector. Business planning and development, involving stakeholders, in the value chain (farmer-to-consumer) need to be institutionalized.
- Awareness of PPP through the creation of an enabling environment, such as process patenting, branding and incentives to both producers and industry, is the need of the hour to promote agroforestry in the region.

References

Albrecht, A. and Kandji, S.T. (2003) Carbon sequestration in tropical agroforestry systems. *Agriculture, Ecosystems and Environment* 99, 15–27.

Dhyani, S.K. and Handa, A.K. (2013) Agroforestry in India and its potential for ecosystem services. In: Dagar, J.C., Singh, A.K. and Arunachalam, A. (eds) *Agroforestry Systems in India: Livelihood Security & Ecosystem Services. Advances in Agroforestry 10*. Springer India, pp. 345–366.

Dhyani, S.K., Sharda, V.N. and Samra, J.S. (2005) Agroforestry for sustainable management for soil, water and environment quality: looking back to think ahead. *Range Management & Agroforestry* 26(1), 71–83.

Dhyani, S.K., Handa, A.K. and Uma (2015) Agroforestry research in India: present status and future perspective. In: Dhyani, S.K., Newaj, R., Alam, B. and Dev, I. (eds) *Organized Agroforestry Research in India: Present Status and Way Forward*. Biotech Book Publishers, Delhi, Ch. 1.

Dhyani, S.K., Ram, A. and Dev, I. (2016) Potential of agroforestry systems in carbon sequestration in India. *Indian Journal of Agricultural Sciences* 86(9), 1103–1112.

ISFR (2013) *India State Forest Report*. Forest Survey of India (Ministry of Environment and Forests and Climate Change), Dehradun, India.

ISFR (2015) *India State Forest Report*. Forest Survey of India (Ministry of Environment and Forests and Climate Change), Dehradun, India.

NAP (National Agroforestry Policy) (2014) Department of Agriculture and Cooperation, Ministry of Agriculture, Government of India, pp. 1–13.

Rao, K.P.C., Verchot, L.V. and Laarman, J. (2007) Adaptation to climate change through sustainable management and development of agroforestry systems. *SAT eJournal* 4, 1–30.

23

The Impact of Global Climate Change on Agricultural Growth

Introduction

Climate change and agriculture are interrelated processes, both of which take place on a global scale. Global warming is projected to have significant impacts on conditions affecting agriculture, including temperature, carbon dioxide, precipitation, sea-level rise, increasing ocean acidification, ultraviolet-B (UV-B) radiation, extreme weather events, glacier retreat and disappearance, and El Niño–Southern Oscillation (ENSO) effect on agriculture, and the interaction of these elements. These conditions determine the carrying capacity of the biosphere to produce enough food for the human population and domesticated animals. The overall effect of climate change on agriculture will depend on the balance of these effects. Assessment of the effects of global climate changes on agriculture might help to properly anticipate and adapt farming to maximize agricultural production. The effects of carbon dioxide (CO_2) enrichment, without associated changes in climate would probably be beneficial for agriculture. However, more severe warming, floods and drought may reduce yields. Higher temperatures, however, could increase the rate of microbial decomposition of organic matter, adversely affecting soil fertility in the long run. An analysis of the biophysical impact of climate changes associated with global warming shows that higher temperatures generally hasten plant maturity in annual species, thus shortening the growth stages of crop plants. Also, studies analysing the effects on pests and diseases suggest that temperature increases may extend the geographic range of some insect pests currently limited by temperature. The effects of increased UV-B radiation reduce yield in certain agricultural crops. Livestock may be at risk, both directly from heat stress and indirectly from reduced quality of their food supply. Fisheries would be affected by changes in water temperature that shift species ranges, make waters more hospitable to invasive species and change life-cycle timing.

In India, the increase in mean annual maximum temperature was 0.76°C and in mean minimum temperature 0.22°C. Increase in annual mean temperature was 0.49°C during the period 1901–2003. In terms of increase in temperature, the west coast of India is warmer, followed by the north-east and the western Himalayas when compared to other regions of the country. The years 2009–2010 were recorded as the warmest in the country since 1901. Increases in temperature and rainfall were noticed in the country in tune with global warming and climate change, though spatial and seasonal differences were evident. At the same time, rainfall during the monsoon season was in deficit in recent years, like 1987, 2002 and 2009, which adversely affected foodgrain production. Climate change impacts agriculture both directly and indirectly. The type and magnitude of impact will vary depending on the degree of change in climate, geographical region and type of production system. Assessment of the impact of climate change is carried out through controlled experimentation and simulation modelling. Experimental results obtained are extrapolated on a regional basis in relation to projected climate change under different scenarios. The key influences are:

- change in productivity, with reference to quantity and quality of crops;
- change in agricultural practices like water use and application of fertilizers, insecticides and herbicides; and
- environmental influences, particularly in relation to the frequency and intensity of soil drainage, which may lead to loss of nitrogen through leaching, soil erosion and reduction of crop diversity.

The major effect on crops is shortening of crop duration, which is related to the thermal environment. Increase in temperature will hasten crop maturity. In annual crops, the shortening of crop duration may vary from two to three weeks, thus adversely impacting productivity. Another direct effect in crops such as rice, wheat, sunflower etc. is on reproduction, pollination and fertilization processes, which are highly sensitive to temperature. The indirect influences operate through changes in water availability due to inadequate or excess rainfall and the effect of increases in temperatures on pest and disease incidence.

The earth's climate has remained dynamic throughout the 4.5 billion years of its history, and climate has periodically changed following a natural cycle. These climatic changes had a profound influence on sea level, rainfall patterns and temperature-related weathering processes. However, temperature increases in the late 20th century seem unique and provide evidence that a GHG effect has already become established, above the level of natural variability of the last 1000 years, and that is greater than the best estimate of global temperature change for the last interglacial (Jat *et al.*, 2016).

Agriculture is the foundation for humanity's basic needs, and it has been satisfying the food, nutrition and livelihood requirements of the ever-growing population for a long time. However, global food security threatened by climate change is one of the most important challenges in the 21st century – supplying sufficient food for the increasing population while sustaining the already stressed natural resources and environment. Climate change is exacerbating the challenges faced by the agricultural sector, negatively affecting both crop and livestock systems in most regions. Climate change has already caused significant impacts on water resources, food security, hydropower and human health across the world. Changes in the frequency and severity of extreme climate events and in the variability of weather patterns will have significant consequences for human and natural systems. The changes in crop production-related climatic variables would possibly have major influences on regional as well as global food production. Increasing frequencies of heat stress, drought and flooding events are projected for the rest of this century, and these are expected to have many adverse effects over and above the impacts due to changes in mean variables alone (IPCC, 2012).

Agriculture is contributing a significant share of GHG emissions that are causing climate change; 17% directly through agricultural activities and an additional 7–14% through land-use changes. During the past two centuries, the world has witnessed a remarkable increase in the atmospheric concentrations of GHGs, namely carbon dioxide (CO_2), methane (CH_4) and nitrous oxide (N_2O), as a result of human activities after 1750 (pre-industrial era). In 1750, the concentrations of these gases were: 280 ppm, 715 ppb and 270 ppb, respectively, which increased to 379 ppm, 1774 ppb and 319 ppb, respectively, in 2005. It showed an increase of 0.23%, 0.96% and 0.12%, annually. The same has further increased to 385 ppm, 1797 ppb and 322 ppb, respectively, in 2008, representing 1.6%, 1.2% and 0.9% increases, respectively, from 2005 levels at an annual increase of 0.53%, 0.43% and 0.31%, annually (IPCC, 2007; WMO, 2009; Jat *et al.*, 2016). GHG emissions have increased four-fold, but emissions per unit of GDP have reduced by three quarters in developing countries. In 2010, the Asia-Pacific region emitted a total of around 20 billion t of GHGs, four times more than it was emitting in 1970. Over those 40 years, Asia-Pacific regional emissions increased from 20% to 40% of the global total. Within the Asia-Pacific region, China is the largest emitter of GHGs and has increased the most in absolute terms and in its relative contribution: from 32% of regional

GHG emissions in 1970 to 56% in 2010. Of this increase, industrialization (fossil-fuel combustion and cement production) contributed 67% and the remaining 33% was through land-use change. The increase in GHGs in the atmosphere is now recognized to contribute to climate change. The projections suggest that global temperatures may rise by 0.6–2.5°C by 2050 and 1.4–5.8°C by 2100 (IPCC, 2007). Studies have shown that there would be at least a 10% increase in irrigation water demand in arid and semi-arid regions of south Asia with a 1°C rise in temperature. Thus, climate change will result in increased demand for irrigation water, further aggravating resource scarcity. Increase in mean temperature; changes in rainfall patterns; increased variability, both in temperature and rainfall patterns; changes in water availability; the frequency and intensity of extreme events; sea-level rises and salinization; and perturbations in ecosystems will all have profound impact on agricultural production. Moreover, climate change can intensify the degradation process of natural resources that are central to meet the increased food demand, while, on the other hand, changing land-use patterns, natural resource degradation (especially land and water), urbanization and increasing pollution will affect the ecosystem in this region directly and indirectly through their impact on climatic variables (Lal, 2016). For example, about 51% of the Indo-Gangetic plains may become unsuitable for wheat crop, a major food security crop of the region, due to increased heat stress by 2050 (Lobell *et al.*, 2012). Therefore, adaptation to climate change is no longer a nebulous option, but a compulsion to minimize the loss due to adverse impacts of climate change and reduced production (IPCC, 2014). Moreover, while maintaining a steady pace of development, the region would also need to reduce its environmental footprint from agriculture.

Climate Change, Population and Food in Asia

Asia is home to more than one half of the world's population living on one third of global land. As per the estimates of the United Nations Census Bureau (UNCB), the present world population has exceeded 7 billion, which is mainly concentrated in the developing world. The population in south Asia has increased about three times, in Africa by more than three times, and in the world as a whole by more than twice during the past 50 years. The rapid and continuing increase in population and economy implies increased demand for food in the region. In order to feed the world's population of 9.6 billion, overall food production needs to be increased by 70% between 2005 and 2050 and this will result in a further 30% increase in global GHG emissions from agriculture by 2050 (Tubiello *et al.*, 2014). This growth in agriculture and associated emissions will mostly occur in Asian and African countries where large numbers of people depend on agriculture and allied sectors for their livelihood. In south Asia, more than 95% of agriculturally suitable land is already under cultivation. Hence there is no scope for horizontal expansion of farming. Whereas, with high risks of climate change-induced extreme weather events, crop yields in the region are predicted to decrease by between 7% and 10 % in the near future. Hence, the world's food situation will be strongly dominated by the changes that would occur in Asia because of its huge population, changes in diet pattern and associated increased demand for food, feed, fibre and fuel. Alleviating poverty and attaining food and nutrition security while protecting natural resources under an adverse environmental scenario due to global climate change and spiralling cost of inputs, as experienced in the recent past, will be a major challenge in the 21st century for most of the countries in the Asian region.

Climate Change and Energy Scenario in Asia

The direction that Asian countries would take to meet their energy needs in the coming 30 years will have profound impacts on global climate change and energy security for the region and the world as a whole. Energy consumption has increased more than four-fold in developing countries in the Asia-Pacific region and is dominated by non-renewable energy sources (UNEP, 2015). The fifth assessment report of the IPCC

(2014) revealed that the energy sector remained the major contributor of GHGs (35% of the total). Asia accounts for about 26% of global CO_2 emissions, and its share of emissions is projected to increase to nearly 50% by 2030. In addition, the burning of coal to meet Asia's energy needs is projected to increase five-fold by 2030, accelerating GHG emissions and further contributing to global climate change. Increasingly, Asian countries are importing fossil fuels to sustain their rapid economic growth, and this is raising concerns for further energy security. By 2030, it is expected that 80% of Asia's oil will be imported from the Middle East. Reserves of natural gas in Asia (a cleaner-burning fossil fuel) are limited, and 40–75% of natural gas will have to be imported by 2030 to satisfy demand. This future dependence on imported fossil fuels raises legitimate concerns for Asian countries about price volatility and shocks, and supply disruptions. Also, the majority of the world's most polluted cities are in Asia and the impact of urban air pollution on health and mortality in Asia is severe. Urban air pollution in Asia is linked to over 500,000 premature deaths every year, accounting for 65% of premature deaths from air pollution worldwide.

Climate Change and the Paris Agreement

More than 190 nations met in Paris in December 2015 and reached a landmark agreement to strengthen the global climate effort. The Paris Agreement commits countries to undertake 'nationally determined contributions' (NDCs) and establishes mechanisms to hold them accountable and to strengthen ambition in the years ahead. The 2015 Paris Agreement (COP 22) of the United Nations Framework Convention on Climate Change (UNFCCC) aims to keep global temperature rises below 2°C above pre-industrial levels. To realize this goal, we need to reduce GHG emissions globally. Among the top three largest emitters of GHGs in the world, two are from Asia. China ranks first and India third after the USA (https://wri.org/blog/2014/11/6-graphs-explain-world%E2%80%99s-top-10-emitters), and therefore the Asian region has a larger role to play in reducing global emissions and determining the

climate of the future. India's Intended Nationally Determined Contributions (INDCs) to UNFCCC pledges to reduce the emission intensity of its GDP by 33–35% by 2030 over 2005 levels (http://www.moef.nic.in/climate-change-docs-and-publications). As one of the major emitters of the world, such commitment could make a noticeable difference to prevent climate extremes globally.

The agriculture sector is responsible for 18% of gross national GHG emissions in India. Livestock, rice cultivation, fertilizer input and burning of crop residues are major emission sources within agriculture. India, at present, is in a phase of rapid economic and demographic transition. Per capita income has been rising steadily since the 1980s. With rapid economic growth and expected population of about 1.71 billion (http://www.populstat.info/Asia/indiac.htm), food demand in India is expected to double by 2050, necessitating an increase in agricultural production. This will exert intense pressure on agroecosystems that are already overburdened. The environmental impacts of meeting the increased food demand will be further intensified by climate change because this accelerates degradation processes in vulnerable environments and leads to unknown interactions and feedback in the complex web of relationships among social, environmental, economic and food systems.

Climate Change Impacts on Agriculture

Climate change is projected to impinge on the sustainable development of most developing countries of Asia as it compounds the pressures on natural resources and the environment associated with rapid urbanization, industrialization and economic development. The impact of climate change on agriculture is now real, and without adequate adaptation and mitigation strategies, food insecurity and loss of livelihoods are likely to be exacerbated in Asia. In this regard, the fifth assessment report of the IPCC, released in 2014, has clearly revealed that increases in the emission of GHGs have resulted in the warming of climate systems by 0.85 [0.65–1.06]°C. It has further projected that temperature increases by the end of this century are likely to be in the range 2–4.5°C. It is expected that future tropical cyclones will become more intense, with larger peak wind speeds and heavier precipitation. Himalayan glaciers and snow cover are projected to contract. It is also very likely that hot extremes, heatwaves and heavy precipitation events will continue to become more frequent. Increases in the amount of precipitation are expected more in high latitudes, while decreases are likely in most subtropical regions. At the same time, the projected sea-level rise by the end of this century is likely to be between 0.18 and 0.59 m. Fresh water availability in central, south, east and south-east Asia, particularly in large river basins, is projected to decrease due to climate change, which, along with population growth and increasing demand arising from higher standards of living, could adversely affect more than a billion people by the 2050s.

Positive Effects of Climate Change

Climatic changes are affecting agriculture through their direct and indirect effects on crops, soils, livestock and pests, and hence global food security. Climate change influences agriculture production differently in different areas, with a decrease of production in tropical and subtropical areas and increase in temperate areas. Thus, climate change is expected to impact more in already vulnerable developing regions with relatively lower technical and economic capacity to respond to these threats. There is evidence of negative impacts of climate change on the yield of cereals and other crops with variable magnitude in diverse ecologies. Studies in India indicate that a moderate increase in temperature will have a substantial impact on rice, wheat and maize yield (Aggarwal and Rani, 2009; Aggarwal et al., 2010). For example, in 2004, due to high temperatures, wheat crop matured by 10–20 days earlier in India, leading to a loss of more than 4 million t of wheat production. The impact may be further worsened by increasing water scarcity, frequent floods, heat stress and droughts, and declining soil organic carbon content. Extreme events including floods, droughts, forest fires and tropical cyclones have already increased in temperate and tropical Asia in the last few decades. Run-off and water availability are projected to decrease in the arid and semi-arid regions of Asia. Sea-level rise and an

increase in the intensity of tropical cyclones are expected to displace tens of millions of people in the low-lying coastal areas of Asia, with expectation of around 17% of land being inundated in Bangladesh alone. On the contrary, the increased intensity of rainfall and contraction of the monsoon period would increase flood risk in temperate and tropical Asia. Lobell et al. (2012), in their study from northern India, show that wheat production will face significant losses due to high temperatures. As stated earlier, it is more likely that more than 50% of the Indo-Gangetic plains may become unsuitable for wheat due to increased heat stress by 2050 (Ortiz et al., 2008).

Increase in CO_2 concentration can cause CO_2 fertilization, and it has been shown to increase crop growth, dry-matter production and yield in specific regions, although this effect is related to frequency of water stress or changes in climatic factors such as temperature or rainfall, and nutrient status. Most crops grown under an enriched CO_2 environment showed increased growth and yield as enhanced CO_2 affects the growth and physiology of crops, enhancing photosynthesis and water-use efficiency. Differences in physiology of C3 and C4 plants make C4 plants more photosynthetically efficient than C3 ones, especially when the level of CO_2 is high. If the direct effect of CO_2 is included, yields are projected to increase for rainfed crops under both the A2 and B2 emissions scenarios (as per IPCC) in the 2080s. The increase is likely to be highest for rainfed maize under the A2 scenario, possibly because the higher CO_2 concentration would boost the yield of rainfed maize under the current water-limited conditions prevalent in some regions of the country. In crop models, it has been seen that yields have slightly increased in maize. Simulated yield of winter maize showed an increase from the baseline in the range of 8.4–18.2%, 14.1–25.4% and 23.6–76.7% for 2020, 2050 and 2080, respectively. Maize with increased CO_2 and concurrent rise in temperature showed a decrease in duration and days to anthesis from the baseline, with total dry-matter production, grain weight and grain number showing an increase from the baseline to 2080. Some of the other positive effects include a shift in the area of cultivation of some crops so as to create new economic and market zones that might benefit the people of the region. This has been particularly observed in some horticultural crops in temperate regions, an example is the shift of apple from lower altitudes to higher altitudes, due to which farmers in the lower elevations have taken up cultivation of fruit crops like pomegranate and kiwi and also have taken up commercial vegetable production and floriculture. Similarly, mango cultivation has been seen to shift to places of slightly lower temperature, thereby contributing to extended availability in the market. One of the other possible positive effects of climate change can be seen in protected cultivation of horticultural crops, which provide unique opportunities for assured, climate-resilient and enhanced production of quality products. The technology is expected to be in greater demand as it can create job opportunities for the unemployed.

Adverse Effects of Climate Change

The recent report of the IPCC has particularly highlighted the vulnerability of developing countries of the Asian region, especially its mega-deltas, to increasing climate change and variability due to its large population, predominance of agriculture, large climatic variability and limited resources to adapt. There are likely to be negative effects also on livestock productivity due to increased heat stress, lower pasture productivity and increased risks due to animal diseases. Increase in sea surface temperature and acidification will also lead to changes in marine species distribution as well as production.

We need to prioritize adaptation options in key sectors, such as stress-tolerant crops, storm-warning systems, water storage and diversion, contingency planning and infrastructure strengthening. Alternative agricultural practices as adaptive measures should emanate from the search for indicative adaptation options with a focus on prevailing farmers' practices in different areas with varying degrees of vulnerability (e.g. water scarcity or aridity) and other environmental constraints. Large areas of rainfed agriculture in this country serve mostly as a sink rather than a source of emissions. Although enteric fermentation remains a major source of GHG emissions, there are large opportunities to reduce these emissions through better feeding and manure management. There is a need to come up with guidelines on crop and animal husbandry practices that ensure reasonably high productivity while minimizing the GHG emissions. Location-specific, usable scientific results will form an important part of strategies to combat climate variability. The focus should be on the dynamics, diversity and flexibility of adaptations, which can harness the opportunities in the changing economic, technological and institutional scenarios. The maximization of agriculture's mitigation potential and adaptation measures will necessitate investments in technological novelty. Increased efficiency of inputs and creation of incentives and monitoring systems that are inclusive of small and marginal farmers will play an important role in the success of both mitigation and adaptation, especially adaptation-led mitigation.

Climate modelling, coupled with socioeconomic scenarios, forms a useful tool for exploring the long-term consequences of climate change and adopting this approach would also help in evaluating the available response options. The use of socioeconomically driven models would allow a cohesive perspective with emphasis on economic viability on mitigation and adaptation options, which is important because the cost of adaptation and mitigation will decide the success of the strategy. Development of shared socioeconomic pathways and integrated socioeconomic scenarios will be a useful focal point for collaborative efforts between integrated assessment and impact, adaptation and vulnerability researches. Models in the future would be gradually refined as the understanding of factors influencing migration behaviour, such as risk-awareness, social networks and labour-market connections, is enhanced. Several adaptation measures have been developed by the NARS, consisting of ICAR institutes and SAUs.

Strategies for Coping with Global Climate Change

The facts emerging from the situation analysed above draw global concern over, and a sense of urgency in addressing, the options by which threats to Asian agriculture due to climate change can be met successfully in the near future. On the positive side, the agriculture sector also provides significant potential for greenhouse gas mitigation and adaptation to climate change. This, however, would demand a reorientation of agricultural research that would comprehensively address all urgent concerns of climatic change through well-defined adaptation and mitigation strategies, which could help maximize food production, minimize environmental degradation and achieve socioeconomic development. Coping with global climate change is a must, and for that there are two strategies: (i) adaptation through learning to live with the new environment (e.g. time of planting, changing varieties, new cropping systems); and (ii) mitigation through offsetting the causative factors, such as reducing the net emission of GHGs.

Adaptation strategies

The potential strategies and actions for adaptation to climate change effects could be as follows:

Climate-resilient genotypes

- Intensify the search for genes for stress tolerance across the plant and animal kingdoms.
- Intensify research efforts on marker-aided selection and transgenic development.
- Develop genotypes for biotic (diseases, insects etc.) and abiotic (drought, flood, heat, cold, salinity) stress management by traditional plant breeding and/or genetic modification.
- Attempt transforming C3 plants to C4 plants.

Science-based resilient land-use systems

- Shifting of cropping zones and cropping systems optimization.
- Critical appraisal of agronomic strategies and evolving new agronomy for climate change scenarios.
- Exploring opportunities for maintenance/restoration/enhancement of soil properties.
- Use of multi-purpose adapted livestock species and breeds.

Value-added weather management services

- Developing spatially differentiated operational contingency plans for temperature and rainfall-related risks, including supply management through market and non-market interventions in the event of adverse supply changes.
- Enhancing research on applications of short-, medium- and long-range weather forecasts for reducing production risks.
- Developing knowledge-based decision-support systems for translating weather information into operational management practices.
- Innovations in risk management like crop insurance, using ICT and weather forecasting; bundling of risk management interventions; PPP models in strengthening big data analysis of climate-specific management options; and initiating climate-sensitive extension services as well as climate site-specific advisory systems for desired impact.
- Developing pest- and disease-forecasting systems covering a range of parameters for contingency planning and effective disease management.

Integrated study of the 'climate change triangle' and 'disease triangle', especially in relation to viruses and their vectors

- Develop dedicated high-tech centres of excellence for advanced studies on the 'climate change triangle'.
- Develop human resource capacity to provide timely and adequate training to all stakeholders to understand the relationship and forecasting of incidence and intensity of viruses and their vectors in relation to climatic risks and extremes.

Documentation of indigenous traditional knowledge (ITK) and exploring opportunities for its utilization

- There is plenty of traditional knowledge and wisdom available on coping with climatic extremes, which should be used effectively in developing technologies and strategies on climate change adaptation.
- Indigenous traditional knowledge should be one of the key components of the national mission on traditional agriculture initiated by the Ministry of Agriculture and Farmers' Welfare.

Reforming the global food system

- Food habits, globally, have undergone a significant transformation, whereas agricultural production systems were designed based on traditional food habits; hence there is a trade-off in demand and production of food systems under local conditions.
- There is a need for significant reform in food systems to meet local food requirements as well as current food systems, in view of their climate sensitivity and adaptability.

Mitigation strategies

The basic strategies for mitigating climate change effects are reducing and sequestering emissions. However, before jumping on bandwagon mitigation strategies, the following points should be considered for effective implementation of mitigation strategies:

- Improve inventories of emission of GHGs using state-of-the-art emission equipment coupled with simulation models and GIS for upscaling.
- Evaluate carbon sequestration potential of different land-use systems including opportunities offered by conservation agriculture and agroforestry.
- Critically evaluate the mitigation potential of biofuels and the genetic improvement and use of engineered microbes.
- Identify cost-effective opportunities for reducing methane generation and emission in ruminants by modification of diet, and in rice paddies by water and nutrient management.
- Renew focus on nitrogen fertilizer-use efficiency with the added dimension of nitrous oxide mitigation.
- Assess biophysical and socioeconomic implications of mitigation of proposed GHG-mitigating interventions before developing a policy for their implementation.
- In order to identify the most cost-effective and feasible mitigation interventions in agriculture without compromising its primary goal of food security, it is important to identify the emission hotspots, understand the major contributors of emissions and

explore possible mitigation options along with the associated cost of adopting such mitigation options. This helps to understand the links between various production practices and GHG emissions, identify mitigation responses and facilitate more informed policy formation that is consistent with the food-security and economic-development priorities of countries.

Reducing emissions

Strategies for reducing emissions include:

- *Avoid deforestation.* Forestry and other land uses contribute nearly 11% of the total GHGs. Hence, afforestation efforts can contribute significantly to GHG reduction.
- *Minimize soil-erosion risks:* Management systems and scientific land-use plans need to be developed and deployed for reducing the emissions contributed by soil erosion.
- *Eliminate biomass burning and incidence of wild fires:* Biomass burning has become one of the major challenges and contributes significantly to air pollution. Efficient management of crop residues has demonstrated ample scope for reducing emissions from agriculture for which well-tested, efficient, viable and scalable technological options (e.g. happy turbo-seeder technology) are available for scaling up, and to avoid rice-straw burning in the Indo-Gangetic plains.
- *Improve input-use efficiency (e.g. fertilizers, energy, water, pesticides):* Of the total environmental footprint of agriculture, nitrogen and water contribute a major share. Hence, precision water and nitrogen management in agriculture can contribute to significant reductions in the environmental footprint of farming.
- *Conservation agriculture:* CA is one of the significant natural resource management (NRM) innovations, and has demonstrated potential benefits not only for adaptation to climatic risks but also for mitigation of GHGs, and hence has recognized climate-smart agriculture (CSA) globally.
- *Solar energy:* Massive efforts are being made to generate solar energy (green energy), which has significant potential for reducing GHG emissions.

Sequestering emissions

Soil is one of the major sinks of CO_2. Soil carbon sequestration has been studied for well over 150 years, and its importance for supporting many agronomic and ecological functions has been recognized. The vital role of soil carbon sequestration in the global carbon cycle, and hence its role in climate change, began attracting interest in the 1980s. Over the past 30–40 years, soil carbon sequestration has gone from an obscure footnote of bio-geochemistry/climate science, to centre-stage in actions to curb climate change, as exemplified by the '4 per 1000' initiative arising from the Paris Climate Agreement. The stored soil carbon is vulnerable to loss through both land-use management change and climate change. There are numerous agricultural sources of GHG emissions with hidden C costs of tillage, fertilizer, pesticide use and irrigation. In general, net C sequestration must take into account these costs. The important strategies of soil C sequestration include restoration of degraded soils and adoption of best management practices (BMPs) of agricultural and forestry soils. For example, in India, the potential of soil C sequestration is estimated at 39–49 (44 ± 5) Tg C p.a., 7–10 Tg C p.a. for restoration of degraded soils and ecosystems, 5–7 Tg C p.a. for erosion control, 6–7 Tg C p.a. for adoption of BMPs on agricultural soils and 22–26 Tg C p.a. for secondary carbonates (Lal, 2004). Therefore, agricultural practices, collectively, can make a significant contribution at low cost to increasing soil carbon sinks and reducing GHG emissions.

A large proportion of the mitigation potential of agriculture (excluding bioenergy) arises from soil carbon sequestration, which has strong synergies with sustainable agriculture and generally reduces vulnerability to climate change. A considerable mitigation potential through sequestration is available from reductions in methane and nitrous oxide emissions in some agricultural systems. However, there is no universally applicable list of mitigation practices and the mitigation through sequestration practices needs to be evaluated for individual agricultural systems and settings (e.g. CA). The biomass from agricultural residues and dedicated energy crops can be an important bioenergy feedstock, but its contribution to climate mitigation to 2030 depends on demand for bioenergy from transport and energy supply, on water availability, and on requirements of land for food and fibre production. Hence, widespread use of agricultural land for biomass production of energy may compete with other land uses and can have positive and negative environmental impacts and implications for food security.

Warning against Complacency and Catalysing Stakeholders

The APAARI, which has been instrumental in promoting regional cooperation for agricultural research in the Asia-Pacific region, has been organizing a series of expert consultations (APAARI, 2009; 2012) for debating on emerging issues vis-à-vis agricultural research and development (ARD) concerns in the Asia-Pacific region. In this endeavour, biofuel and climate change were identified as major themes during the expert consultation on Research Need Assessment organized by APAARI during 2006. Accordingly, the issue of climate change and its imperatives for agricultural research in the Asia-Pacific region were deliberated in an international symposium organized jointly by APAARI and Japan International Research Center for Agricultural Sciences (JIRCAS). Participants representing NARS, CGIAR, IARCs, Global Forum on Agricultural Research (GFAR), Australian Centre for International Agricultural Research (ACIAR), Advanced Research Institutes (ARIs), universities and regional forums from 30 countries came out with agricultural research priorities for adapting agriculture to climate change in the form of the Tsukuba Declaration on Adapting Agriculture to Climate Change (APAARI, 2009).

Tsukuba Declaration on Adapting Agriculture to Climate Change

- The Asia-Pacific region sustains almost half the global population, with high rates of population growth and poverty. Agriculture continues to play a critical role in terms of employment and livelihood security in all countries of the region. At the same time, this region has the largest concentration of hungry and malnourished people in the world. Droughts, floods, heatwaves and cyclones occur regularly. Climate change is likely to raise regional temperatures and lead to decline in fresh water availability, sea-level rise and glacial melting in the Himalayas. The IPCC has considered the developing countries of the Asia-Pacific region, especially the mega-deltas of Asia, as very vulnerable to climate change.

- Attainment of SDGs, particularly alleviating poverty, assuring food security and environmental sustainability against the background of declining natural resources, together with a changing climate scenario, presents a major challenge to most of the countries in the Asia-Pacific region during the 21st century.

- Water is a key constraint in the region for attaining food production targets and will remain so in future. Steps are therefore needed by all stakeholders to prioritize enhancing water-use efficiency. In addition, measures for water storage using proven approaches such as small on-farm ponds, large reservoirs, groundwater recharge and storage, and watershed approach managed by the farming communities, require attention.

- It was fully recognized that increasing food production locally will be the best option to reduce poor people's vulnerability to climate change variations. Available agricultural technologies can help increase the yield potential of crops that has not yet been tapped in many countries of the Asia-Pacific region. Hence, concerted efforts, backed by policy makers at the national level, would be the key to enhancing food security as well as ensuring agricultural sustainability.

- New genotypes tolerant to multiple stresses – drought, flood, heat, salinity, pests and disease – will help further increase food production. This would require substantial breeding and biotechnology- (including genetically modified varieties) related efforts based on collection, characterization, conservation and utilization of new genetic resources that have not been studied and used. CGIAR centres, ARIs and NARS of the region have a major role to play in this context. This will require substantial support in terms of institutional infrastructure, human resource capacity and the required political will to take up associated agricultural reforms. Therefore, there is a need to fervently call upon national policy makers, overseas development agencies, other donor communities and the private sector to increase funding support for agricultural research and development in the Asia-Pacific region.

- It was also recognized that a reliable and timely early warning system of impending climatic risks could help determination of the potential food-insecure areas and communities. Such a system could be based on using modern tools of information and space technologies and is especially critical for monitoring cyclones, floods, drought and the movements of insects and pathogens. Advanced Research Institutions, such as JIRCAS, could take the lead in establishing an Advanced Centre for Agricultural Research and Information on Global Climate Change for serving the Asia-Pacific region.

- The increasing probability of floods, droughts and other climatic uncertainties may seriously increase the vulnerability of resource-poor farmers of the Asia-Pacific region to global climate change. Policies and institutions are needed that assist in containing the risk and providing protection against natural calamities, especially for small farmers. Weather crop/livestock insurance, coupled with standardized weather data collection, can greatly help in providing alternative options for adapting agriculture to increased climatic risks.

- Governments of the region should collaborate on priorities to secure effective adaptation and mitigation strategies and their effective implementation through creation of a regional fund for improving climatic services and for effective implementation of weather-related risk-management programmes. Active participation of young professionals is also called for.

- It was recognized that there are several possible approaches to enhancing carbon sequestration in the soils of the Asia-Pacific region such as greater adoption of scientific soil- and crop-management practices, improving degraded lands, enhancing fertilizer-use efficiency and large-scale adoption of CA. To be effective, these would require simultaneously improved use of inputs such as fertilizers, crop residues, labour and time. This soil carbon sequestration has the added potential advantage of enhancing food security at the national/regional level. We urge the global community to ensure appropriate pricing of soil carbon and related ecosystem/environmental services in order to motivate small farmers to adopt new management practices that are linked to proper incentives and rewards.

- APAARI has been instrumental in stimulating regional cooperation for agricultural research in the Asia-Pacific region. Global climate change and its implications for agriculture underline the need for such an organization to become even more active at this juncture. APAARI, in collaboration with its stakeholders, especially CGIAR centres, ARIs, GFAR and other regional forums, should continue facilitating regional collaboration in a consortium mode and take advantage of new initiatives such as the Challenge Programme on Climate Change

for building required capability to adapt and mitigate the effects of climate change and ensure future sustainability of all concerned in the region. The deliberations also led to identification of research priorities and both adaptation and mitigation strategies to deal with the challenge of climate change.

Epilogue

The impact of climate change on agricultural production in the Asia region is real. Hence, immediate action at sub-national, national and regional level to understand and address the issues of climate change becomes a priority. Strategy around both adaptation and mitigation is urgently called for, which would require research reorientation and major policy interventions to increase investment. Regional and global collaboration would help in effectively addressing these concerns and building both institutional and human resource capabilities, the two cradles for sustainable agriculture.

References

Aggarwal, P.K. and Rani, D.N.S. (2009) Assessment of climate change impacts on wheat production in India. In: Aggarwal, P.K. (ed.) *Global Climate Change and Indian Agriculture: Case Studies from the ICAR Network Project*. Indian Agricultural Research Institute, New Delhi, pp. 5–10.

Aggarwal, P.K., Kumar, S.N. and Pathak, H. (2010) *Impacts of Climate Change on Growth and Yield of Rice and Wheat in the Upper Ganga Basin*. Worldwide Fund for Nature (WWF) and Indian Agricultural Research Institute (IARI), New Delhi.

APAARI (2009) International Symposium on Global Climate Change: Imperatives for Agricultural Research in the Asia-Pacific: Proceedings and Recommendations. Tsukuba, Japan, 21–22 October 2008.

APAARI (2012) Workshop on Climate Smart Agriculture in Asia: Research and Development Priorities: Proceedings and Recommendations. Bangkok, 11–12 April.

IPCC (2007) Intergovernmental Panel on Climate Change (IPCC). *Fourth Assessment Report: Climate Change*. IPCC, Geneva, Switzerland.

IPCC (2012) Managing the risks of extreme events and disasters to advance climate change adaptation. In: Field, C.B., Barros, V., Stocker, T.F. *et al* (eds) *A Special Report of Working Groups I and II of the Intergovernmental Panel on Climate Change*. Cambridge University Press, Cambridge, UK.

IPCC (2014) Summary for policymakers. In: Edenhofer, O., Pichs-Madruga, R., Sokona, Y., Farahani, E., Kadner, S. *et al.* (eds) *Climate Change 2014: Mitigation of Climate Change*. Contribution of Working Group III to the Fifth Assessment Report of the Intergovernmental Panel on Climate Change. Cambridge University Press, Cambridge, UK.

Jat, M.L., Dagar, J.C., Sapkota, T.B., Singh, Y., Govaerts, B. *et al.* (2016) Climate change and agriculture: adaptation strategies and mitigation opportunities for food security in South Asia and Latin America. *Advances in Agronomy* 137, 127–235.

Lal, R. (2004) Soil carbon sequestration: impacts on global climate change and food security. *Science* 304, 1623–1627.

Lal, R. (2016) Feeding 11 billion on 0.5 billion hectare of area under cereal crops. *Food and Energy Security* 5(4), 239–251.

Lobell, D.B., Sibley, A. and Ivan Ortiz-Monasterio, J. (2012) Extreme heat effects on wheat senescence in India. *Nature Climate Change* 2, 186–189.

Ortiz, R., Sayre, K.D., Govaerts, B., Gupta, R., Subbarao, G.V. *et al.* (2008) Climate change: Can wheat beat the heat? *Agriculture, Ecosystems and Environment* 126, 46–58.

Tubiello, F.N., Salvatore, M., Cóndor Golec, R.D., Ferrara, A., Rossi, S. *et al.* (2014) *Agriculture, Forestry and Other Land Use Emissions by Sources and Removals by Sinks*. Working Paper Series. Food Agriculture Organization of the United Nations. DOI: 10.13140/2.1.4143.4245.

UNEP (2015) *Indicators for a Resource Efficient and Green Asia and the Pacific – Measuring Progress of Sustainable Consumption and Production, Green Economy and Resource Efficiency Policies in the Asia-Pacific Region*. United Nations Environment Programme, Bangkok.

WMO (2009) *Greenhouse Gas Bulletin* no. 5. World Meteorological Organization, Geneva, Switzerland.

24

Towards Climate-smart Agriculture

The Asia-Pacific region is agriculturally vibrant. However, climate change is a major challenge as well as a threat to food security, sustainable livelihood and biodiversity. Weather variability influences agriculture in all its dimensions, adversely – biologically, physically and chemically, which are critical elements for production, productivity and profitability on farms (including livestock and fisheries). The predicted climate change by 2050 would further reduce agricultural production by 10–20%, while demand would increase by 70% (FAO, 2009; Nelson *et al.*, 2009). The existing scenario and models indicating increased frequency of droughts, floods and temperature-related events of climate change undoubtedly would have serious implications for future food availability as well as global nutritional, environmental and political security. Climate change would intensify degradation processes of natural resources (central to meeting increased food demand), and changing land-use patterns, natural-resource degradation (especially land and water), urbanization and increasing pollution would affect the ecosystem of the region directly as well as indirectly. Moreover, the region's agrarian landscape is predominantly one of smallholder farmers. At present, more than 650 million people, half the world's poor (income US$1/day), are hungry and poor. It is also estimated that by 2020, foodgrain requirement in the Asia-Pacific region would be between 30% and 50% more than for 2012. Moreover,

agriculture, forestry and land use account for ~26% of global GHGs. Hence, it is critical that all-out efforts are made now to adopt CSA in order to reduce GHG emissions. Agriculture must become more productive, more resilient and more climate-friendly.

The ever-increasing population, interlinked with natural-resource degradation, can potentially add further to adversity due to climatic risks, leading to a large number of people becoming vulnerable to climate change. During the past half-century (1965–2015), in the process of achieving multi-fold increases in food production, inefficient use and inappropriate management of resources (water, energy, agrochemicals), mainly to achieve the Green Revolution, have impacted vastly the quality of natural resources, and contributed to climatic variability affecting farming adversely. Increasing climatic variability also affects most of the biological, physical and chemical processes involved in the productivity enhancement of agricultural systems, including livestock and fisheries.

Adapting to Climate Change

Adaptation to climate change is no longer an option, rather a compulsion to minimize losses due to adverse impacts of climate change and to reduce vulnerability. Agricultural technologies that promote sustainable intensification and adaptation to emerging climatic variability along with mitigating GHG emissions are known as climate-smart agriculture (CSA). Hence, the scientific R&D initiatives for both adaptation and mitigation must now be prioritized. To ensure global peace, equity and prosperity under the growing challenges of climate change, increased investment to promote CSA becomes critical at national, regional and global level. This would require effective participation of all stakeholders: farmers, policy makers, processors, private sector and consumers, with special emphasis on the following.

Adaptation and contingency planning

As a strategy, both adaptation and mitigation are the twin approaches to address the adverse impacts of climate change. Of the two, the former is an easy option to be adopted on a large scale in different agro-ecologies. Breeding of crop varieties and hybrids that are early in maturity, tolerant to both biotic and abiotic stresses, nutritionally superior and having consumer acceptability will ensure greater adaptation to varying climate uncertainties. By reducing the maturity period of chickpea from 150–180 days to as little as 75–80 days enabled its cultivation to spread from north to south India. Similarly, nobalization of sugarcane, leading to shorter duration and drought resistance, enabled it to grow even in the sub-temperate regions of north India. The same is true for pigeonpea and many other crops. In future, concerted efforts to breed varieties that are drought-, salt- and heat-tolerant would enable farmers to cope well with the ill effects of climate change. Also, greater emphasis needs to be given to short-duration crops that can fit well into diverse cropping systems in case the monsoon is either delayed or insufficient. Establishment of seed banks with the option of growing short-duration crops if the monsoon is late could be an effective contingency plan. Similarly, CA, micro-irrigation, fertigation and sub-surface drip-irrigation systems are the available options for CSA. CSA would certainly need advance weather forecasting, advance crop planning, stocking of seeds in seed banks and climate-smart tools and techniques for need-based use of fertilizers, chemicals and energy for agricultural operation. In this context, better knowledge dissemination, timely availability of inputs, credit and insurance against climatic risks would be necessary to support farmers, mainly in high-climate risk-prone areas.

Making 'grey' areas 'green'

To combat the twin problems of meeting food, nutrition and energy needs, increasing population and depleting natural resources, there is an urgency to focus on CA, diversification and resilience through good agronomic practices to attain the Evergreen Revolution. Time is ripe for laying greater emphasis on rainfed agriculture (covering still more than 55% of total area), which is critical for sustainability, improving livelihood and enhancing the income of resource-poor farmers. For better risk management, diversified agriculture, silvipastoral approaches through crop–livestock integration, arid horticulture and agroforestry have to be adopted. Accordingly, a paradigm shift is needed in rainfed agriculture towards integrated natural resource management (INRM). Soil and water conservation, precision in weather forecasting, resilient crop varieties/breeds and a farming-systems approach based on scientific land-use planning are approaches around CSA that will enable making 'grey' areas 'green'.

Outscaling innovations for resilience

Significant efforts have also been made to develop location-specific CSA practices through large national and international initiatives and programmes like the CGIAR research programme on Climate Change, Agriculture and Food Security (CCAFS), and the National Innovation on Climate Resilient Agriculture (NICRA) of the Indian Council of Agricultural Research (ICAR). The CSA technologies and learning from such programmes need to be replicated through enabling policies and increased investments. Efforts need to be made to capture farmer-led innovations on climate-smart agricultural practices and blend them with modern science. The learning through community-based approaches in climate-smart villages (CSVs) are good examples of blending science and society for participatory learning and building evidence on CSA (Jat, 2017), which needs to be replicated over large areas. For addressing issues of resource fatigue and bridging existing yield gaps, the recommendation domain of the best-bet CSA practices, resource mapping and characterization using new tools and techniques like remote sensing and GIS would help considerably. Documenting success stories of potential climate-smart technologies like stress-tolerant genotypes/hybrids, better feeding management of locally well-adapted livestock breeds, CA, laser land-levelling, micro-irrigation systems, use of customized fertilizer nutrients and replicating them in similar ecologies, production systems and farm conditions, through functional regional networks, should receive high priority.

Knowledge-sharing for CSA

Climate change is a complex issue and the solutions to it are not straightforward and are rather knowledge-intensive. As R&D efforts on CSA are relatively new, special focus on knowledge sharing and capacity development of a range of stakeholders is urgently needed. There have been efforts by several organizations and countries to develop, validate and promote CSA technologies. There are also several success stories from different production ecologies in the region that need to be shared. Also, there is a need to learn from each other's successes as well as failures so that farmers need not 'reinvent the wheel'. On the contrary, they should build on the existing knowledge base to take advantage of CSA innovations that have already made an impact at scale. In this context, greater awareness of the use of decision-support systems like green seeker, leaf-colour chart, CA using zero till, laser land-levelling, micro-irrigation, sub-surface drip irrigation, fertigation and diversification of cropping systems will help in faster adoption of new technologies to combat climate change.

Future Road Map

Based on several consultations and stakeholder engagements, some declarations such as: the Tsukuba Declaration on Adapting Agriculture to Climate Change (APAARI, 2008), the Suwon Agrobiodiversity Framework (Bioversity International, 2010), the Bangkok Declaration on Reorienting Agriculture Research for Development (APAARI, 2009), the Delhi Declaration on

Agrobiodiversity (ISPGR, 2016) and the Inter-Drought-V Hyderabad Declaration on Climate-Smart Agriculture have been endorsed to promote technologies to counter the effects of climate change on agriculture, farmers' livelihoods and national food and nutritional security. Recognizing the SDGs to end hunger, achieve food security, improve nutrition and make farming resilient and sustainable, the following road map is being proposed:

- Innovations addressing the food-water-energy nexus and regional cooperation and integration are required to address various issues relating to CSA. The key areas of intervention should be climate-resilient food-value chains, rural market infrastructure, use of renewable energy and improving rural livelihoods. For these, a multi-sectoral approach is needed with a focus on value addition and partnership.

- Concerted efforts are needed to make best use of available knowledge and technologies through defining recommendation domains (technology targeting), increased investments (double) in managing land and water resources and strengthening input delivery as well as market-linkage mechanisms.

- Efforts are needed to manage current climatic risks for poverty alleviation and equitable development through effective use of climate-related services, ranging from satellites to mobile phones; use of cloud sourcing and cloud computing for weather forecasting (short-, medium- and long-range) and early warning systems, and linking these to real-time agro-advisories and input-output markets. Innovations are needed in risk management, like crop insurance using ICT and weather forecasting; bundling of risk management interventions; PPP models in strengthening big-data analysis of climate-specific management options; initiating climate-sensitive extension services and climate-site-specific advisory systems for desired impact. Promotion of climate-smart technologies is critical, like improved varieties/hybrids/breeds of crop and livestock, and management practices (tillage, residues, water, nutrient, machinery, housing) adapted to diversity of production systems and farmers' local conditions, rather than *ad hoc* recommendations across wider geography/landscape.

- There is an urgent need to identify and exploit potential benefits of climate change, as it may not be all negative in its impact. For example, change in temperature zones, increase in rainfall and shorter crop duration would have adaptation domains that can be utilized using science-based evidence.

- Improved targeting of technologies and policies are the missing links for contingency planning/climate risk management. For example, establishment of seed banks to manage possible climatic risks and provision of crop insurance in high-climate risk-prone areas.

- Maximization of synergies among interventions is urgently required by developing farming-systems approaches, rather than single-technology-/commodity-centric initiatives as in the past. There is a need to develop CSA-enabled development plans and prioritize interventions using the following:
  - Build biophysical and socioeconomic datasets from the bottom up.
  - Use an integrated modelling framework for making interventions spatially explicit.
  - Address climatic and socioeconomic scenarios holistically.
  - Support multi-objective trade-off analyses.
  - Support more informed decision-making for: (i) what crops are to be cultivated; (ii) which CSA technologies and practices are to be followed; (iii) where to target future investments; and (iv) when and how those investments are to be made.

- There is need to prioritize CSA interventions that have multiple wins for addressing simultaneously poverty, governance, institutions and human capital, being critical for inclusive agricultural growth.

- For making climate-smart agricultural practices and innovations relevant to the smallholder farmers under their local circumstances, systems research through participatory approaches needs to be emphasized, whereby cropping/farming system modelling extends and adds value to traditional agronomic practices.

- Spatial data infrastructure is required to support integration of data assets to target technologies, track progress and integrate development sectors to ensure economic, social and environmental sustainability against the backdrop of increased inter-annual weather variability. Availability of spatial and temporal data assets should be shared under the Global Open Data for Agriculture and Nutrition (GODAN) and to similar open-data initiatives.

- Concerted efforts are needed for strengthening private–public producer partnerships to research, in collaboration with IT and finance sectors, genotype × environment × management (GEM) in relation to adaptation to weather risks and raising productivity and profitability for farmers.

- Non-linear models and approaches of technology delivery including technology-led business and service should complement the existing public extension system to accelerate last-mile delivery of CSA innovations and their impact at scale. Accordingly, there is an urgent need to institutionalize new business/extension models for scaling CSA at the local level, involving village-governance systems, including women and youth. This can be reinforced by appropriate incentives based on performance and outcomes based initiatives. The role of all stakeholders, especially NGOs and the private sector, is critical and needs to be encouraged through an enabling policy environment.

- Special efforts are also needed to strengthen existing extension services, including those relating to gender issues. There is also a need to build the capacity of tomorrow's farmers through regional knowledge platforms on CSA, ICT-based dissemination of knowledge, participatory videos, dedicated television channels on agriculture, social media and community radio featuring local content and demonstrations. Emphasis on vocational training to build a new cadre of technology agents to provide knowledge without dissemination loss and services on a custom-hire basis will be critical for desired success.

- There is a need for establishment of ecoregion-specific platforms for innovation, communication, cross-learning and capacity development on CSA as climate change is more of a transboundary issue and does not remain confined to geographical/political boundaries.

- Mainstreaming CSA in the basic education system, from school to university would help build much-needed general public awareness.

- There is an urgent need to adopt an aggressive approach for policy and institutional reforms necessary for scaling innovations associated with CSA through appropriate incentive mechanisms for carbon trading by smallholder farmers as a reward for their environmental services in the national interest.

- There is an urgent need to document scalable evidence and impact of CSA approaches/technologies as success stories for wider dissemination as well as for benefits across similar ecologies and landscapes.
- Targeting/implementing scale for CSA should be a climate-smart village, which would be helpful not only for implementation but also for monitoring, evaluation and learning about CSA as well as linking community to input and output markets, a must for adopting CSA.

The Way Forward

Agriculture in the Asia-Pacific must liberate the region from the twin scourge of hunger and poverty and of malnutrition of children and women. Accelerated science and innovation-led agricultural growth must be inclusive in the region and should address the needs and aspirations of resource-poor smallholders. Under the growing challenges of resource degradation, escalating input crises and costs, with the overarching effects of global climate change, the gains in foodgrain production would largely depend on a paradigm shift from integrated germplasm improvement to integrated natural resource management with the focus on ecoregional/farmer typology-specific climate-smart agricultural practices (CSAPs), climate services (especially weather- and market intelligence-related capacity and knowledge) and an enabling policy environment. Future AR4D efforts by the NARS should be reorientated towards a farming-systems approach, CA for sustainable intensification (CASI), CSA, and linking farmers to market. More importantly, it must bridge the income divide between farmers and non-farmers and benefit, equally, the consumers. Developing countries in the Asia-Pacific region would have to increase their investment in agricultural research and innovation for development (ARI4D) to address future challenges and to ensure food, nutrition and environmental security for all in the region.

References

APAARI (2008) Tsukuba Declaration on Adapting Agriculture to Climate Change. Symposium on Global Climate Change: Imperatives for Agricultural Research in Asia-Pacific, 21–22 October, International Congress Center, Tsukuba, Japan.

APAARI (2009) Bangkok Declaration on Reorienting Agricultural Research for Development in Asia-Pacific Region. Regional Consultation on Agricultural Research for Development (AR4D) in the Asia-Pacific, Bangkok, 30–31 October.

Bioversity International (2010) The Suwon Agrobiodiversity Framework: The Way Forward for Managing Agrobiodiversity for Sustainable Agriculture in the Asia-Pacific Region. International Symposium on Sustainable Agriculture Development and Use of Agrobiodiversity in the Asia-Pacific Region, 13–15 October, Suwon, Republic of Korea.

ISPGR (2016) Delhi Declaration on Agrobiodiversity Management. 1st International Agrobiodiversity Congress 2016, 6–9 November, New Delhi.

FAO (2009) How to Feed the World 2050. High-level expert forum on climate change and bioenergy challenges for food and agriculture. Food and Agriculture Organization of the United Nations. Available at: http://www.fao.org/fileadmin/templates/wsfs/docs/Issues_papers/HLEF2050_Climate.pdf (accessed 22 June 2018).

Jat, M.L. (2017) Climate smart agriculture in intensive cereal based systems: scalable evidence from the Indo-Gangetic plains. In: Belavadi, V.V., Nataraja Karaba, N. and Gangadharappa, N.R. (eds) *Agriculture under Climate Change: Threats, Strategies and Policies*. Allied Publishers Pvt Ltd, India, pp. 147–154.

Nelson, G.C., Rosegrant, M.W., Koo, J., Robertson, R., Sulser, T. *et al.* (2009) Climate Change: Impact on Agriculture and Costs of Adaptation. Food Policy Report. International Food Policy Research Institute, Washington, DC.

25

Linking Research with Extension for Accelerated Agricultural Growth

The Asian region is rich in natural resources, human capital and indigenous knowledge, and much faster progress can be achieved if innovations are outscaled on farmers' fields. This chapter draws attention to issues concerning the need for linking research with extension for faster agricultural growth in Asia. The Asian region is agriculturally vibrant. With 38% of the world's total agricultural land, it houses 80% of smallholder farmers supporting 74% of the world's agricultural population. The region encompasses 39 countries, including 19 commonwealth members with two of the world's most populous countries, China (1.41 billion) and India (1.34 billion). With 3.5 billion people, the region accounts for 58% of the world's population (7.6 billion) (http://www.worldometers.info). Agriculture (crops, livestock, fishery, forestry, and the associated natural resources endowments) is the main source of livelihood for nearly 2 billion people. The region is the largest supplier of the world's food and agricultural products, and has witnessed several innovations in agricultural development. It is evident that the Green Revolution was brought out by a science-led synergistic extension approach capitalizing genetic potential, irrigation, fertilizer, appropriate policies and farmers' hard work. This led to an unprecedented transformation in food security and rural development in the region. Since the mid-1960s, Asian cereal production has almost doubled, reaching nearly 1 billion t, recording an annual growth rate of 3%. Increased agricultural productivity, rapid industrial growth and expansion of the non-formal rural economy resulted in quadrupling per capita GDP, almost halving poverty in the region. However, continuing such gains is becoming a major challenge, especially in the context of declining factor productivity, deteriorating natural resources, impact of global climate change and, above all, fatigue in the existing extension system, which is largely in the public sector. Hence, impact of innovations around natural resource management technologies is not all that evident, as was the case with miracle wheat and rice seeds during the Green Revolution in south Asia in the mid-1960s and 1970s.

The Challenges Ahead

Food demand versus small farm holdings

Food insecurity and poverty, accounting for two thirds of the world's hungry and poor, exacerbated by soaring food and fuel prices, global economic downturn and volatile markets, have surfaced as the major development-related concerns in the region. The problem has intensified further with a sharp rise in the cost of food and energy, depleting water resources, diversion of human capital from agriculture, shrinking farm size, soil degradation, indiscriminate and imbalanced use of chemical inputs and the overarching effects of changing climate. The per capita land availability for agriculture in the region (0.3 ha) is one fifth of that in the rest of the world (1.4 ha). The region's agrarian landscape is predominantly smallholder farmers (~80% of the world's small and marginal farmers). It is estimated that by 2050, foodgrain requirement in the Asia region would be around 70% more than the current demand. A low investment in agricultural research for development complicates the problem further. Future dependence on imported fossil fuels raises concerns regarding price volatility and shocks, and supply disruptions in agriculture production. Therefore, ensuring availability of and economic access to food, in both quantity and quality (nutrition), for the poorest of the poor in the developing countries remains a daunting challenge. To this end, the Global Conference on Agricultural Research for Development (GCARD) road map, developed through the interaction of diverse stakeholders from around the world in Montpellier, France, in 2010, highlights urgent changes required in AR4D globally, especially to address the needs of resource-poor smallholder farmers and consumers. It envisages a major paradigm shift towards farming systems research with greater thrust on 'innovations for greater impacts on small holder farmers', requiring partnerships among stakeholders and their capacity-building. To meet future food demands and to achieve successfully the MDGs, especially in the context of the world leaders' meeting, Rio+20, it has been stressed that improving the efficiency and resilience of agriculture around farming systems in the developing countries would be the only way to move forward (Alarcón and Bodouroglou, 2011).

Poverty and malnourishment

According to the FAO, the number of undernourished people in the world has increased during the last decade, and for the first time the number of hungry people crossed the 1 billion mark. Almost two thirds of the world's hungry (642 million) and 67% of the world's poor have homes in this region. The gains made in the 1980s and early 1990s in reducing chronic hunger have been lost, and the hunger reduction target of 50% by 2015 under the MDGs was not attained. Despite the fall in international food and fuel prices since late 2008, prices in the domestic markets in Asia have invariably remained 15–25% higher in real terms than the trend level, resulting in further distress for the poor. Besides poverty, the region is at present home to 70% of the world's undernourished children and women. The number remains stubbornly high, and lately has shown a rise. In the past year the number of hungry, especially in the south Asia sub-region, has increased by 10.5%, derailing all progress made after the Green Revolution. Currently, lack of economic access, not physical access, is a major challenge before society, especially for policy makers, planners, scientists and those engaged in advisory services in the region.

Natural resource degradation and climate change

The fast-declining and degrading land, water, biodiversity, environment and other natural resources are three to five times more stressed due to population and economic and political pressures in Asia as compared to the rest of the world. The region has already reached limits of land available for agriculture, and there remains no further scope for horizontal expansion. Inefficient use and mismanagement of production resources – especially land, water, energy and agrochemicals – has vastly reduced fertility and damaged soil health. At present, soils are both hungry and thirsty. To a greater extent, lack of political will and appeasement policies to provide free or relatively cheap inputs like seeds, fertilizers, water and energy have exacerbated the problem. Moreover, while maintaining a steady pace of development, the region has to reduce its environmental footprint from agriculture. The reduction in water availability and increased animal and plant diseases will primarily affect poor countries and small island states having limited capacity to respond and adapt to such negative impacts. Regrettably, man-made disasters in some countries, due to lack of political will and over-exploitation of natural resources (like the drying of the Aral Sea in Uzbekistan due to overuse of water upstream by central Asian countries), can result in increased misery for the people.

Opportunities to Harness

Genetic resource management

Agricultural biodiversity is a key resource for achieving food and nutritional security. The Asia region has a rich diversity of fauna and flora, including agroforestry species, and is the centre of origin of many important crops, livestock and forest-tree species. This resource can serve as a goldmine for specific/unique traits to be harnessed for germplasm improvement, through breeding and biotechnology applications for developing varieties/breeds possessing high productivity, better nutritional quality, resistance to biotic (diseases and insect pests) and abiotic (drought, frost, flood, salinity) stresses and high adaptation to climatic change. The Green Revolution was mainly due to exploitation of dwarfing and photo-insensitive genes in wheat and rice. Germplasm conservation through use can significantly help in achieving sustainable agricultural growth and development in Asia. It is, therefore, necessary that each country builds an effective NARS by involving all stakeholders, and has a national action plan to conserve scientifically in national gene banks all their valuable genetic resources for posterity. New science such as biotechnology, including GM crops, ICT, nanotechnology etc. offers ample opportunities to benefit the farming community. Fortunately, new innovations in agricultural science, like single-cross maize hybrids, quality protein maize (QPM), hybrid rice, hybrid sorghum, hybrid pearl millet and other crops, Bt cotton, GM technology in corn, rice, canola, soybean, brinjal etc., are available, which need to be outscaled.

Innovations in natural resource management

One of the main reasons for slow growth in agriculture is relatively poor dissemination of emerging technologies relevant to the needs of smallholder farmers. Innovations are needed to meet the major challenge of consistently increasing resource scarcities and to bring in structural transformation in the socioeconomic context to reduce the cost of inputs and for improving the livelihoods of resource-poor smallholder farmers. To liberate nations from hunger and poverty, while sustaining existing natural resources, policy makers have to have a renewed thrust and commitment to additional funding for AR4D. Without this, the task of achieving inclusive growth will remain elusive. Innovations around good agricultural practices such as CA, balanced use of fertilizers, small-farm mechanization for resilience, micro-irrigation, IPM and scientific land use for crop diversification would contribute considerably to arresting natural resource degradation, climate change adaptation and mitigation as well as increased farm productivity and profitability. One such successful example in the region is of CA in the Indo-Gangetic plains, led by the regional NARS (Bangladesh, India, Nepal and Pakistan) and facilitated by the International Maize and Wheat Improvement Center (CIMMYT), which led to a cost–benefit ratio of 1:19 (investments of US$3.5 million led to an output equivalent of US$64 million) through adoption of zero tillage for wheat planting over 2.5 million ha. The area under CA can be easily increased almost four-fold (10 million ha), provided concerted efforts are made in a mission-mode approach to outscale this innovation. Another successful example is of laser land-levelling, adopted recently over 2.5 million ha in northwest India, primarily due to custom-hire service windows. In Haryana alone, it led to saving 1 billion m^3 of water annually. Similarly, direct-seeded rice (DSR) in Basmati varieties has picked up fast in the last two to three years. Replication of such success stories can help farmers in similar ecological situations in other countries without 'reinventing the wheel', provided knowledge is shared through effective extension systems.

Recently, the International Food Policy Research Institute (IFPRI) has brought out an interesting publication entitled *Millions Fed*, covering around 20 success stories from around the world, half of which are from Asia alone. Similarly, the Asia-Pacific Association of Agricultural Research Institutions (APAARI) has published around 45 success stories from the Asia-Pacific region, indicating how developing countries have made faster progress through outscaling new innovations. Learning from them, many NARS have gained equally well, like adoption of hybrid rice in India, the Philippines and Vietnam, based on spectacular results in China. Similarly, the story of baby corn in Thailand could be repeated in India. Success of IPM in rice in Indonesia has also been repeated in many countries. GM cotton technology has also been adopted fast in China, India and Pakistan.

Other potential sectors

Horticulture

This region also has huge potential to promote horticulture. Most of the countries in the region have not paid due attention to this sector. At present, the need is to diversify the food basket by producing more vegetables and fruits. The post-harvest losses happen to be high, ranging from 10% to 30%. This situation needs to be changed through the application of processing technology, value addition, cold storage and rapid transportation.

Livestock

Besides crop productivity enhancements, strategies are needed to usher in 'White' and 'Blue' Revolutions in this region. This would need new production models to enhance contribution from the livestock and fishery sectors. Mechanization and automation of dairy farms, measures to provide good-quality feed and fodder, provision of improved seed varieties for fodder crops, and value addition of milk and meat products are some of the measures for enhancing the livestock industry.

Fishery

Fishery is another potential sector that can help achieve food and nutritional security. The inland fish farms, with adoption of modern technologies, managed by skilled human resource, can make all the difference. The success story of tilapia fish farming in the Philippines, sea bass in Israel and king prawn in Thailand and India are some examples worth emulating. In fact, willingness to adopt new ideas has transformed typical fish farmers and even young professionals in the Philippines, Thailand, India and elsewhere into thriving entrepreneurs.

Strengthening Collaboration and Partnerships

The Green Revolution was the outcome of partnership between NARS, international centres like CIMMYT, the International Rice Research Institute (IRRI) and the extension system, including progressive farmers. Regional and global networks and partnerships for knowledge-sharing and enhanced capacity development of different stakeholders are a must for outscaling innovations in similar ecologies. It has been increasingly realized that under the changing scenario of production to consumption, a linear approach in technology development and deployment would not serve the purpose of meeting MDGs. For inclusive growth in agriculture through large-scale uptake of new technologies, a major paradigm shift in approach is needed from R&D to AR4D, involving greater participation of all stakeholders. Past experiences from regional organizations/programmes like the Asia-Pacific Association of Agricultural Research Institutions (APAARI), the South Asian Association for Regional Cooperation (SAARC), the Association of Southeast Asian Nations (ASEAN), the Rice-Wheat Consortium (RWC) and the Cereal Systems Initiative for South Asia (CSISA) have revealed that regional partnerships are important to catalyse faster adoption of new technologies, mainly through sharing success stories around good agricultural practices (SAARC, 2016).

APAARI has been instrumental, since its inception in 1990, in promoting regional cooperation for agricultural research, and has organized a series of expert consultations on emerging issues concerning AR4D. Some of these were on: food crisis and biofuel; productivity enhancement; biotechnology and biosafety; post-harvest management; CA; climate change; and women and youth. From Sri Lanka, the Council for Agricultural Research Policy (CARP) has been an active member of APAARI from the very beginning in most of these initiatives. A similar partnership with the Department of Agriculture (DOA) in Thailand would prove beneficial.

Knowledge-sharing and Capacity-building

APAARI is supporting a major programme known as the Asia-Pacific Agricultural Research Information System (APARIS), under which more than 45 success stories from the region, beside proceedings and recommendations of several expert consultations and workshops, have been published and widely disseminated. Details of these success stories can be accessed from the APAARI website (www.apaari.org). APAARI has also proclaimed some important regional declarations relating to AR4D such as the Tsukuba Declaration on Climate Change, the Suwon Framework on Agrobiodiversity, the Bangkok Declaration on Strengthening Agriculture Research for Development etc. All these have received considerable attention of policy makers and planners in many countries towards reshaping/reorientating their research and extension agenda for the benefit of resource-poor farmers who are in dire need of technical backstopping knowledge.

Strategy for Linking Research with Extension

Research should be sensitive to local needs and meet the aspirations of farmers and consumers, and there should be closer working relations between research and extension organizations. The scientists involved in basic, strategic, applied and adaptive research, together with subject matter specialists, extension workers and farmers, should be seen as an integral component of the knowledge dissemination and agricultural advisory system. The interface between research and technology transfer is indeed very critical for converting outputs into outcomes. In fact, we need to link 'land to lab' and 'village to institution'. This would require a paradigm shift from a top-down to bottom-up approach for technology generation, refinement and adoption. In the present context, the agriculture sector has to be more scientifically oriented and technology-driven. As stated earlier, almost all problems of contemporary Asia require interdisciplinary, inter-institutional and regional, rather than national, solutions. Furthermore, the research agenda of the institutions could

be better organized for technology development and dissemination. In all institutions, technology transfer programmes need to be an integral part of technology development to empower farmers with proper knowledge. Farmers' participatory research has to receive major attention (FAO, 2013).

The Indian Extension System – an Example

Research and extension to work hand in hand has been a challenge over the past 50 years. From time to time, several experiments were conducted to make the extension system vibrant, effective and meaningful. Agriculture in India is a state subject. Accordingly, states follow the central government schemes launched from time to time. Built on the foundations of the Community Development Programme (CDP), started in 1952, public sector extension followed an evolutionary pathway. The Intensive Agriculture District Programme (IADP) (1961–1962), the Intensive Agriculture Area Programme (IAAP) (1964–1965), the High Yielding Varieties Programme (HYVP) (1966–1967) and the Farmers' Training and Education Programme (FTEP) (1966–1967) are some of the significant developments leading to the growth of agricultural extension in India. Undoubtedly, these programmes created awareness and paved the way for acceptance and application of genetic resource technologies. These, however, were ineffective in serving the needs and aspirations of small and marginal farmers. This weakness, inherent in the early technology transfer system during the Green Revolution period, is believed to have widened the gulf between the resource-rich and resource-poor farmers (Paroda, 2014).

In India, agricultural institutions (ICAR institutes and SAUs), in collaboration with the Department of Agriculture, generate technologies and transfer them to stakeholders. In the 1960s, the agricultural production situation was so critical that intensification of agriculture with the use of HYVs became unavoidable. The programmes, such as the Integrated Agriculture Development Programme (IADP), the Intensive Agriculture Area Programme (IAAP), the National Demonstration (ND) and the High Yielding Variety Programme

(HYVP), gained momentum. The sole purpose of these programmes was for increasing crop yields by using modern means of production. This approach, though paying good dividends, generally failed to help poor farm households. The emphasis was broadened from agricultural development to rural development, and programmes like the Small Farmers Development Agency (SFDA), the Marginal Farmers and Agricultural Labor Development Agency (MFALDA), the Drought Prone Area Programme (DPAP) and the Integrated Rural Development Programme (IRDP) were launched in the 1970s (Paroda, 2014).

The most significant development took place under the World Bank project and the Training and Visit (T&V) extension programme, started in the mid-1970s. The emphasis was on efficient technology transfer using promising research results. The system, however, proved to be of little help to small farmers, especially those in rainfed areas. To bridge this gap and increase the reach of extension services, the ICAR launched its frontline demonstration programmes such as the Operational Research Project (ORP), Lab-to-Land and National Demonstration. The ICAR established Krishi Vigyan Kendras (KVKs), which are also known as district science centres. This is an institutional mechanism for a front-line extension approach, and so far about 700 KVKs have been established all over India. The need was also felt for technology appraisal, refinement and transfer, and the Institution Village Linkage Programme (IVLP), based on participatory methodology, was launched in selected locations in 1998 under the World Bank's National Agricultural Technology Project (NATP) (Paroda, 2014).

The approach has been reversed from 'top down' to 'bottom up'. New institutional arrangements for technology dissemination through establishment of the Agricultural Technology Management Agency (ATMA) at district level have enabled better coordination and convergence of all rural advisory programmes. Under the project, a State-level Agricultural Management and Extension Training Institute (SAMETI) has been created to provide training to state-extension functionaries. Later, other projects, like the National Agricultural Innovation Project (NAIP), were implemented to give agricultural research/technology generation systems an explicit development- and business-oriented perspective through innovative partnership models.

Application of ICT has also been promoted in agriculture. The Indian Tobacco Company (ITC) has spearheaded an Integrated Rural Development Programme (IRDP) to empower farmers and raise rural incomes. The strategy for this is broadly centred around information and knowledge dissemination, access to quality inputs and markets, generation of supplementary incomes and natural resource augmentation. Farmers are given critical information and relevant knowledge on farm productivity, prices and markets through ITC's e-Choupal. This platform enables access to quality inputs for better productivity besides expanding reach to markets. Dedicated radio and TV channels are also in the offing. The private sector has also come up to support farmers by empowering them with better technology and providing them with quality inputs. Also, use of smart phones has become popular to access knowledge (Paroda, 2014).

Outscaling Farmer-led Innovations

In the pursuit to enhance both agricultural production and income, farmers do consistently try to make agriculture efficient and cost-effective. In the process, they have come out with numerous innovations around improved farming practices and livelihoods. These innovations have supported food security. Farmers identified a number of new/indigenous traditional crops and developed varieties with enhanced productivity and better quality through selection. They also identified livestock breeds and developed technologies for low-cost animal and fish rearing and processing, efficient horticultural practices, value addition and better marketability of farm products. In addition, a number of farm implements and tools have been designed and manufactured by farmers to increase operational efficiency and productivity. Commendable work has been done by women farmers, especially in germplasm conservation, post-harvest management and value addition, which helped to enhance farm income. In fact, farmers are silently innovating, adopting new practices and continuously improving them. Unfortunately, these farmer-led innovations, over generations, have not been recognized and documented. Also, the IPR on the innovations made by farmers and their documentation often lacked in the past. Value of traditional knowledge and its documentation has also remained unnoticed by scientists. As a result, the advantages of many technologies developed by innovative farmers have not been reaped by other farmers. Efforts are needed to record farmer-led innovations in agricultural practices and blend them with modern science through refinement and validation in a participatory mode. The innovative farmers need encouragement and financial support for their creativity. Accordingly, an Agriculture Innovation Fund/Board needs to be created at the national level to supplement the efforts of such farmers by awards/rewards and by providing them with some financial assistance (APAARI, 2011; Dar, 2014).

Linking Farmers to Market

Agriculture is the only enterprise where cost is determined by others, rather than by the producers. To ensure competitive price of produce, the role of middlemen has to be minimized and market forecasting systems have to be strengthened so that farmers can take the right decisions on crop planning, production and sale of produce. Recent studies by the US Agency for International Development (USAID) have indicated that 50–70% of smallholders are in transition from subsistence to commercial farming in many countries of Africa and Asia. In most of the south-Asian countries, urbanization and industrialization are not creating sufficient numbers of off-farm jobs to accelerate agricultural commercialization. Overcoming the commercialization barrier requires an upgrading process around investment in local infrastructure, strengthening of business services and improvement in farmer's skills through efficient extension systems, which are not visible. Also, in view of the considerable decline in public extension services over the last two to three decades, farmers were not able to access vital technologies and services. Studies show convincingly that income growth generated by agriculture would be up to four times more effective in reducing poverty

than growth in other sectors (Growth Commission, 2008). Therefore, income growth in agriculture needs to be stimulated further by linking farmers to markets. There is a need to develop a sustainable model for marketing that should allow farmers to sell directly to consumers. Hence, value-chain development involving farmers, direct sales by farmers, contract farming, organized retailing by farmers and the establishment of farmers' associations, self-help groups and/or companies would go a long way in achieving these goals (APAARI, 2011).

Micro-financing

Providing effective and efficient financial services in the agriculture sector continues to be a challenge. The FAO argues that poorly functioning financial markets make farmers reluctant to adopt new practices and technologies and also reduces their risk-taking abilities. Therefore, objectives of micro-financing cannot be overlooked. Some flagship institutions in Asia such as Bank for Agriculture and Agricultural Cooperatives (BAAC) in Thailand; village banks (Unit Desas) of Bank Rakyat in Indonesia (BRI-UD); and Grameen Bank (GB) in Bangladesh have demonstrated how to successfully supply loans and other financial services in rural areas. Such institutions need to be created on a large scale. There is a need to establish a closer relationship between finance and production, income distribution, empowerment and welfare. The happy situation is that several innovations are being examined to make financial support available to farmers. In India, Kisan (farmer) Credit Cards (KCC) are being issued to all farmers to avail themselves of credit at low interest rates (APAARI, 2011).

Policy support

The appropriate policies on provisions of subsidies on key inputs; promotion of efficient technologies such as CA; innovations and improved varieties; and creation of institutions such as farmers' cooperatives, self-help groups, farmers' clubs and farmers' companies need to be inculcated in agriculture development plans. In the coming years, south-Asian countries would need to foster long-term productivity policies by investing heavily in agricultural R&D, while introducing institutional reforms to create an environment facilitating adoption of new technologies. Emphasis needs to be given to look again at domestic agricultural policy to make it more effective for infrastructure development, risk management and easy credit availability.

Empowering Women for Inclusive Growth

Globally, about 43% of women are engaged in agriculture. In India, 60% of farming operations are performed by women. Therefore, agriculture can be a primary driver for empowering women. Innovations would improve their work efficiency and would also ensure overall household development and nutrition security. However, women are invariably deprived of access to agricultural knowledge, credit, technology to overcome their drudgery and market-related services. Often they are deprived of their rights to land and resources. All these adversely impact their performance. *The State of Food and Agriculture Report of 2010–11* by the FAO indicated that reducing the gender gap between male and female farmers would raise yields on farms by 20–30%. As a consequence, this would lead to reduction of undernourished people, globally, by 12–17%. This, in turn, would translate into 100–150 million fewer hungry people. Hence, technology generation relevant to women farmers and its adoption should become an important agenda for future agricultural growth (ICAR, 2012).

Retaining Youth in Agriculture

Asia can reap demographic dividends if attention is paid to creating more and better jobs, improving the technical skills and education of youth and providing efficient matching of labour supply and demand through regulations and mobility. The ageing population of farmers and declining interest among rural youth in taking up agriculture as a profession are challenges for agricultural sustainability in India and also in other countries of the region. A large section of youth invariably prefers to migrate to cities to seek employment, especially government jobs. Hence, a major challenge today is how to retain youth in agriculture. This certainly cannot be left unaddressed. The declining interest of rural youth is directly related to existing poor physical amenities, socioeconomic conditions and lack of an enabling environment. Economic factors such as low-paid employment, inadequate credit facilities, low profit margins and lack of insurance against crop failure are also discouraging other factors. Social factors include public perception about farming, especially parental desire that their children should opt out of agriculture. Environmental issues include poor soil health, non-availability of water for irrigation and climate change. Proper incentives for their involvement in agricultural education, research and extension and linking them to expanding markets would have positive effects in attracting youth in agriculture (YPARD, 2012; MSSRF, 2014).

Earlier, seed, pesticide, fertilizer and farm machinery were the only potential sectors to employ agricultural graduates/rural youth. Lately, new opportunities are emerging in IT-linked agri-extension, seed technology, biotechnology, food processing, cold storage, packaging, supply-chain management, insurance and farm credit. The private sector and NGOs are also engaging rural youth. In this context, we now need greater thrust on vocational training of youth (including females) for relevant skill acquisition and greater confidence-building to serve as 'technology agents' as well as efficient knowledge/service providers on a custom-hire basis. It is high time that all-out efforts are made at all levels to engage youth in multifarious activities around 'plough-to-plate', so as to make farming an attractive as well as lucrative profession. Knowledge-based agriculture around secondary and speciality agriculture can enhance opportunities for additional income for youth (YPARD, 2012).

The Future Road Map: The Need for a Paradigm Shift

The success of the Green Revolution was mainly due to a holy alliance between researchers, extension specialists and farmers. The technology dissemination approach adopted was top-down and centred around individual farmers. Faster adoption of technology was also on account of miracle seeds of wheat and rice, promoted largely by the public extension system, which, over the years, has become relatively weak. On the contrary, new innovations around natural resource management require a bottom-up approach, involving farmers' participation, while ensuring confidence-building among farming communities to take risks and make agriculture more scientific and resilient. In the process, sharing of knowledge on good agricultural practices, without dissemination loss, and incentives for critical inputs become highly crucial to be successful. Also, partnership among key stakeholders becomes essential to promote growth in agriculture. In the process, care is also needed to overcome complacency that has crept into the public extension/advisory services. A paradigm shift is needed from the present NARI system to NARES. This would require active involvement of stakeholders such as farmers, NGOs, the private sector, scientists and policy makers. Another shift has to be in the extension approach towards translational research to ensure outscaling of innovations for greater impact on productivity and income.

In this context, an extension approach has now to be around farming communities rather than individual farmers. Also, NRM-related innovations would require more lead time to assess impact on farmers' fields, unlike the impact of HYVs on crop productivity. This throws a new institutional challenge for needed reforms in existing extension systems, which mostly depend on public organizations. The role of the private sector, especially through involvement of youth and gender in agriculture, becomes most relevant in the present situation. Hence, empowering youth (both men and women) through vocational training and building a cadre of technology agents to provide technical backstopping as well as custom-hire services to smallholder farmers would go a long way in linking research with extension for accelerating agricultural growth. We need to link 'land with lab', 'village with institute' and 'scientists with society' to ensure faster adoption of resource-saving technologies that would benefit producers and consumers. In the process, the agriculture technology agents would become job creators and not job seekers and provide, on farmers' doorsteps, the best technologies as well as quality inputs. Another strategy could be to create 'agri-clinics' where technology agents could join hands to ensure a single-window system of advisory services to farmers so that they need not run from 'pillar to post'. In fact, a good farmer is knowledge-hungry and does not want to depend only on government subsidy.

The Way Forward

Agriculture in Asia must liberate the region from the twin scourges of hunger and poverty and of malnutrition in children and women. The region must continue to feed the world with adequate food. Accelerated science and innovation-led agricultural growth must be inclusive and should address the needs and aspirations of resource-poor smallholder farmers in the Asia-Pacific region. Under the growing challenges of resource degradation, escalating input crisis and costs, and with the overarching effects of global climate change, the major gains in foodgrain production will largely depend on a paradigm shift from integrated germplasm improvement to integrated NRM. The future AR4D efforts by NARS must now be reoriented towards a farming systems approach involving farmers' participatory approach. We need to employ innovative ways for effective dissemination of knowledge and lay greater emphasis on outscaling innovations for needed impact on the livelihood of smallholder farmers. Henceforth, 'farmer first' should be the goal of all NARES to bridge the income divide between farmers and non-farmers, and it should benefit equally producers and consumers. To ensure this, developing countries in the Asia-Pacific region must enhance their investments (almost triple) in AR4D to address effectively the emerging challenges, thus ensuring food, nutrition and environment security for all.

References

Alarcón, D. and Bodouroglou, C. (2011) Agricultural innovation for food security and environmental sustainability in the context of the recent economic crisis: Why a gender perspective? Draft paper. *World Economic and Social Survey 2011*. Available at: http://www.un.org/en/development/desa/policy/wess/wess_bg_papers/bp_wess2011_alarcon.pdf (accessed 13 June 2018).

APAARI (2011) *Proceedings of the International Conference on Innovative Approaches for Agricultural Knowledge Management: Global Extension Experiences*, 9–12 November. Asia-Pacific Association of Agricultural Research Institutions, Bangkok.

Dar, W.D. (2014) Leading to Serve and Empower the Poor: Transformational Leadership in Agricultural Research for Development. International Crops Research Institute for the Semi-Arid Tropics. ICRISAT, Hyderabad, India.

FAO (2013) Expert consultation on strengthening linkages between research and extension to promote food and nutrition security, organized by the Food and Agriculture Organization's Economic and Social Commission for Asia and the Pacific (ESCAP) and the Center for Alleviation of Poverty through Sustainable Agriculture (CAPSA), 11–12 December, Bangkok.

Growth Commission (2008) *The Growth Report: Strategies for Sustained Growth and Inclusive Development.* Commission on Growth and Development, World Bank, Washington DC.

ICAR (2012) *Proceedings of the First Global Conference on Women in Agriculture (GCWA).* Indian Council of Agricultural Research, New Delhi and Asia-Pacific Association of Agricultural Research Institutions, Bangkok, New Delhi, 13–15 March.

MSSRF (2014) *Attracting and Retention of Youth in Agriculture: Workshop Phase II.* 5–6 April 2014. Organized by M.S. Swaminathan Research Foundation and Tamil Nadu Agricultural University (TNAU), Coimbatore, India.

Paroda, R.S. (2014) Need for Linking Research with Extension for Accelerated Agricultural Growth in Asia. Strategy paper. Trust for Advancement of Agricultural Sciences, Pusa Campus, New Delhi.

SAARC (2016) *Proceedings of the Regional Workshop on Research-Extension Linkages for Effective Delivery of Agricultural Technologies in SAARC Countries.* Jointly organized by SAARC Agricultural Information Centre, BARC Campus, Dhaka, Bangladesh (http://www.saic-dhaka.org) and National Academy of Agricultural Research Management, Rajendranagar, India (http://www.icar.naarm.ernet.in), 20–22 November. SAARC Agricultural Information Centre (SAIC), Dhaka, Bangladesh.

YPARD (2012) ICAR to Focus on Retaining Rural Youth in Agriculture. Young Professionals for Agricultural Development. 14 September. Available at: https://ypard.net/news/icar-focus-retaining-rural-youth-agriculture (accessed 13 June 2018).

26

Empowering Farmers through Innovative Extension Systems

Agriculture must liberate India from the twin scourges of hunger and poverty while ensuring sustainability of natural resources. It must also address effectively the concerns of malnutrition among children and empowerment of women; being important SDGs. To ensure these, needs and aspirations of resource-poor smallholder farmers must be addressed through innovation-led, accelerated and sustainable agricultural growth. Historically, adoption of high-yielding dwarf varieties of wheat and rice during the Green Revolution era addressed both hunger and poverty. Of late, however, the yield gaps in agriculture, and income divide, in the farm and non-farm sectors have widened, primarily due to gaps in knowledge and skills and lack of timely access to improved technologies. Outscaling of appropriate technologies to reach farmers has emerged as a complex issue. Why farmers are not able to access or adopt new technologies are the major issues that create problems for the development officials and scientists alike. Further, growing challenges of natural resource degradation, escalating input costs, market volatility and, above all, the effects of global climate change contribute to declines in yield as well as farm income, thus making agriculture both non-profitable and unattractive. Thus, it is crucial to ensure inclusive growth in agriculture through innovative and synergistic approaches for achieving sustainable food and nutritional security. Therefore, 'agriculture research for development (AR4D)' requires a paradigm shift to 'agricultural research and innovation for development (ARI4D)' (TAAS, 2015).

Changing Paradigm in Extension

Agricultural extension in India and elsewhere consistently requires transformation. The current transitional phase also needs a 'renewed interest' and 'policy attention'. Public extension systems played a vital role during the Green Revolution but were confined mainly to the irrigated areas. This success was also due to a holy alliance among researchers, extension specialists, farmers and policy makers. At that time, the technology dissemination approach remained top-down, focusing on individual farmers. The current scenario of Indian agriculture is confronted with multifaceted challenges arising out of inefficient management of natural resources (soils, water, agrobiodiversity). All these have led to a decline in factor productivity and farm profitability. Apparently, this complexity cannot be overcome by routine transfer of technologies; rather, efforts would be needed towards translational research, requiring outscaling of innovations through 'outside the box' extension systems. Also, conscious deployment of rural youth, women and progressive farmers would help in a speedy transfer of technology and the needed impact on the livelihood of smallholder farmers. Farmers' welfare needs to be ensured through a 'farmer first' approach to benefit equally producers and consumers. Further, in view of diverse demands for new innovation, new products, new information and new extension services, there is a need to shift from a top-down to a bottom-up approach, involving farmers' participation at the grassroots level, while ensuring confidence-building among farming communities to take risks and adopt more scientific and resilient agriculture. In the process, knowledge sharing on good agricultural practices, without dissemination loss, and incentives for critical inputs become highly critical to achieve future development successes in the agricultural sector. At the same time, partnerships among key stakeholders become vital for promoting agricultural growth. In the process, care is also needed to overcome complacency that has crept into the public extension system. Hence, this necessitates greater vibrancy in the NARES; requiring active involvement of stakeholders (farmers, NGOs, private sector, scientists and policy makers) and a shift in the extension approach towards outscaling of innovations for greater impact on smallholder farmers through higher productivity and income (TAAS, 2015).

In this context, the extension approach has to focus on farming communities rather than on the individual farming household approach. With the increasing challenges of addressing land degradation, soil quality and water-use efficiency, the NRM-related innovations, unlike the adoption of HYVs, showing immediate impact on crop productivity, need much more lead time to translate and assess impact on farmers' fields. This obviously throws a new institutional challenge for needed reforms in the existing extension system, which mostly depend on public organizations; and the role of the private sector becomes highly relevant, especially for involving rural youth, including women, in agricultural extension.

Involving Youth in Agriculture

Empowering youth through vocational training and building a cadre of 'technology agents' to provide technical backstopping as well as custom-hire services to smallholder farmers would go a long way in linking research with extension, and thereby accelerating agricultural growth. There is also a need to link 'land with lab', 'village with institute' and 'scientists with

society' to ensure faster adoption of efficient resource-utilization technologies, which would benefit producers and consumers. In the transformation process, the agricultural technology agents need to become job creators and not job seekers and provide best technologies as well as quality inputs on farmers' doorsteps. Another strategy could be to establish 'agri-clinics', where technology agents can join hands to ensure a single-window system of advisory services to farmers (YPARD, 2012).

Lately, in the changing socioeconomic environment, there has been steady improvement in the use of ICT, rural infrastructure and literacy standards in farming communities. Public sector institutions such as the central and state government departments of agriculture, horticulture, livestock, including fisheries; the central and SAUs; a network of Krishi Vigyan Kendras (KVKs); and the Agricultural Technology Management Agency (ATMA) are empowering farmers. But efficient agro-advisory in the wake of increasing demand for quality agricultural knowledge, together with input support, can be best delivered through pluralistic agricultural extension, i.e. a mix of public–private sector participation. Emergence of private sector institutions such as corporate organizations, community-based organizations, farmer associations, farmer cooperatives, self-help groups, watershed and water-user associations, farmer-producer companies, NGOs; input providers for seeds, nutrients and pesticides; and service providers for small tools and implements; para-professionals (Kisan Mitras etc.); input producers; the private corporate sector; fertilizer companies; marketing firms; and processing enterprises should be encouraged to bring in much-needed complementarities in providing agro-knowledge advice to farmers (Saravanan, 2010; Singh *et al.*, 2011; YPARD, 2012).

Empowering progressive farmers is also a must. Their farm-led innovations can be recognized and promoted for further scaling and refinement to make them efficient, cost-effective and adapted to local situations. Also, information acquisition by smallholders from other progressive farmers is invariably more effective. The information demand by farmers is often changing as they intend to make farming more diversified and resilient. Thus, a demand-driven extension approach around integrated farming systems should be addressed while promoting secondary and speciality agriculture.

Also, there is a need for convergence among different government-sponsored programmes. Accordingly, concerns for collaboration, convergence and synergy need to be addressed along with issues of optimizing institutional arrangements of prevailing pluralistic agricultural extension and farm advisory.

New Institutional Mechanisms

Institutional reforms (extension and non-market) emphasize stakeholder participation and have the potential to improve efficiency and effectiveness of extension efforts. They emphasize how agricultural extension cannot operate in isolation but as part of a broader information system, the agricultural knowledge system (AKS), which comprises three pillars: research, extension and agricultural higher education. The three pillars involve complementary investments that should be planned and sequenced as a system rather than separate entities. Success is only possible with strong cross-institutional linkages between AKS systems and their clientele. Agricultural extension needs to expand its focus on non-farm micro-enterprise development initiatives as a way of improving livelihoods, because most rural people depend upon multiple sources of income. Agricultural extension services should go beyond providing technical support, and provide market extension and advice on the importance of farmers organizing into farmers' associations. Approaches to extension should change in response to the changes in the global environment through participatory learning and action (PLA), including participatory rural appraisal (PRA), rapid rural appraisal (RRA), participatory learning methods (PALM), participatory action research (PAR) and farming systems research (FSR). There are four basic themes: (i) collaboration through participation; (ii) acquisition of knowledge; (iii) social change; and (iv) empowerment of participants. Action researchers are responsible for developing a learning environment that challenges the status quo and for generating better alternatives to improve their future. The CRASP definition of action research has

been used: *critical* collaborative enquiry by *reflective* practitioners, who are *accountable* in making the results of their enquiry public, *self-evaluative* of their practice, and engaged in *participative* problem-solving and continuing professional development (Rasheed, 2012; *The Hans India*, 2017).

ICT for Knowledge Dissemination

Farming and ICT seem to be the most distantly placed knowledge sets; farming being the most primitive and basic and IT being the most advanced and modern. However, ICT plays a significant role for the betterment of farming. The information related to policies and programmes of government, schemes for farmers, institutions through which these schemes are implemented, new innovations in agriculture, good agricultural practices, institutions providing new agricultural inputs (high-yielding seeds, new fertilizers) and training in new techniques are disseminated to farmers through the use of information technology to ensure inclusiveness and to avoid a digital divide. Access to price information, agriculture information, national and international markets, increasing production efficiency and creating a 'conducive policy environment' are the beneficial outcomes of eAgriculture, which enhance the quality of life of farmers. The management of soil, water, seed, fertilizers, pests, harvest and post-harvest is the important component of eAgriculture where technology aids farmers with better information and alternatives. It uses a host of technologies like remote sensing, computer simulation, assessment of speed and direction of wind, soil-quality assays, crop yield predictions and marketing using IT. In India, there have been several initiatives by state and central governments to meet various challenges facing the agriculture sector in the country. The eAgriculture is part of Mission Mode Project, which has been included in the NeGP (National e-Governance Plan) in an effort to consolidate learning from the past, integrate all the diverse and disparate efforts currently underway, and upscale them to cover the entire country (eAgriculture, 2012; Uphoff, 2012; Singh *et al.*, 2015; Karubanga *et al.*, 2016; Mamur *et al.*, 2016; *The Hans India*,

---

Box 26.1. eAgriculture system. (From: https://www.intel.in; eAgriculture video, youtube.com; Grameen Intel Social Business (www.Grameen-Intel.com) and Intel World Ahead Program (www. intel.com/worldahead).)

- eAgriculture initiatives: these bring together a wide array of local and regional stakeholders to form a mutually beneficial value chain.
- Grameen Intel and other social businesses: information and expertise, consulting services, technology and programmes to reach rural and impoverished markets.
- Governments and multilateral development agencies: programme support to enable and increase rural outreach, improve food security, create jobs and develop partnerships with local businesses and community organizations.
- Banks and other financial institutions: credit, capital and other financial instruments (crop insurance, subsidies) for entrepreneurs and farmers.
- Universities and agriculture extension systems: technology to strengthen extension systems; advice and technical support for farming communities; training and capacity-building for entrepreneurs; research and development projects designed to solve problems faced by farming communities.
- Supply chain (e.g. suppliers, commodity markets, aggregators): best-of-class products and services for farmers that improve returns to all stakeholders, including farmers.
- Technology companies: internet connectivity, hardware and software solutions that create access to new markets, value chains and business models.
- Community organizations (e.g. farmer cooperatives, rural telecentres, government- and NGO-run agriculture service centres): help entrepreneurs, provide grassroots agriculture domain and business support, and enable programmes to scale efficiently.

2017 (http://www.thehansindia.com)). The highlights of eAgriculture are given in Box 26.1.

National Dialogue on Extension

In view of the above, a National Dialogue on 'Innovation Extension Systems for Farmers' Empowerment and Welfare' was organized jointly by TAAS and ICAR in New Delhi from 17–19 December 2015, in which 242 stakeholders participated from all over India. The discussions centred around issues such as current status and challenges, the need to revisit existing extension systems, farmers' perception and need, role of media and communication systems, empowerment and involvement of women and youth, role of NGOs and the private sector, policy and institutional reforms, and urgent need for effective coordination and convergence. Participants were unanimous in the need for renewed thrust to transform the present agricultural system to make it more meaningful, relevant and effective by involving new actors of extension such as youth and women, NGOs and progressive farmers (YPARD, 2012; TAAS, 2015). In general, the following points are important for immediate attention:

- Effective and efficient agricultural extension and advisory services are critical to achieve higher productivity, promote agricultural trade to help raise farmers' income, while achieving a national target of 4% growth in agriculture.
- The scope of agricultural extension has undergone certain fundamental changes with a growing number and diversity of extension service providers.
- The public extension system caters merely to 15%, whereas such services provided by others like the private sector, NGOs, farmers and social media are yet to be optimally organized and mainstreamed.
- A real transformation in the existing agricultural extension requires demand-driven, multidimensional, multi-agency, market-oriented, pluralistic and outside-the-box approaches.
- Empowerment of women and youth for agricultural extension and farmers' welfare is critical for large-scale adoption of highly scientific, resilient, productive and remunerative secondary and speciality agriculture by farming communities.
- Knowledge sharing on good agricultural practices, without dissemination loss, is indeed

critical to achieve better results in the agriculture sector; for which the role of print, social media, like radio and TV, and ICT (especially mobile phones) is considered essential.

- Innovations in agricultural extension would henceforth demand 'paid extension' services, especially when there is a scope to increase farmers' income, for which an enabling policy environment is now emerging for the private extension system through small-scale entrepreneurs as technology agents and input providers.

Need to Reorient the Extension System

To overcome the multiplicity and increasing complexity of problems being faced by farmers, there is a need to adopt (TAAS, 2015) the following strategies:

- A 'farmer first' approach needs to be promoted with twin objectives. On the one hand, to better understand critical needs of farmers, and on the other, to identify options that can address these needs in a manner benefitting all involved in the agricultural value chain. To ensure this, a National Mission on Agricultural Extension needs to be established as a priority by the Ministry of Agriculture and Farmers' Welfare to plan, undertake and promote collaborative extension interventions by public, private, NGOs and progressive farmers and to give a modern extension thrust across the board, optimizing effective coordination and evolving efficient convergence mechanisms. The new national mission may also oversee coordination and convergence of various state- and district-level extension activities by the KVKs, ATMA, private sector, NGOs and progressive/innovative farmers. Initially, an annual budgetary provision of around Rs 15 billion could be made for implementing the much-needed mission-mode approach in agricultural extension.
- Multidisciplinary, inter-institutional efforts towards translational research must be accelerated with required policy and financial support, especially to outscale innovations after validation and needed refinements.

- Conscious deployment of rural youth, women, farmer professors and authorized/trained/certified input providers are to be ensured through innovative approaches, such as the formation of farmers' self-help groups, farmers' cooperatives, farmer-producer companies, farmer-to-farmer training, agri-clinics etc. to catalyse speedy technology transfer and diffusion.

Foresight Approach for a Paradigm Shift

A foresight approach to ensure a paradigm shift from top-down to bottom-up needs to be adopted to meet new demands for innovations, products, information and extension services (Singh *et al.*, 2015; TAAS, 2015), such as:

- Ensuring farmers' participation at grassroots level and confidence-building among the farming community to take risks and adopt more scientific and resilient farming technologies. Simultaneously, provide policy incentives for critical inputs as well as farmers' participatory activities by all stakeholders and market players.
- Encouraging farming systems' extension by interdisciplinary, inter-institutional extension teams, comprising subject matter experts, as was envisioned under the earlier institution–village linkage programme (IVLP) for effective agricultural extension.
- Promote knowledge-sharing on good agricultural practices aimed at minimizing dissemination loss for services relating to inputs, technologies, insurance, processing, value addition, markets etc.
- Encouraging required partnerships among key stakeholders to promote demand-driven, multi-stakeholder-oriented agricultural extension around integrated farming systems. This should be ensured through in-built incentives to adopt innovative technologies that optimize the use of natural resources, though requiring more adoption time to assess, refine and diffuse NRM-related technologies on farmers' fields.
- Providing innovative alternate knowledge/information dissemination systems with

authentic content in farmer-friendly communication mode such as the Kisan TV channel, ICT, smart phones, print media and radio, to ensure their distant reach and effectiveness.

- Emphasizing linking farmers to market is a key step towards inclusive market-oriented development (IMOD) for smallholder farmers. Also, focusing on designing women- and youth-centric programmes for their active role in market-oriented agri-food value chains with provision of timely incentives.

- Stimulating the national agricultural extension system beyond free extension; paid extension services through agri-clinics are to be encouraged with an in-built safeguard mechanism in place.

- Private sector participation in the national agricultural extension system is to be encouraged through corporate social responsibility (CSR), and also through much-needed PPP, supported by an enabling environment.

- Emphasis shall be laid on documentation and wider dissemination of successful extension models under diverse agro-ecologies and farming situations. Similarly, lessons learnt from failures can be assessed to take corrective measures elsewhere.

- Extension research should go beyond production to post-production extension. As such, higher emphasis needs to be placed now on innovation, growth and development.

- Communication systems need to be enhanced in rural areas in order to play a more proactive role in effectively reaching farming communities through excellent linkages with agricultural universities/colleges, ICAR institutes, NGOs, private companies and other key R&D players.

A Road Map for Innovative Extension

Considering the emerging challenges before Indian agriculture – existing constraints for technology transfer and options for scaling innovations for improving productivity and good agricultural practices around integrated NRM, opportunities for agricultural diversification, secondary and speciality agriculture, and options for linking farmers to market – a new road map for innovative extension systems is urgently needed to ensure faster scaling of innovations for greater impact. The following action points need urgent consideration:

- Establish agri-clinics, by encouraging well-trained individuals as small-scale private entrepreneurs, or by a group/club/association of progressive farmers. At least one agri-clinic/10,000–20,000 farm families needs to be established under the national mission, with funding provision of around Rs 5 million each (preferably on a 50-50 basis). Accordingly, to cover the existing 140 million farm families, 14,000 agri-clinics would be needed for which a budgetary requirement of around Rs 35 billion is to be met from the overall budget of the proposed National Mission on Agricultural Extension. Moreover, all agri-clinics may not be established in one go, and hence can be taken up in a phased manner over five years (needing around Rs 7 billion each year), based on the well-defined accreditation/recognition process.

- Induct farmer professors to facilitate farmer-to-farmer knowledge extension and skill transfer without dissemination loss, to provide vocational trainings for rural youth and farm women for 'Skill-up India' and 'Stand-up India' initiatives, build capability of Panchayats and ensure better support of existing institutions for technology/input delivery, credit, subsidy, insurance, value addition and marketing. To begin with, around five to ten farmer professors can be inducted in each district, for which budgetary provision of approximately Rs 500 million to 1 billion may be kept in the mission's overall budget.

- Establish a National Farmers' Innovation Fund (NFIF) of about Rs 1 billion with the support of both government and the private sector to encourage and involve progressive and innovative farmers to promote farmer-to-farmer extension and to support needed initiatives to build farmer–scientist links for outscaling innovations through testing, refinement and adoption on a large scale. It should also provide incentives and rewards in different forms to innovative farmers.

- A cabinet committee on farmers' welfare needs to be constituted to meet the aspirations of Indian farmers as well as those who are contributing to sustainable development and growth of agriculture. In particular, this committee has to ensure much-needed coordination and convergence for cohesive implementation of agriculture- and rural development-related programmes by different union ministries and government departments.

- Without further delay, concerted efforts need to be made for implementation of the recommendations of the High-Power Committee on the Management of Krishi Vigyan Kendras (KVKs), to ensure improved efficiency, effective monitoring and required relevance of farmer–science connections.

- Emphasis should be given to strengthening, coordination and modernization of KVKs rather than their further multiplication. For sector-wise strengthening of much-needed site-specific programmes/activities, there is a need to revisit the enhanced cadre strength of ten scientists/KVKs and to redeploy some subject matter specialists to take care of diversified/relevant areas such as horticulture, agroforestry, animal science, fisheries, post-harvest processing and social science.

- To establish Agricultural Technology Information Centres (ATICs) in all KVKs to promote 'land–lab' linkages and to reap benefits of research through promoting new innovations. There is a need to revisit existing ATMA-KVK convergence models and to bring in needed reforms concerning allocation of resources to meet contingent and exigency needs for training and knowledge/information sharing related to agriculture with local farmers through KVKs, and to shed redundancy and improve efficiency in all district-/local-level agricultural extension matters (Saravanan, 2010).

- To ensure expansion of scope of the proposed National Agricultural Education Project (NAEP), being funded by the World Bank and implemented by the ICAR, to address much-needed reforms in the public extension system and to strengthen capacity-development activities through informal training of private entrepreneurs to act more effectively as technology agents. The plan should be revised and implemented as the National Agricultural Education and Extension Project (NAEEP). This would trigger innovations by creative and skilled young minds for serving society and the agriculture sector with a human 'face'.

- Kisan Aayog (Farmers' Commission), on the pattern of Punjab and Haryana, needs to be established across the country in each state to facilitate the required transformation in agricultural extension, to promote both the national and local sustainable agricultural development agenda and to assist/advise the states in promoting relevant farmers' welfare-related policies and programmes based on well-defined and formally adopted state agriculture policies.

- Revamp agricultural extension-related education by initiating new courses on rural entrepreneurship, agricultural journalism, agri-business management etc., to bring innovative concepts and new economic options for rural youth. Also, there is an urgent need to teach agriculture as a subject for science students in the high schools to generate much-needed awareness of the role of agriculture in household/national food and nutritional security.

Conclusion

Extension should respond to both external and internal forces. Shifting from a traditional, top-down approach towards a more participatory approach is, therefore, the need of the hour. The challenges are technology and its transfer, and the process of problem-solving capacity-building. The extension workers must transform themselves from messengers to facilitators, for success in the changed scenario. The concept of participatory innovation development and extension is based on dialogue, farmer experimentation and strengthening of the organizational capacities of rural communities. The key steps are adopting the right kind of tools, like participatory rapid appraisal (PRA) tools, to obtain better results and introducing training programmes based on raising awareness through participatory, dialogue-based education. Newer extension methods including ICT and eAgriculture systems

27

Women's Empowerment
for Agricultural Development

Preamble

Agriculture is the backbone of the Indian economy. Women play a crucial role in building this economy (FAO, 2010–11). Over the years, there has been a gradual realization of the key role of women in agricultural development and their important contribution in the field of agriculture, food and nutritional security, horticulture, livestock, fisheries, processing, sericulture and other allied sectors. Rural women are thus the most productive workforce in the economy of developing nations like India (Kokate *et al.*, 2012). Their activities typically include producing agricultural crops, tending animals, processing and preparing food, working in agricultural and allied rural enterprises, collecting fuel and water, engaging in trade and marketing, caring for family members and maintaining their homes. Many of these activities are not defined as 'economically active employment' in the national context but they are critical for the well-being of rural households. Statistical data are available regarding their participation in the agricultural sector and allied activities but their impact on the home environment has not been accounted for (Gates, 2014). Variations in women's participation in agricultural work depend on supply-and-demand factors linked to economic growth and agricultural modernization. Farm women do have an impact on their children's education, as they often encourage them to be educated to have a better life.

Multi-dimensional Role of Women

It is a well-recognized fact that it was women who first domesticated crop plants and thereby led to the settlement of people and also the art of farming. While men went out hunting, women started gathering seeds and began cultivating them for food, feed, fodder, fibre and fuel. Women have played, and continue to play, a key role in conservation of basic life support systems such as land, water, flora and fauna (Kokate *et al.*, 2012). They have protected not only the health of soil, through organic recycling, but also accelerated the process of genetic diversity and conservation of crop plants. The multi-dimensional role of women includes: (i) agricultural activities like sowing, transplanting, weeding, irrigation, fertilizer application, plant protection, harvesting, winnowing, safe storage etc.; (ii) domestic activities like cooking, child-rearing, water collection, fuel-wood gathering, household maintenance etc.; and (iii) allied activities, which include cattle management, fodder collection, milking etc. Despite women's extensive and varied participation in agriculture, they continue to have less access to modern agricultural technologies. As a result, their intensive efforts invariably yield meagre economic returns (Kokate *et al.*, 2012).

Rural Indian women are extensively involved in agricultural activities. However, the nature and extent of their involvement differs with varied agricultural production systems. The mode of female participation in agricultural production varies with the landowning status of farm households. Their role ranges from managers to landless labourers. In the overall farm production chain, women's average contribution is estimated at 55–66%, with percentages much higher in certain regions. It has been observed that in the Himalayas, a pair of bullocks works 1064 hours, a man 1212 hours and a woman 3485 hours in a year on a 1 ha farm. This illustrates women's significant contribution to agricultural production (Kokate *et al.*, 2012). Women farmers, a quarter of the world's population, produce over 50% of the world's food and share 43% of the agricultural labour force. Women invest ten times more of their earnings than men on the well-being of the family, including family health, child health, education and nutrition (Akter *et al.*, 2017), yet they have less access than men to agricultural-related assets, inputs and services. Equal access and participation not only can help reduce gender inequality but also boost crop productivity by 20–30%, and raise overall agricultural output in developing countries by 2–4%. This gain in production could lessen the number of hungry people in the world by 12–17% besides increasing women's income. Women's empowerment thus has a direct impact on agricultural productivity and household food, and, more so, nutritional security. Similar sentiments have been echoed through the SDGs, adopted during the UN Conference on Sustainable Development in September 2015. Out of the 17 SDGs, one goal (5) is to 'achieve gender equality and empower all women and girls' (United Nations, 2015).

Limited Role of Women in Decision-making

In the developing world, women constitute the backbone of the agricultural and rural economy through their involvement in diversified activities, yet they remain one of the most vulnerable

and deprived groups. Women shoulder the entire burden of looking after livestock, bringing up children and performing other household chores. In India, women shoulder the most strenuous activities in farming, such as almost 50% transplanting and threshing, 27–30% harvesting, most fodder and livestock management activities and more than 60% in post-harvest operations. The nature and extent of their involvement differs with the variations in agro-production systems. However, the exact contribution, both in terms of magnitude and of nature, is often difficult to assess, and shows a high degree of variation across countries and regions (Kokate *et al.*, 2012).

Secondly, their access to material and social resources, as well as involvement in decision-making, is limited. In India, land ownership most often is with men, and they influence and dictate decisions concerning farming and family affairs. The agricultural produce marketing activities are often controlled by men, providing them with complete control over financial resources. The advent of the industrial revolution has significantly influenced societal set-up in the rural world, where more and more women are venturing into farming as men are migrating to urban areas for work. Women's hard work has not only been unrecognized but has also remained mostly unpaid. They are invariably paid lower wages than men for the same agricultural work, often not getting access to credit, as they are not involved in marketing activities and do not own land. On average, only 11% of women have access to land holdings, mostly as small and marginal farmers (World Bank and IFPRI, 2010). With grossly inadequate access to education and technology, a host of other socioeconomic factors have adverse impacts on the lives of women farmers. This gender inequality comes at a huge cost, not just for women but to society as a whole. Even in ancestral properties there is a taboo that the successor would generally be the son(s) and not daughter(s), although Indian law provides equal status in such cases. Strengthening women's ownership and use of rights, although vital, are not enough, as often men control women's agricultural property as well as the products of their work, much of which is carried out as unpaid labour. If the products of women's labour are sold, the men often control the income. Women for societal/cultural reasons are less involved in decision-making. Technology generation and dissemination in agriculture are often gender-blind, not addressing specific needs and constraints of women. Lately, owing to these reasons, there is a growing appreciation of the widening gap between the potential outcomes that could be achieved if women's empowerment in agriculture were to be adequately addressed. The thought process is, therefore, changing. Discrimination against women can undermine the overall economic development of a nation.

Women's Role in Innovation Systems

Unless innovation sensitively responds to the constraints faced by women, success may remain elusive for women. Women and men must be equitably educated and resourced. Historically, women have not been equipped to engage in the innovation-development process. Mechanisms for enabling women to exert influence over the setting of agendas in the innovation process, consequently, are of crucial importance. Women's role in post-harvest management, processing/storage and marketing are increasingly important in tying production to nutrition and income-generating outcomes. For example, in the dairy sector, innovations can bring refrigerated trucks, cooling tanks for safe storage as well as equipment for value-added products such as yoghurt, ghee and cheese. All these require a serious rethink about women's role for the future growth of agriculture. Farming systems have to be considered in a more inclusive and holistic way, covering the broad range of issues that would enable women-led innovations for accelerated rural development. These include women's role in the household, in particular, for child nutrition, patterns of household food security and consumption, contributions to rural income and emphasis on children's education, especially girls, thus addressing indirectly the check on population growth.

Changes in rural–urban dynamics are crucial, given the impact of seasonal and long-term migration, remittances and the growth of peri-urban and urban agriculture. The systematic and meaningful involvement of women in knowledge generation and innovation systems is essential, so that new innovations adequately reflect their needs. This implies that organizations, enterprises and individuals that conceptualize, invent or adopt new products, services, technologies and tools should involve women across the entire agricultural value chain. At the same time, biological and cultural factors can put women and girls at particular risk of undernutrition, malnutrition and poor health, especially during their reproductive period. Good agriculture, nutrition and health-related programmes must, therefore, take account of gender issues at all stages, making women both creators and beneficiaries of new policies and investment initiatives (Kokate *et al.*, 2012).

It is also a well-known fact that despite their eagerness, women have often not been able to take full advantage of opportunities arising from new technologies, innovations and markets. The constraints and opportunities that women face in agriculture vary across different agro-ecological and geographical regions of the country, depending upon, among others, socioeconomic and cultural contexts.

Empowering Women in Agriculture

Recognizing that empowerment of women is necessary to change the face of agriculture and the rural sector, there is a growing realization and commitment of the global community to achieve more sustainable and broad-based agricultural growth by addressing gender-related issues in agriculture through national, regional and global initiatives and partnerships (World Bank, 2012). There is also a greater degree of coordination, consultation and convergence of initiatives undertaken by the international institutions – Consultative Group on International Agricultural Research (CGIAR), Global Forum on Agricultural Research (GFAR), Food and Agriculture Organization of the United Nations (FAO), regional forums and many national agricultural research systems. One of the important global initiatives for transforming agriculture to empower women and deliver nutrition and income security was taken by the GFAR, called Gender in Agriculture Partnership (GAP). The GAP highlights the role of men and women as producers, developing participatory processes, addressing social norms and power relations in creating disparities, and puts a spotlight on women farmers as the backbone of agricultural and rural sustainability. The GFAR, among its many roles, works to highlight the important work of women farmers – as food producers, nutrition providers and caretakers, as scientists, innovators and teachers in villages and cities, in governments and in leadership roles in NGOs and as progressive farmers. Globally, gender equality has become an integral part of the R&D programmes being implemented by international organizations, NGOs, government agencies and International Agricultural Research Centres (IARCs) through the CGIAR Research Programme (CRP) on Gender.

India is at the forefront in acknowledging the role of women in agriculture. It established the world's first National Research Centre for Women in Agriculture (now ICAR-Central Institute for Women in Agriculture) in Bhubaneswar in 1996. The centre has been engaged in developing methodologies for identification of gender implications in farming systems approaches and developing women-friendly technologies under different production systems. Empowerment processes are strengthened through educational interventions, transfer of technologies, feasibility trials and knowledge-sharing. The centre also emphasizes undertaking vocational training to impart skills necessary to undertake different vocations and relieve women from drudgery by providing time- and labour-saving tools and equipment. Empirical evidence suggests that women have moved from beneficiaries to active partners in shaping empowerment. Recognizing the role of women in agriculture, Dr M.S. Swaminathan had proposed to move the 'Women Farmers Entitlement Bill' 2011 in the Rajya Sabha, which seeks, inter alia, access to water, credit, inputs and land ownership for women farmers as a policy reform to create an enabling environment.

Linking women, agriculture and nutrition requires multisectoral thinking and action to address major nutritional deficiencies that continue to hamper children's development around the world. Concurrently, it requires institutionalization of research and extension through joint decision-making that involves women themselves in participatory approaches. This needs to be incorporated during the initial design, which needs to be flexible so that it can be adapted to build on or address unintended agricultural consequences (positive or negative). It is

complex and difficult to intervene in a number of policy areas simultaneously and the actions need to be coherent at local, national, regional and global levels. Understanding how policies contribute (how are they working and why) requires evidence to share lessons and to learn about their effectiveness in different contexts. The efforts to monitor and track these impacts must be accompanied by appropriate indicators.

First Global Conference on Women in Agriculture

The government of India was instrumental in hosting a global partnership programme with the expectation that it will inspire other governments to follow. The ICAR, in partnership with the Global Forum on Agricultural Research (GFAR) and the Asia-Pacific Association of Agricultural Research Institutions (APAARI), helped bring together women farmers, policy makers and leaders from more than 50 countries at the first Global Conference on Women in Agriculture (GCWA) in New Delhi, 13–15 March 2012, wherein the importance of reducing the gender gap in agriculture was highlighted to ensure that men and women are equal partners in food and nutritional security. The then President of India, Smt. Pratibha Patil, in her valedictory address, emphasized the need to empower women with new knowledge and skills to bring them into the mainstream of agricultural development and reduce gender disparity, and corroborated that much of scientific knowledge and technologies do not reach rural women for various reasons. This needs a stock exercise to be undertaken in a holistic manner by drawing on the existing evidences of the impact of policies, institutions and programmes to empower farm women and to learn lessons for the future so as to ensure higher economic growth in the agricultural and rural sector. It was also stressed that research systems must seek the input of women as they have, historically, been the source of much traditional knowledge and innovation. While appreciating the efforts of NARES for bringing women to the forefront of agricultural R&D, she suggested forming Mahila Kisan Mandals (women farmers' cooperatives) in every village to educate women on the different aspects of agriculture and related activities, including their future predominant role in agricultural marketing to benefit producers and consumers.

The need for an Action Plan was emphasized to integrate and empower women for inclusive growth and development through an enduring global partnership programme on gender in agriculture. Such an action plan needs emphatic interventions by national and international agencies to ensure enhanced involvement and access to resources by women. It was felt that considering the urgency of addressing gender-related issues in agriculture, globally, a common knowledge-sharing platform on gender – Gender in Agriculture Platform for Gender in Agriculture Partnership (GAP4GAP) – was needed, which could help in collaborative working at the national, regional and global level. The platform should involve partnership from R&D organizations, national governments, regional and global forums, multilateral development agencies and donors and should act as a knowledge repository and provide space for both policy research and advocacy on gender-related issues in farming systems and rural ecologies. The GAP-4GAP can provide technical backstopping, a guide for future investments and facilitate effective networking and collaboration among partners and stakeholders. These gender-related initiatives would need generation and documentation of gender-segregated data, linking women's role to health and nutritional security at household level, enhanced visibility for the role of women, generation of knowledge and evidence for support and contextualization of global issues to suit local needs. Such new programmes on gender empowerment would require adequate resources for mobilizing women forming groups, improving capacity and capability in the technical, organizational and commercial (business micro-enterprises) sector, and support systems (credit, inputs, markets). These should be prepared jointly in consultation with women and other relevant organizations (public, private and voluntary), which can, potentially, complement and supplement the efforts of other stakeholders.

Empowering women and girls deserves a high priority in the developmental agenda but it needs sustenance by radically reorientating the agricultural research agenda to overcome existing gaps and to face emerging challenges of sustainable development and livelihood of resource-poor smallholder farmers, especially women farmers. In the past decade there has been a significant growth in women's self-help groups (SHGs) and enrollment of girls in the agricultural education system. These two significant socioeconomic changes in the developing countries can play a pivotal role in empowering women and transforming rural areas. There is an urgent need to support women's education and SHGs to develop future professionals, entrepreneurs and farmers. In India, about 40% of girl students are currently studying in varied courses of the SAUs. They should be trained and motivated to act as facilitators to empower women in agriculture. A special fund, the Women's Empowerment Fund, must be created at the national level to support gender-specific welfare-associated programmes. Banks and micro-credit services can play a pivotal role in such initiatives.

Women and Household Nutritional Security

Empirically, a strong linkage among agriculture, nutrition and empowerment of women is well-established. Malnutrition is a big problem in developing countries, especially in girls in rural areas. Nutritional insecurity is a complex issue and involves a multi-sectoral solution. Control of women over household income is linked invariably with improved nutrition, health and education of children. For household nutritional security, efforts are needed to integrate scientific and socioeconomic aspects to empower women and form a 'nutrition umbrella base', which can help in developing an integrated strategy. Enhanced government investment, awareness, capacity-building and micro-enterprises should supplement these endeavours. Scientific institutions should produce effective technologies, database, knowledge on nutrition-rich food, and value addition by involvement of women's groups for nutritional security. AR4D systems have to move towards innovations not only in nutritional aspects but also in increasing women's farm work efficiency and reducing drudgery in farm operations. This process would require reorientation towards more gender-sensitive innovations with emphasis on empowerment of farm women, including their financial empowerment. The new business models and agricultural marketing strategies should encourage women members as part of producer and marketing associations. Such efforts should be backed by an overall strategy to improve market access through development of market infrastructure and better access to information through ICT. For this, there is a need to revisit our agricultural education system, to encourage innovations in research outscaling/marketing pathways and to augment the role of women in policy planning and decision-making.

Overall, the gender issues are dynamic and so are agricultural production systems, socio-economic aspects and environmental factors. Already, climate change and weather-related aberrations are influencing adversely farmers' livelihood and agricultural sustainability. These aberrations will have direct and indirect effects on farm women. The future strategy should thus include vulnerability assessment of farm women to climate-related risks and pathways to participate in opportunities such as adaptation and mitigation options. While providing compensation for environmental services, farm women should also be considered as beneficiaries for their role. All stakeholders must come together to understand gender issues better and share their experiences as to what works and what does not for women's empowerment in agriculture. Considering the urgency of addressing all gender-related issues in agriculture across the world, more policy support and institutional mechanisms are required for achieving desired results. Collective action is obviously required for empowerment of women so that they come together on a single platform to address concerns.

Ensuring Visibility of Gender

Despite growing evidence of the substantial role of women in agriculture and household food and nutritional security, many policy makers, agricultural scientists and development professionals are yet to recognize their important role in agriculture. As a result, agricultural policies and R&D programmes in many countries continue to be gender-blind, ignoring the importance of women's work and the complexity and sensitivity of many of the barriers that constrain farm women's ability to perform and contribute effectively to the economic status of their families and society. Ironically, most rural women are not so conscious of the economic and social importance of their work, and are hesitant to demand recognition or rights.

Rural women's invisibility has, in fact, been caused by society's neglect, in general, and by agricultural policy makers and professionals in particular. This is a cause for the under-performance of women in the agriculture sector. Hence, to address these concerns, complementary strategies and mechanisms are needed that can help increase women's visibility and roles in: (i) agricultural value chains (crops, horticulture, livestock, forestry, fisheries); (ii) household food, nutrition and health security; and (iii) research, education, extension and policy-making organizations. Also, we need to demonstrate, using justified quantitative and qualitative data, the value of women's contribution, and let their voice be heard in future decision-making at various levels, by promoting women's leadership in building the social fabric of society at all levels (IFAD, 2013).

The Way Forward

Women play a vital role in a wide range of agricultural activities, thus contributing to sustainable agricultural development. To achieve inclusive agricultural growth, it is necessary to empower women, address gender issues, drudgery of women and their health and nutritional status. Further, these issues are to be addressed through gender-friendly technology assessment, refinement and extension methodologies. If we look at women's role in food production, we can notice enormous discrimination; women in the sector receive less than 10% of credit offered to small-scale farmers. The FAO has estimated that if women farmers had the same access as men, agricultural output in 34 developing countries would have risen by an estimated average of up to 4%. This would have reduced the number of undernourished people in those countries by as much as 17%, translating to 150 million fewer hungry people. Thus, investments in women and overcoming their drudgery are perhaps the best actions for future development. The evidence is clear: when women farmers have the opportunity to earn and control the home income, they are more likely to spend on their children's nutrition, education and health. Improving the knowledge and status of women would, therefore, deliver significant outcomes in terms of agricultural production, food security, child nutrition and health and education, thus contributing significantly towards SDGs. In view of this, urgent action is required at the national, regional and international levels on the following:

- There is a need for collective advocacy to raise awareness of women's needs in agriculture and to ensure their visibility in terms of their valuable contribution towards agricultural development.
- Women need to be educated and empowered to make their own choices for better farming options and for responding to new opportunities for diversified agriculture and better living.
- Women's abilities needs to be enhanced in order for them to actively participate in the development processes by changing their perceptions and increasing their opportunities for greater social responsibility.

- There is a need to encourage collective action and leadership among women to develop programmes that directly address women's needs and make agricultural support systems gender-sensitive.
- Sincere efforts need to be made for removing drudgery from farm women by ensuring access to new tools and implements that increase efficiency and higher productivity. Also, the AR4D agenda needs to be made gender-sensitive and pro-women.
- Urgent attention is needed to address discrimination through appropriate policies, legislation, enforcement mechanisms and establishing women's rights (e.g. access to markets, ownership of land etc.).
- It must be ensured that institutions and legal support mechanisms are in place to promote women's ownership and control of resources (e.g. land, bank accounts, farm implements).
- Social, educational and cultural institutions must change to create an environment where women realize their full potential. For this, investment in women's human capital through education and training for skills development is critical to exploit their abilities, time and energy.

References

Akter, S., Rutsaert, P., Luis, J., Htwe, N.M., San, S.S., Raharjo, B. and Pustika, A. (2017) Women empowerment and gender equity in agriculture: a different perspective from Southeast Asia. *Food Policy* 69, 270–279.

FAO (2010–11) Women in agriculture: closing the gender gap. In: *The State of Food and Agriculture*. Electronic Publishing Policy and Support Branch, Office of Knowledge Exchange, Research and Extension, Food and Agriculture Organization of the United Nations, Rome, pp. 1–149.

Gates, M.F. (2014) Putting women and girls at the center of development. *Science* 345, 1273–1275.

IFAD (International Fund for Agricultural Development) (2013) *Gender Equality and Women's Empowerment*. Rome. Available at: www.ifad.org/gender/policy/gender_e.pdf (accessed 15 June 2018).

Kokate, K.D., Srinath, K., Mal., B., Adhiguru, P., Dash, H.K., Chahal, V.P. *et al.* (eds) (2012) *First Global Conference on Women in Agriculture (GCWA): Proceedings*. New Delhi, 13–15 March. India Council of Agricultural Research (ICAR), New Delhi, and Asia Pacific Association of Agricultural Research Institutions (APAARI), Bangkok. Available at: http://www.taas.in/documents/pub30.pdf (accessed 15 June 2018).

United Nations (2015) *Transforming Our World: The 2030 Agenda for Sustainable Development*. United Nations, New York. Available at: https://sustainabledevelopment.un.org/post2015/transformingourworld (accessed 15 June 2018).

World Bank (2012) *World Development Report 2012: Gender Equality and Development*. World Bank, Washington, DC.

World Bank and IFPRI (2010) *Gender and Governance in Rural Services*. World Bank, Washington, DC.

28

Attracting and Retaining Youth in Agriculture

Introduction

The global population may reach 9 billion by 2050, and youth would represent around 20% (FAO, 2014). Most young people (around 85%) live in the developing countries (UNDESA, 2011). India has a comparative advantage over other countries in terms of the distribution of its young population. As per India's census, the total youth population increased from 168 million in 1971 to 422 million in 2011. In 2017, it reduced to 356 million (10–24-year-olds), against China's 269 million. India's population has been observed to remain young longer than China's and Indonesia's, the two major countries, along with India, that determine the demographic features of the Asian continent (CSA, 2017). India also enjoys a demographic dividend with more than 60% of its population of working age. According to a World Bank report, in India, the working-age population will outnumber the dependent population for at least three decades (until 2040). As per the National Higher Education Commission (NHEC) estimates, the average age of the Indian population in 2020 will be 29, as against 40 in the USA, 46 in Europe and 47 in Japan (British Council, 2014). Agriculture still remains the key sector, providing livelihood and employment opportunities to more than 60% of India's population living in rural areas. Overall, in the developing world, youth and agriculture are the twin pillars of progress and prosperity, keys to achieving global SDGs (Paroda et al., 2014).

The progress and prosperity of a nation depend, to a large extent, on its well-trained, enlightened and disciplined youth. Indeed, young people are a major resource and agents of change for overall growth and development, as they possess tremendous enthusiasm, creativity, energy, imagination and dedication. The energy and passion of youth, if harnessed properly, can bring significant positive change in all sectors, including agriculture and society as a whole (Saharawat et al., 2013). Young people are creative digital innovators and active citizens eager to contribute positively towards SDGs. While the world's youth cohort is expected to grow, employment and entrepreneurial opportunities for youth, particularly those living in developing countries' economically stagnant rural areas, remain limited, poorly remunerated and of poor quality (Percy-Smith and Akkermans, 2011–12). Therefore, it is vital that young people are brought into the mainstream of agriculture.

Major Challenges

In the recent past, retaining youth in agriculture has been one of the major challenges in the developing world. The principal challenges in retaining youth in agriculture include: insufficient access to knowledge, information and education; limited access to land; inadequate access to financial services; lack of formal and informal on-the-job training; limited access to markets; and limited involvement in decision-making and policy dialogues (Saharawat *et al.*, 2013). Over the years, the community has become gradually poorer due to small land holdings, which comprise over 80% of total farm households. Multiple risks associated with agriculture intensify the challenges owing to over-exploitation of natural resources linked with rapidly increasing globalization, soaring fuel and food prices, volatile markets and growing climatic volatility. Youth is a great resource, to be used for agricultural development. In the past few decades, because of rapid industrialization and urbanization, youth and agriculture are experiencing unprecedented transformation. Another major dilemma in the developing world is the poor social image of agriculture and, hence, rural youth are moving towards the urban sector, looking for alternative and better opportunities (Paroda *et al.*, 2014). It is evident through successful business models of leading public and private sector organizations, as well as multinational companies (e.g. the IT sector), that youth are more innovative and productive as well as receptive to new technologies. On the contrary, in the agriculture sector there is a wide gap between energy (youth) and experience (older people), which is a cause of the backward nature of farming and the slow adoption of innovations and new technology. These are huge losses in the technology dissemination process, delinking science with society and making farming non-remunerative, non-resilient and unattractive to youth (Saharawat *et al.*, 2013).

Under the above scenario, agriculture is not a remunerative and respectable profession, particularly for youth, and is not a sustainable pathway to meet food, nutrition and livelihood security. The challenges are complex and interwoven. Therefore, youth has to be motivated through advances in innovation, capacity development, partnership and a participatory approach, through enhanced skills and a positive attitude towards their role in the overall agricultural and rural development of the country.

Role of Youth in Agriculture

The challenge to retain youth in agriculture has been recognized globally. It first figured prominently in 2006 during the global conference organized by the Global Forum on Agricultural Research (GFAR) in New Delhi. The deliberations resulted in an agreement to form a youth-led international forum, which led eventually to the formation of Young Professionals in Agricultural Research for Development (YPARD) and the first Global Conference on Agricultural Research for Development (GCARD 1), held at Montpellier, France, in 2010. The importance of youth in agriculture was further emphasized and structurally debated during GCARD 2, organized at Punta del Este, Uruguay, in 2012. GCARD 2 had put forth 'Youth and Agriculture' as one of the topics for focal discussions. The chair of GCARD 2's organizing committee emphasized that, globally, agriculture is considered an ageing and undervalued profession and youth needs special encouragement in all aspects of AR4D. As a follow-up to the GCARD 2 discussions, ICAR, in association with APAARI and TAAS, organized a national workshop on 'Foresight and Future Pathways of Agricultural Research through Involvement of Youth in India', in March 2013, at the National Agriculture Science Centre (NASC) complex, New Delhi. About 300 participants from different ICAR institutes and agricultural universities, including young farmers, students, private sector representatives and senior mentors, attended the workshop. The workshop was to debate the role of youth, being an important critical mass in ICAR, in meeting agricultural R&D needs. Currently, the country has around 7000 agricultural scientists in India's public sector, NARS, of which more than 35% are below the age of 40. The two days of deliberations covered a wide range of disciplines and issues related to Indian agriculture, natural resource management, crop improvement and protection, horticulture, post-harvest technology, livestock and fisheries development, agricultural engineering and implements, ICT and socioeconomics. The deliberations identified research needs across disciplines and regions where youth can play a prominent role. The key recommendations of the deliberations include: the urgent need to reorientate agricultural research towards a farming systems mode by ensuring interinstitutional and interdisciplinary collaboration; creating state-of-the-art research facilities; undertaking joint research with the private sector and international/advanced research centres through the creation of excellent research infrastructure; provision of a seed grant (Rs 1-1.5 million); encouraging scientists to initiate research; short- to long-term training for young scientists at advanced research institutions; emphasizing greater involvement of women in decision-making bodies; and greater emphasis on human resource development through special allocation of funds for skills development (Saharawat *et al.*, 2013).

Retaining Youth in Agriculture

The Attracting and Retaining of Youth in Agriculture (ARYA) programme was initiated by the ICAR after deliberations in the workshop, and is being implemented successfully by the Krishi Vigyan Kendras (KVKs) in different states of India. Overall, the deliberations led to the development of a road map to define and delineate pathways for developing and nurturing a new generation of young agricultural professionals and entrepreneurs, with greater emphasis on technical capacity development, institutional arrangements, innovative networking, appropriate investments and harnessing the full potential of youth, in order to realize a qualitative change in their lives.

The government formulated a National Policy for Skill Development and Entrepreneurship in 2015 to provide an umbrella framework for all skill development activities carried out within the country, to align them to common standards and to link with demand centres. More than 50% of the Indian population is involved in the agricultural sector but hardly 5% of rural youth are involved in agriculture as a profession. Rural youth are an important means to achieve accelerated agricultural and rural development. Accordingly, effective channelling of this resource to constructive activities can contribute to increased prosperity for all. On the contrary, the current developmental models spur migration of educated and skilled youth away from agriculture, leaving a scarcity of skilled and progressive farmers/entrepreneurs in the rural and agricultural sector. Rural youth has been deprived of minimum facilities, needed opportunities and encouragement in innovative farming over time. Thus, most of the youth who remain in agriculture have limited knowledge and skills and are being forced to find new opportunities in other sectors. As a result, there is an ongoing exodus of rural young men and women from villages to towns and cities, affecting, adversely, rural development and agricultural growth. Considering the huge knowledge and skills gap in the agricultural sector, there is an urgent need to assess skills required within the sector to make it sustainable, entrepreneurial and attractive to youth. The skill development and entrepreneurship programme thus needs greater emphasis on vocational training of rural youth.

In the Asia-Pacific region, the challenges and opportunities for youth in the agricultural profession do not differ much. Different countries are tackling the issue of involving agri-professionals in the farming sector. There are several youth-led successful models for transforming agriculture in the countries. However, these models lack an appropriate mechanism for regional and cross-border learning from different countries' experiences. Keeping these challenges and opportunities in view, a regional workshop on 'Youth and Agriculture: Challenges and Opportunities' was organized jointly by APAARI and the Pakistan Agricultural Research Council (PARC) at Islamabad, Pakistan, in October 2013, in collaboration with GFAR, CIMMYT, the International Center for Agricultural Research in the Dry Areas (ICARDA), the International Center for Research in the Semi-Arid Tropics (ICRISAT), the International Food Policy Research Institute (IFPRI) and Bioversity International. The deliberations highlighted the emerging phenomena of over-urbanization and growing youth unemployment, which are leading to social disparity, on the one hand, and global food insecurity on the other. Prioritizing investment for attracting youth is, therefore, crucial for future agricultural development. Greater and active involvement of youth in farm advisory, empowering them with knowledge to serve society through the creation of technology-led business models and providing value-added services and creating employment opportunities, is the way forward for enhancing agricultural productivity for a food-secure society. This needs a paradigm shift in our approach and policy focused on youth to transform youth from job seekers to job creators.

Capacity development of youth through informal and vocational training and creating awareness of new opportunities in agriculture, including secondary and speciality agriculture, would attract youth in agriculture, help bridge the gap between rural and urban and boost rural economies in the region. The local institutional, national and regional leaderships in the Asia-Pacific region, therefore, need to take initiatives for greater involvement of youth in policy planning and prioritization of investment for shaping their future in farming and preparing them professionally for tomorrow's agriculture and the task of feeding, sustainably, a projected global population of 9.2 billion by 2050. The key points to emerge from the regional consultation include: (i) reorientation of agriculture to agricultural research for results (AR4R) by promoting agri-innovation; (ii) agri-business and entrepreneurship through involvement of youth at national, regional and international levels; (iii) urgently linking agriculture with health, environment, nutrition and other basic science disciplines to address challenges by young professionals; (iv) focusing attention on capacity development of youth through vocational training; (v) inclusion of agricultural education in the school curriculum and farmers' participatory approach to technology generation; and (vi) transfer and adoption to ensure faster growth in agriculture. Innovative approaches to developing and transferring technologies, efficient funding mechanisms, openness in knowledge-sharing, much-required marketing reforms and partnership at national and regional level are important areas to pursue; and to make agriculture intellectually rewarding for youth, special emphasis is needed on secondary agriculture, diversification, protected cultivation, crop intensification and use of ICT (TAAS, 2017).

Agriculture is one of the largest employment-generating sectors. Therefore, there is a need to create awareness among youth regarding emerging opportunities. In south Asian countries, existing administrative structures, lack of prioritization of R&D, fragmentation along disciplinary lines, poor coordination and volatile public funding are some of the real impediments that need to be overcome soon through proper policy advocacy and public-awareness mechanisms. Also, there is an urgent need for strong political will and an enabling policy environment for greater involvement of youth in AR4D initiatives. For this, there is a need to focus more on foresight, research partnership and capacity development. A regional network is urgently needed in the overall interest of future agricultural growth for sharing knowledge, innovations and expertise in similar target environments and socioeconomic settings. For this, international organizations, namely the FAO, IFAD, the World Bank, the Asian Development Bank, CG centres and regional organizations like APAARI, the Association of South East Asian Nations (ASEAN) and the South Asia Association for Regional Cooperation (SAARC) need to devise appropriate mechanisms involving the NARS of the region. The way forward, therefore, warrants providing unblemished and tangible pathways for engaging youth in agriculture through developing and practising farm youth- and gender-friendly agricultural technologies, practices and policies.

In view of the current agricultural challenges, increasing youth population and rapid globalization, developing world agriculture would require a paradigm shift in the mindset, from traditional agriculture as the means of livelihood to a business-oriented, specialized agriculture involving skilled youth in rural areas. It is obvious that empowering youth in agriculture would be an important vehicle for change. The current agricultural occupation scenario has to be made remunerative through scaling new innovations and entrepreneurships. It is clear that quality/skilled youth can only be attracted and retained in farming if it becomes economically rewarding and intellectually satisfying, associated with improved rural infrastructure and better educational and primary healthcare facilities. The comprehensive strategies for plausible transformation in future would demand more rewarding jobs in all agro-based and agro-related activities with equal opportunities and facilities in rural and urban areas, better options for public–private sector investments in agriculture and rural-sector infrastructure, and promotion of small agri-firms and producer companies to promote agri-food and value-chain systems (GLF, 2014). To empower rural youth, including women, there is an urgent need to transform the extension system into an innovation extension platform that delivers technology-orientated knowledge, inputs and value-added services. The extension approach would have to focus around farming communities rather than an individual farm household approach, as was the case in the past.

Looking Ahead

The present situation demands skill development of rural youth through vocational training and building a cadre of technology agents to provide technical backstopping as well as custom-hire services to smallholder farmers. Another strategy could be to create agri-clinics, where technology agents can join hands to ensure a single-window system of advisory services. In future, efficient agro-advisory in the wake of increasing demand for quality and new agricultural knowledge, together with input support, can be best delivered through pluralistic agricultural extension, i.e. a mix of public and private sector involving participation of youth, in particular. The emergence of private sector institutions such as corporate organizations, community-based organizations, young farmers' associations, farmers' cooperatives, self-help groups, watershed and water-user associations, producer companies, non-governmental organizations, farmer producers, input providers, service providers, para-professionals (Kisan Mitras etc.), input producers, the corporate sector, organic and inorganic mix fertilizer companies and rural-based, low-cost primary processing enterprises can all be encouraged to save smallholder farmers in India. These specialized agri-knowledge services would help promote speciality, secondary, diversified, value-added and entrepreneurial agricultural systems. Such entrepreneurship platforms would not only empower youth to become knowledge agents but would also attract and retain them in agriculture. Overall, these endeavours would certainly enable agriculture to become a reputable profession.

References

British Council (2014) Understanding India: the Future of Higher Education and Opportunities for International Cooperation. Available at: https://www.britishcouncil.org/sites/default/files/understanding_india_report.pdf (accessed 16 June 2018).

CSA (2017) *Youth in India*. Social Statistics Division, Central Statistics Office, Ministry of Statistics and Programme Implementation. Available at: http://mospi.nic.in/sites/default/files/publication_reports/Youth_in_India-2017.pdf (accessed 16 June 2018).

FAO (2014) Youth and Agriculture: Key Challenges and Concrete Solutions. Working paper by FAO, IFAD and CTA. Available at: http://www.fao.org/3/a-i3947e.pdf (accessed 16 June 2018).

GLF (2014) Building Cross-cutting Skills and Landscapes Knowledge for Effective Youth Leadership. On the sidelines of 20th conference of the parties to the UN Framework Convention on Climate Change (COP20), Lima, Peru, 5–7 December.

Paroda, R., Ahmad, I., Bhag, Mal, Saharawat, Y.S. and Jat, M.L. (eds) (2014) Regional workshop on youth and agriculture: challenges and opportunities. Proceedings and Recommendations, Islamabad, 23–24 October 2013.

Percy-Smith, A. and Akkermans, L. (2011–12) *Working Towards a New Generation of Young Professionals in ARD*. YPARD (Young Professionals' Platform for Agricultural Research for Development), Rome.

Saharawat, Y.S., Dhillion, M.K., Bhattacharyya, R., Jat, M.L., Dadlani, M., Gupta, H.S., Ayyappan, S. and Paroda, R.S. (2013) Foresight and future pathways of agricultural research through youth. *Proceedings and Recommendations*. NASC complex Pusa, New Delhi, 1–2 March.

TAAS (2017) Scaling conservation agriculture for sustainable intensification in South Asia – a regional policy dialogue. *Proceedings and Recommendations*. Trust for Advancement of Agricultural Sciences, New Delhi.

UNDESA (2011) World Population Prospects: the 2010 Revision, Highlights and Advance Tables. Working Paper ESA/WP 220.

29

Revitalizing the Indian Agricultural Education System

Agriculture is an integral part of the socio-economic fabric of India, sustaining the livelihoods of over 60% of rural households and providing employment to nearly the same percentage of the population. The sharp rise in India's post-independence population has been matched by a commendable rise in foodgrain production, starting with the Green Revolution of the early 1970s. This production touched a record high of 277.49 million t during 2017–18, with a remarkable increase of 23% (around 4 million t) in pulse production. The agricultural research system comprising researchers, teachers and extension workers spread all over the country had been the backbone of this growth. However, continuing to achieve such production gains to ensure future food security for an ever-increasing population is likely to be a challenging task. By 2030, we would need to produce 70% more foodgrains than we are producing today; that in the face of multiple challenges like climate uncertainties, depleting natural resources, shrinking farm sizes and indiscriminate and imbalanced use of chemical fertilizers and pesticides. The need to strengthen the agricultural research system, including education, is, therefore, critical to build capable human resources that are vital for future growth.

largely devoted to teaching, research and extension carried out by the state departments (Randhawa, 1968). The first agricultural college was established in 1877 at Saidapet, which was later shifted to Coimbatore. Agricultural education was initiated in Bengal Engineering College in 1898. The Imperial Agricultural Research Institute was established at Pusa, north Bihar, in 1905, followed by the establishment of agricultural colleges at Coimbatore, Nagpur, Kanpur, Pune and Sabour during 1906–1908.

During the early post-independence period, a large number of agricultural colleges affiliated to traditional universities were opened, but due to lack of resources, the quality of education remained poor and unable to produce appropriate human capital to address the growing demands for food and other agricultural commodities. Recognizing this deficiency in the education system, the University Education Commission (1948) recommended the establishment of 'rural universities', followed by the report of the Joint Indo-American Teams in 1955 and 1959. This led to the establishment of the concept of modern agricultural universities in India based on the US land grant system, beginning in 1960 with the establishment of the first SAU at Pantnagar, Nainital, Uttar Pradesh.

Agricultural Education

At the time of independence, India had 17 agricultural and veterinary colleges, which were

US Land Grant System

The US land grant system of agricultural education took shape with the enactment of the

Morrill Act in 1862, which enabled US states to provide public land, and using the proceeds of these lands to establish colleges that would teach agriculture and mechanical arts (Committee on the Future of the Colleges of Agriculture in the Land Grant, University System, National Research Council, 1996). The Second Morrill Act (1890) provided for annual appropriation to the states to support their land grant colleges, most of which were subsequently transformed into fully-fledged universities. While the Morrill Act of 1862 gave the mandate to teach, the mandate for research was granted to these colleges by the Hatch Act of 1887, as a consequence of which state agricultural experimental stations (SAESs) were established. The SAESs later evolved into on-campus and off-campus establishments as well as branch stations. The faculty of these stations is entitled to access Hatch research funds, which are administered by the US Department of Agriculture (USDA). The responsibility for 'extension' was given to the land grant colleges by the 1914 Smith Lever Act, which was to be carried out through cooperative activity between the states and the federal government through USDA and the land grant colleges. Subsequent legislation incorporated: emphasis on systems role, funding for forestry research, formula funds for research in animal sciences, a competitive grants programme and a new research initiative on environmental studies. Overall, teaching, research and extension have been three basic operational modes of the land grant system. The land grant system has served the US exceedingly well with some of the universities like Wisconsin and California having grown into world-class institutions of education and research not only for agricultural but also for several other disciplines.

The Modern Indian Agricultural Education System

The above-mentioned joint Indo-American teams, in a report submitted in 1960, recommended that assistance in the establishment of agricultural universities should only be granted when these commit to basic principles such as: (i) autonomous status; (ii) location of agricultural, veterinary, animal husbandry, home science, technological and science colleges on the same campus; (iii) integration of teaching by offering courses in any of these institutions to provide a composite course; and (iv) integration of education, research and extension functions.

In 1960, the government constituted a committee headed by Dr R.W. Cummings to advise the state governments on the establishment of agricultural universities. On the basis of the recommendations of this committee, the ICAR developed a model Act to be adopted by the universities. The first proposal for opening an agricultural university on the above pattern was submitted by the government of Uttar Pradesh, which culminated in the establishment of Uttar Pradesh (now G.B. Pant) Agricultural University at Pantnagar, District Nainital, in 1960. In the initial stage, the US Agency for International Development (USAID) supported one-to-one hand-holding of Indian agricultural universities by US agricultural universities. For example, Uttar Pradesh Agricultural University and Jawaharlal Nehru Krishi Vishwa Vidvalaya established links with the University of Illinois; Punjab Agricultural University and the University of Udaipur with the Ohio State University; and Orissa University of Agriculture and Technology with the University of Missouri. The assistance included training of an Indian faculty in US universities and participation of US specialists in teaching, research and extension activities in these universities in India.

Presently, the agricultural education system being operated under the Indian Council of Agricultural Research comprises SAUs (62), deemed-to-be universities (5), central agricultural universities (2) and central universities (4) with agriculture faculties. The staff and student positions of these universities are summarized in Table 29.1. Within the ICAR, the Agricultural Education Division has the mandate: (i) to plan, promote and coordinate agricultural education in the country; (ii) to enhance the quality and relevance of higher agricultural education in the country; and (iii) to strengthen the agricultural university system for developing quality human resources in agriculture and allied sciences. This is proposed to be achieved through appropriate planning, development, coordination and quality assurance in higher agricultural education in India. Thus, ICAR has the mandate on a par with the University Grants Commission (UGC) to coordinate functions related to agricultural education and its federal funding and monitoring of

Table 29.1. Staff and students in two agricultural universities with maximum and minimum strengths along with the average of all universities (2012–13). (From: http://www.iauaindia.org/introduction.htm). The analysis is based on 56 universities for which complete information was available.

University	Staff			Students		
	Approved	Actual	%	Approved	Actual	%
KU, Gandhinagar	22	28	136.4	107	103	96.3
PAU, Ludhiana	1387	792	57.1	1175	1072	91.2
Average of all universities	475	321	64.8	1321	896	89.8

educational quality standards through a process of accreditation and support for new infrastructure and programmes.

Ranking of agricultural universities

The Ministry of Human Resources Development (MHRD), Government of India, under its National Institute Ranking Framework (NIRF), ranks all the Indian universities based on: teaching, learning and resources; research and professional practice; graduation outcome; outreach and inclusivity; and perception. Among the universities in the NIRF 2017 list of 100 top universities are listed six agricultural universities: Indian Agricultural Research Institute (23), Tamil Nadu Agricultural University (28), Punjab Agricultural University (40), Tamil Nadu Veterinary & Animal Science University (60), Anand Agricultural University (62) and Dr Y.S. Parmar University of Horticulture & Forestry (84) (NIRF, 2017). Obviously, most other universities would need to make greater improvements in required parameters to find a place among the top institutions.

Lately, the Indian Council of Agricultural Research has also evolved a system of ranking of the country's agricultural universities. Table 29.2 lists the top ten universities in terms of their ranking for the year 2016–17 (Agricultural Education Division, 2017).

Changing Needs of Agricultural Education

While the country has made commendable progress in agricultural production, the challenges of meeting the food and nutritional security for the ever increasing population call for accelerated efforts towards enhancing the quantity

Table 29.2. Ten top universities in ICAR ranking (2016–2017). (From: Agricultural Education Division, 2017; https://icar.gov.in/files/071715062804_0au-ranking-2017.pdf)

Rank	Name of agricultural university
1	ICAR-National Diary Research Institute, Karnal
2	ICAR-Indian Agricultural Research Institute, New Delhi
3	Punjab Agricultural University, Ludhiana
4	Chaudhary Charan Singh Haryana Agricultural University, Hisar
5	ICAR-Indian Veterinary Research Institute, Bareilly
6	University of Agricultural Sciences, Bangalore
7	Tamil Nadu Agricultural University, Coimbatore
8	G.B. Pant University of Agriculture & Technology, Pantnagar
9	Guru Angad Dev Veterinary and Animal Sciences University, Chennai
10	Tamil Nadu Veterinary & Animal Sciences University, Chennai

and quality of food so produced. As already stated, by 2030, the country would need to produce 70% more foodgrains than currently, in the face of multiple challenges. To achieve this, the country will need to strengthen and vastly improve its agricultural research and education system to develop capable human resources, so critical for making faster agricultural growth.

While the country possesses a vast network of agricultural education institutions, it is widely acknowledged that the quality and kind of education has not kept pace with the rapid advances being made in agriculture related research and development around the world. Available statistics give a clue to the reasons for this shortfall.

Table 29.1 reveals large discrepancies among universities in staff and student numbers and the large shortfall in the required numbers of staff compared to students. Public sector spending in agricultural research has shown a decline in real terms after an initial flip during 2000–2012 (Stads, 2016; Fig. 29.1). Invariably, more than 70% of the budget is being spent on salaries and the total number of available researchers (research and teaching staff) has fallen considerably since 2000 (Stads, 2016; Fig 29.2).

Revisiting the Land Grant System

As mentioned earlier, starting with the first SAU at Pantnagar in 1960, there has been phenomenal growth in universities, now numbering 73. With more than 57 years of experience, a need to revisit the agricultural education system is being realized to ensure needed reforms and improvements. In the past, however, periodic efforts were made to review the agricultural education and research system. Table 29.3 provides details of such reviews undertaken for the needed reforms from time to time. However, a holistic view of the needed midcourse corrections is urgently needed.

A comprehensive dialogue on the Indian agricultural education system was held by the National Academy of Agricultural Sciences (NAAS) during its Agricultural Science Congress in Bhubaneshwar in 2013, in which many leaders from around the world participated and

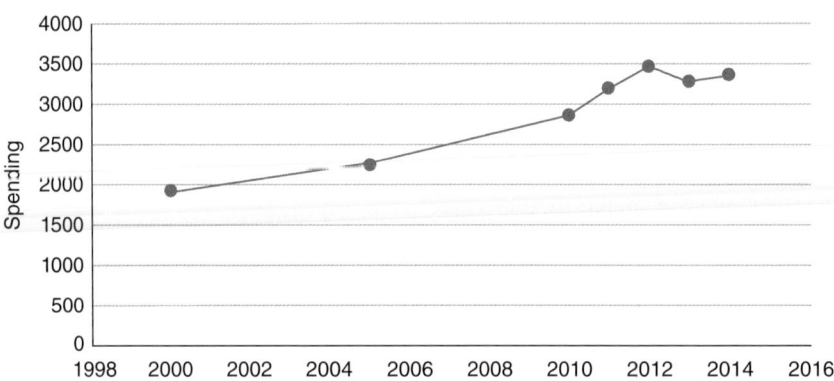

Fig. 29.1. Public Sector spending on agricultural research in India (in millions of PPP (purchase power parity) $). (From: Stads, 2016)

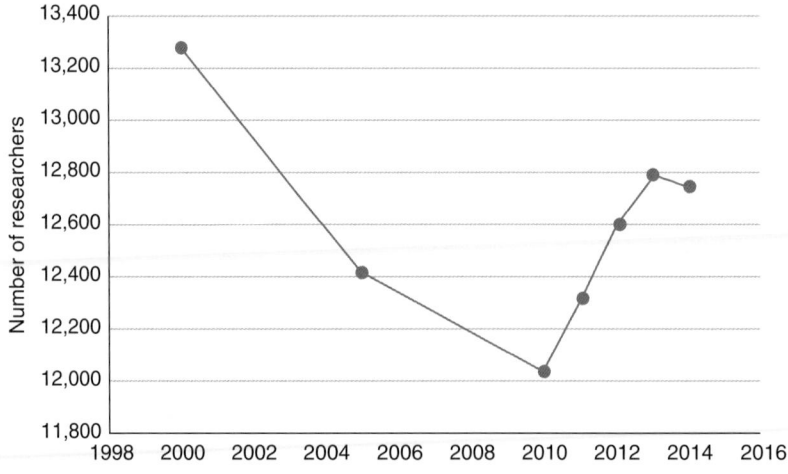

Fig. 29.2. Total number of agricultural researchers (in full-time equivalent). (Source: Stads, 2016)

Table 29.3. Review of the Indian agricultural education system. (From: Makwana, 2013; Singh, 2014)

Year	Event
1966	ICAR developed Model Act for agricultural universities
1974	Standing committee on agricultural education replaced by Norms and Accreditation Committee (NAC)
1994	National Assessment and Accreditation Council (NAAC) established to assess and accredit higher education institutions
1988	G.V.K. Rao Committee recommended revamping of ICAR including agricultural education
1996	NAC replaced by Accreditation Board
2007	Revision of UG course curricula and syllabuses, and norms
2009	Revision of PG course curricula and syllabuses, and norms
2013	Bhubaneswar Declaration on Transforming Agricultural Education for Reshaping India's Future

presented valuable experiences from all over the world and suggested several reforms as well as changes in the system (Singh, 2014). In this conference, as well as through other reviews, some issues constraining the desired growth of agricultural universities have been identified:

- proliferation of universities, particularly private agricultural colleges/universities established with inadequate planning, infrastructure and meagre resource allocation;
- bifurcation of SAUs into different disciplines like veterinary, fishery and horticulture is a retrograde step, which must be discouraged;
- no quality control and rather inadequate accreditation system;
- considerable political interference on policy issues, recruitment and day-to-day functioning of the universities;
- excessive internal bureaucracy stifling innovation and reward system for researchers and educational staff;
- practical abandoning of merit-based promotional system in favour of time-linked seniority-based promotions leading to mediocrity and nepotism;
- limited allocation for working capital for research, with most of budget spent on salaries;
- lack of a robust review and assessment system that rewards merit and excellence in science, especially linked to innovation;
- increasing in-breeding, with students receiving degrees from, and spending their entire working careers in, the same institution;
- outdated course curricula that do not keep pace with the scientific and technological advances being made around the world, changing social

needs and required flexibility that encourages learning of concepts and skills;
- limited flexibility in choice of courses and education programmes being offered to students;
- lack of student-based teacher-evaluation system linked to incentives and rewards for excellence.

The Way Ahead

The national and global challenges of growing population; food and nutrition for all; increasing consumption of processed meat and meat-based products; and the reduction and deterioration of natural resources including agricultural land and water necessitate a major readjustment in research and educational objectives and approaches. The UN SDGs (United Nations, 2015) obligate countries to take necessary measures to end poverty, hunger and malnutrition, and to achieve this, accelerating and improving agricultural education is imperative. In addition, harnessing the benefits of new and rapidly advancing sciences like biotechnology, nanotechnology, GIS and remote sensing and ICT requires a review of course curricula and flexibility in the education system. As mentioned above, earlier readjustments in R&D and educational priorities, and adoption of new approaches, have yielded rich dividends in terms of expected outputs and social and economic outcomes. The following recommendations have been made from time to time to improve the national agricultural education scenario and deliver the expected outputs:

- There is a need to have a re-look at the current status of the Land Grant System as a whole, to revitalize the Indian agricultural education system and restore and revitalize its basic concepts and principles while ensuring much-needed reforms for a change for good.
- The agricultural education system should be broad-based and must fulfil the needs of all stakeholders associated with agricultural production, processing and the marketing value chain.
- The universities must have full autonomy and be free of political interference and nepotism, with the established preeminence of the Vice-Chancellor's position.
- There is a need to establish an Agricultural Education Commission on the lines of the Farmers' Commission, which was headed by Dr M.S. Swaminathan, to review and suggest needed reforms at the national level.
- A statutory body like the Agricultural Education Council, on the lines of the Indian Veterinary Council, should be established urgently under the Department of Agricultural Research and Education (DARE), with an effective functioning of the accreditation system linked with federal funding.
- Boards of governors and university-level academic committees should include international experts to ensure excellence in research and education.
- Restructuring and amalgamation of agricultural education institutions, especially sub-discipline-based bifurcated universities, are urgently required to improve their relevance and quality of education in a holistic manner.
- Universities must be sufficiently funded to enable high-quality and advanced research, teaching and communication facilities.
- Besides salaries, operational funds in the ratio 60:40 should be ensured for all universities through provision of a competitive research grant system. Faculties should be encouraged to generate resources from outside sources/organizations.

- The recruitment and promotion system needs a thorough review with merit and excellence given prime importance in selection as well as career advancement.
- Agriculture must find a prominent place in the course content at primary and secondary school levels, especially to generate awareness and attract youth to agriculture.
- The course content of agricultural universities should be relevant to the evolving needs of the country taking into account the rapid global scientific and economic advancements. Courses in new emerging fields such as nanotechnology, biotechnology, ICT, GIS, post-harvest technology, agri-business management and market intelligence should be included in the revised course curricula.
- The education system should be accessible to all stakeholders offering theoretical and practical training of different levels and durations, including informal vocational training programmes.
- Due emphasis needs to be given to interdisciplinary courses that are highly relevant to the current agricultural production as well as the farming system's needs. Also, inter-institutional collaboration must be encouraged, both for teaching and research.

Conclusion

For accelerated agricultural growth, so critical to address SDGs, reforms in existing research and the education system will be necessary. Also, to address successfully emerging challenges, and to embrace secondary and speciality agriculture, capable human resources and reformed educational systems will be critical to improve the livelihood of smallholder farmers and to make agriculture an attractive and economically rewarding profession. Hence, higher investment in the existing agricultural education system, associated with needed reforms, is extremely important and fully justified. Agricultural education needs to be given high priority attention for soon achieving an Evergreen Revolution.

References

Agricultural Education Division (2017) Ranking of Agricultural Universities for the Year 2016–17. ICAR, New Delhi. Available at: https://icar.gov.in/files/071715062804_0au-ranking-2017.pdf (accessed 18 June 2018).

Committee on the Future of the Colleges of Agriculture in the Land Grant, University System, National Research Council (1996) *Colleges of Agriculture at the Land Grant Universities: Public Service and Public Policy.* National Academies Press, Washington, DC.

Indian Agricultural University Association (IAUA). Strength of Staff and Students of IAUA Member Universities. Available at: http://www.iauaindia.org/introduction.htm (accessed 17 June 2018).

NIRF (National Institutional Ranking Framework) (2017) Indian Rankings 2017. Ministry of Human Resource Development, Government of India. Available at: https://www.nirfindia.org/OverallRanking.html (accessed 18 June 2018).

Makwana, A.K. (2013) Agricultural education in India: challenges and prospects. *Agricultural Education* 2, 90–94.

Randhawa, M.S. (1968) Agricultural Universities in India – Progress and Problems. Lecture delivered at the Agricultural Division of the National Association of State Universities and Land Grant Colleges, Washington, DC, 12 November.

Singh, R.B. (2014) Transforming agricultural education for reshaping India's future. *XI Agricultural Science Congress Proceedings.* National Academy of Agricultural Sciences, New Delhi.

Stads, G.-J. (2016) A snapshot of agricultural research investment and capacity in Asia. In: Karihaloo, J.L., Mal, B. and Ghodake, R. (eds) *High Level Policy Dialogue on Investment in Agricultural Research for Sustainable Development in Asia and the Pacific.* Papers presented at Asia-Pacific Association of Agricultural Research Institutions (APAARI), Bangkok, pp. 8–20.

United Nations (2015) Resolution 70/1. *Transforming Our World: The 2030 Agenda for Sustainable Development.* United Nations, New York.

30

A Strategy for Doubling Farmers' Income

All the nations facing problems of poverty, hunger and malnutrition will need to accelerate their agricultural growth for achieving SDGs, especially while aiming at no poverty, zero hunger and a safe environment for all (Paroda, 2017). The Green Revolution not only led to food self-sufficiency but also helped to reduce poverty and hunger. And yet, despite a five-fold increase in foodgrain production, as against a four-fold increase in population, India still has around 250 million people who live in poverty and about 45 million children below age 5 who are malnourished. Moreover, after 50 years of the Green Revolution, India is also facing second-generation challenges like decline in factor productivity growth, poor soil health, loss of soil organic carbon, ground and surface water pollution, water-related stress, increased incidence of pests and diseases, increased cost of inputs, decline in farm profits and the adverse impact of climate change. On the demographic front, India adds annually almost one Australia (about 15–16 million) to its population. Thus, any progress gets nullified by an overall increase in population. Also, around 48% of the population is currently dependent on agriculture and allied fields and the agriculture sector contributes around 17% to national GDP. Moreover, public sector capital investment in agriculture and rural development has declined from almost 20% during the Green Revolution period to currently less than 10%. In the process, many states have remained deprived of growth and development. As a result, most farmers are not benefitted, especially since the majority of them are smallholders and find agriculture not profitable any more.

Why Double Farmers' Income?

Today, around 138 million Indian farmers' main concern is about declining farm income on the one hand and the increasing cost of inputs on the other. A recent study by the National Institute of Agricultural Economics and Policy Research (NIAP) has shown that around 70% of farmers in the country have annual per capita income of less than Rs 15,000 (around US$250). Birthal *et al.* (2017) have further analysed the situation and found that their geographical distribution is widespread, but mostly concentrated in Uttar Pradesh (27.4%), Bihar (11.4%), West Bengal (9.9%), Odisha (6.3%), Rajasthan (5.8%), Madhya Pradesh (5.3%), Maharashtra (4.9%), Assam (3.9%) and Jharkhand (3.2%). Most of these states lack the required infrastructure for agricultural income growth. Moreover, around 70% of farmers are marginal (owning less than 1 ha), and 77% of them earn a meagre income of Rs 6067 per capita p.a. Further, about 40 million farmers have around just 500 sq. m of land, which is not sustainable. Accordingly, the distress of small and marginal farmers has drawn specific attention of policy makers lately. The Prime Minister, considering this as a national priority, rightly called for doubling farmers' income by 2022. It is often argued that the Green Revolution mainly helped the country to achieve national-level food self-sufficiency, whereas it seems to have bypassed the majority (almost 86%) of smallholder farmers having less than 2 ha. Further, besides the second-generation problems of the Green Revolution, farmers are now faced with twin global challenges: (i) global climate change; and (ii) globalization of agriculture. The average land holding is around 1.1 ha, whereas many have much less than even 1 ha, which is not sustainable for a farm family. To make farming profitable, these farmers require both new technologies that can save cost of agricultural inputs while increasing productivity, and the policy support for getting credit at low interest and also higher income by linking them directly to the markets.

Farmers' Income Trend

It is argued that to achieve the set goal, a holistic approach would be needed to reap the benefits from all possible sources of growth, both from agriculture and outside the agriculture sector. Doubling farmers' income by 2022 would require some specific policy and institutional reforms that take into account identification and targeting of low-income farmers, particularly from the regions that were bypassed by the Green Revolution, like eastern, north-eastern and western regions of the country where the capital investment somehow was not made to build the required infrastructure for overall agricultural development. Further, it is also argued that the information on farmers' income, being so crucial to understand the income dynamics of farm households and to devise strategies to improve farmers' income, is not available, except the two surveys in the past – one in 2002/03 and another in 2012/13 conducted by the National Sample Survey Office (NSSO). Chand (2017) has provided estimates of the total income and per cultivator farm income (not farmer's income) for the period 1983/84–2011/12. According to him, farm income was reported to be inadequate to escape poverty for 53% of farm households who operated on less than 0.63 ha of land holdings. As per estimates, between

1993/94 and 2015/16, real farm income had only doubled (Table 30.1) and farm income per cultivator saw a slightly higher increase mainly due to a decline in the number of cultivators after 2004/05, since the younger generation seems to have opted out of agriculture and in to employment in urban areas.

Further, the low income of farmers compared to non-agricultural workers (almost 50%) is one of the reasons for agrarian distress. The low and highly fluctuating farm income is detrimental to investment and forces the cultivators, particularly the youth, to leave farming. Even the labour cost for cultivation has gone up considerably since the implementation of the scheme under the Mahatma Gandhi National Rural Employment Guarantee Act (MGNREGA).

In view of the above, the government's intention to double farmers' incomes by 2022 is indeed laudable. Once achieved, it would reduce agrarian distress and bring in parity between income of farmers and those in the non-agricultural sectors, thus possibly arresting or reversing the current migration trend. The target period to double farmers' income in real terms has been fixed as seven years, i.e. from 2015 to 2022. Hence, considering the past trend, it will require a minimum annual growth rate of 10.4%. Again, it is important to know what is to be doubled; is it the income of farmers or the output/income of the sector or the value-added or GDP of the agriculture sector? If the technology, input prices, wages and labour used could result in per-unit cost savings then farmers' income would possibly rise at a faster rate than the output. In this context, the doubling of farmers' incomes has to be viewed differently to the doubling of farm output.

Table 30.1. Trend of farmers' income in India (1993/94–2015/16). (From: Chand *et al.*, 2015)

Year	Total real farm income of all farmers (Rs 10 million)	Real farm income per cultivator (Rs)
1993–94	303,814	21,110
1999–00	372,923	26,875
2004–05	434,160	26,146
2011–12	632,514	43,258
2012–13	596,695	41,553
2013–14	602,922	42,760
2014–15	597,020	43,106
2015–16	598,764	44,027

It is also argued that if inflation in agricultural commodities is high, farmers' income in nominal terms can be doubled in a much shorter period, but the government's intention appears to be to double the real income of farmers. Unfortunately, the latest data on the number of cultivators is available only up to 2011/12. Therefore, while calculating per cultivator income, it is assumed that farmers would continue their withdrawal from agriculture at the rate observed during 2004/05–2011/12. It is rather contradictory that on the one hand we want farmers' income to be doubled so that they find agriculture attractive, and on the other, economists and policy makers expect them to withdraw from agriculture. This process should remain evolutionary and not be made revolutionary. The real strength of Indian agriculture lies in the fact that it currently sustains around 48% of the population of India.

Initiatives by the Government

For quite some time now, the distress of small and marginal farmers has been drawing the attention of policy makers. In 2004, the government had set up a National Commission on Farmers, headed by Dr M.S. Swaminathan. The Commission had submitted a report in 2006 (Government of India, 2006) aiming at 'faster and more inclusive growth'. It came out with several useful recommendations to revitalize agriculture and protect farmers from the vagaries of nature and price volatility. The key recommendations were: (i) improving farmers' income from farm and non-farm sources; (ii) enhancing efficiency in the use of resources; (iii) minimizing expenditures on non-renewable inputs; and (iv) remunerative prices to farmers at 50% higher than the minimum support price (MSP). Somehow, the last recommendation, which is directly linked to farmers' income, has not yet been implemented. On the contrary, the price fluctuations in the market of farmers' produce and the higher cost of inputs have caused widespread discontent among farmers, resulting in protests and even suicides, thus drawing urgent attention of the policy makers to the need to draw up a strategy for doubling farmers' real income.

As a first step, the government changed the name of the ministry to Ministry of Agriculture and Farmers' Welfare. It also initiated programmes like Attracting Rural Youth in Agriculture (ARYA), Mera Gaon Mera Gaurav, National Skill Qualification Framework, Skill Training, Value Addition and Technology Incubation Centres in Agriculture (VATICA), Knowledge Systems and Homestead Agricultural Management in Tribal Areas, Nutri-sensitive Agricultural Resources and Innovations (NARI), Climate-Smart Villages, and web and mobile advisory services. The potential role of farmer-producer organizations (FPOs) in innovation and scaling for increasing overall income has also been given due importance.

The present government has taken many new initiatives for increasing farmers' income such as: (i) 'per drop, more crop'; (ii) availability of quality seeds; (iii) soil test-based nutrient management distribution of soil health cards; (iv) post-harvest crop losses – large investment in warehousing and cold chains; (v) value addition by the farmers; (vi) creation of a national agricultural market by removing distortions and having e-markets to link farmers to market; (vii) Pradhan Mantri Fasal Bima Yojana; (viii) high priority to diversification towards high-value activities – horticulture, dairying, food processing, poultry, sericulture, bee-keeping and fisheries.

Also, the government, in its budget of 2014–15, had established a National Adaptation Fund for Climate Change, a long-term Rural Credit Fund, provision of financial assistance of Rs 5,00,000 for Bhoomi Heen Kisan (landless formers) through National Bank for Agriculture and Rural Development (NABARD), launching of soil health cards, Pradhan Mantri Krishi Sinchayee Yojana (PMKSY) and the Agri-Tech Infrastructure Fund. In its budget of 2015–16, the government had emphasized rural infrastructure development and created a Long-term Credit Fund, Short-term Cooperative Rural Credits Refinance Fund and Paramparagat Krishi Vikas Yojana (PKVY) to promote organic farming. Further, in the budget of 2016–17, a provision for a Long-term Irrigation Fund was made and the Union Budget of 2017–18 made some special provisions: (i) Rs 10 trillion allotted to ensure adequate flow of credit to under-serviced areas; (ii) Rs 90 billion allotted to increase the coverage under Pradhan Mantri Fasal Bima Yojana (PMFBY); (iii) contract farming emphasized for strengthening and linking the horticulture sector and agro-processing units; and (iv) Rs 20 billion allotted for dairy processing and infrastructure development to NABARD for modernizing milk-processing units. Besides these, several other measures were taken in the past to promote agriculture and farmers' income such as MGNREGA, Rashtriya Krishi Vikas Yojana (RKVY) etc.

The resources of NABARD are also being augmented substantially following the Parliament's nod to a six-fold increase in its authorized share capital to Rs 300 billion. The Development Financial Institution (DFI) is eyeing a balance sheet size of Rs 7 trillion by 2023 as against Rs 3.9 trillion as at present. The rural India-focused DFI plans to achieve this balance sheet by stepping up focus on providing support to irrigation projects, dairy farming, improving market infrastructure in rural areas (so that farmers get good prices for their produce), enhancing credit flow to deprived areas such as central and eastern states, and support for rural housing.

Despite these initiatives, the agricultural economists have differing views; some have even expressed doubts and consider the goal unrealistic and unachievable since there is negligible information available on farmers' income and there is no clarity as to how to double their income (Gulati and Saini, 2016). This is because the real income in the past has increased by only 5.2% p.a. between 2002/03 and 2012/13. At this rate, it may take at least a decade to double the real income of farmers, unless a new and dynamic strategy is put in place and implemented in a mission mode to achieve higher than 10% income p.a., which appears to be a gigantic task. NITI Aayog has indicated that doubling farmers' income may take a little longer than the target year of 2022, unless needed reforms are expedited (Chand, 2017). Also, the combined effect of growth was found to be 75.1% in seven years and 107.5% in ten years. According to him, if the farmers' income growth is considered to rise at the same rate as experienced between 2001 and 2014 (except price factor), income will rise by 66% by 2022/23 and will possibly double in ten years, i.e. by 2025/26.

Strategy for Faster Agricultural Growth

It is quite clear that 'business as usual' will not achieve the target of doubling farmers' income;

nor the suggestion by some to take farmers out of farming. What would farmers do without the new skills and where would they find employment? Instead, it is better to retain farmers in agriculture by making the profession more attractive and rewarding through diversified options, including post-production management and value addition-related activities. Obviously, out-of-the-box thinking with focused efforts on outscaling innovations linked to higher productivity, sustainability and profitability through the most appropriate diversified, secondary and speciality agriculture linked to post-harvest management, especially around proper storage, value addition and better access to market, would help achieve the goal of doubling farmers' income.

It has also been established from past trends that to achieve 8% growth in GDP, a minimum of 4% growth in the agriculture sector is a must. Hence, there is no room for complacency just because India had achieved Green, White and Blue Revolutions in the past and the problem of food scarcity has been resolved. On the contrary, the problems of smallholder farmers have magnified and real income has declined. To reverse this trend, we need a clear strategy, including a road map, that can lead us to sustainable and profitable farming using innovative approaches to harness opportunities. Also, as stated earlier, accelerating agricultural growth is critical for achieving the SDGs, especially to remove poverty, have zero hunger and ensure environmental security. Moreover, the greater the emphasis on agricultural research for innovation, the higher will be the growth of agricultural GDP (Pratt and Fan, 2010). In fact, the Green Revolution in itself was an innovation-led initiative around use of high-yielding dwarf wheat and rice varieties that responded favourably to higher inputs leading to a quantum jump in productivity. The cradles of success were: (i) political will; (ii) good institutions and human resources; (iii) availability of critical inputs (seeds, water, fertilizer etc.); (iv) enlightened extension workers and hardworking farmers; and (v) partnership at the global level.

Considering the current challenges of factor productivity growth decline, depleting natural resources, increasing cost of inputs, higher incidence of diseases and pests, higher cost of inputs, less profit to farmers and, above all, the adverse impact of climate change, the task of increasing income, especially of 86% of farmers who are small and marginal (Government of India, 2018), would require technologies by which they can save costs on inputs and have more income by higher productivity and by linking themselves to markets. Therefore, the strategy to double incomes would demand sustainable intensification, diversification, improved resource-use efficiency and resilience in farming that is economically rewarding. In this regard, the following three-pronged strategy needs to be pursued:

- improved productivity and production efficiency;
- agricultural diversification including secondary and speciality agriculture; and
- policy support and linking farmers to market.

Improving Productivity and Production Efficiency

Bridging the yield gap

India's cropped area has been stagnant at around 141 million ha for over a decade, whereas net irrigated area is currently 65.3 million ha and the gross cropped area is 195 million ha with cropping intensity of 135%. Of this, almost 55% is still rainfed. Since there is no scope for horizontal expansion, vertical expansion through increased productivity is the only way forward, for which considerable scope exists. In this context, a clear strategy was suggested for productivity enhancement state-wise/crop-wise, projecting an increase of 80 million t of foodgrains (Hooda Committee Report, 2010). Some states have productivity less than the national average, whereas some can achieve yet higher productivity in view of rich resources and availability of technological options.

The existing yield gaps can also be bridged by increasing seed replacement rates/the area under seeds of improved varieties, especially hybrids, by adopting large-scale use of biotechnology, including the use of GM food crops and by adopting good agronomic practices that are based on natural resource conservation and both water- and nutrient-use efficiency.

Globally, the use of GM crops has benefitted farmers in reducing costs on pesticide use and for increased productivity. More than 189.8 million ha was cultivated, globally, in 2017 under GM crops, whereas India has, so far, released only cotton, covering around 11 million ha, with considerable benefits to millions of smallholder farmers. Moreover, it has reduced the use of pesticides by almost 40% and has increased both production and productivity of cotton leading to exports worth around US$3 billion annually. Thus, the government must come out with a clear strategy in support of using these innovations in crops like maize, soybean, canola, rice and brinjal, which can help farmers to raise their incomes while reducing costs on inputs and getting higher productivity.

Conservation agriculture

In addition, there is a possibility of increasing cropping intensity through efficient water use. Also, there are options for improved input-use efficiency, especially of fertilizers, pesticides and energy to ensure resilience in agriculture. For this, conscious efforts are needed to swap unsustainable elements of the conventional tillage-based monoculture production practice with temporally and spatially highly productive, profitable and sustainable intensification through large-scale adoption of CA as a vehicle of change. It is well-established, globally, that over 180 million ha, CA helps in achieving sustainable and profitable agriculture through three principles – minimal soil disturbance, permanent soil cover and proper crop rotation. CA-based management practices also help in adapting climatic risks and in lowering environmental footprint. CA technologies have been developed, adapted and promoted over the past two decades, primarily to conserve resources and increase farm income. The CA-based management optimization in the cereal-based cropping systems in south Asia have helped in increasing crop productivity, input-use efficiency with economic returns, improving soil health, increased adaptive capacity of production systems to climate risks, reducing emissions and enhancing soil-carbon sequestration (Jat *et al.*, 2016).

Conceptually, CA-based sustainable intensification (CASI) is not a single technology; it is an innovation for sustainable farming, assimilating effective germplasm/crops, integrated nutrient/pest management, minimal and efficient farm mechanization and efficient soil and water management practices. Therefore, it requires application of farming systems-related coherent interventions that would increase both income and adaptive capacity of farmers for diversified as well as resilient agriculture. Additionally, its infusion is seen to sustain ecological services and provide greater environmental benefits to a landscape (TAAS, 2017).

Scaling innovations

There are some major innovations that currently need to be outscaled as a matter of priority, keeping in view the expected impacts on production and productivity. These are: (i) hybrid rice – the current area coverage (over the last two decades) is only around 2.5 million ha, whereas scope exists for covering at least 10 million ha in the next decade; (ii) single-cross maize hybrids – the area covered under these hybrids is less than 60%, whereas scope exists to double maize production in the next decade provided more than 90% of maize area is brought under promising single-cross hybrids; (iii) the area under CA in rice-wheat cropping systems in the Indo-Gangetic plains is about 3.5 million ha, whereas scope exists for almost 10 million ha. CA innovation also has vast scope under rainfed farming covering around 55% of the total 141 million ha of cultivable area in India; (iv) protected cultivation – the current area under protected cultivation in India is only around 50,000 ha, compared to more than 2 million ha in China; (v) micro-irrigation – out of a total irrigated area of 64.7 million ha, the area so far covered under micro-irrigation is around 8.6 million ha, which can certainly be doubled by 2022 provided direct subsidy support to the farmers is enhanced for adopting practices such as: drip, sprinkler, laser levelling, plastic mulching, raised-bed planting and direct seeding of rice. Also, the current initiatives by the government to augment and complete irrigation schemes may add an additional 2 million ha area under irrigation. However, for more efficient water use, both free supply of water and flood-irrigation practices

will have to be stopped as a matter of national policy. It will also be a bold decision if water is brought under concurrent list (like Israel), to resolve inter-state disputes and enhance water productivity in the larger national interest, and to bring more area under irrigation.

Increasing nutrient-use efficiency

One of the reasons for higher productivity in irrigated areas has been the increased use of chemical fertilizers. Today, India uses, on average, around 105 kg/ha of nutrients and total consumption of chemical fertilizers is around 32 million t, of which nitrogenous fertilizers are around 25 million t. On the contrary, nutrient-use efficiency (NUE) is not more than 30%. Thus, increasing fertilizer-use efficiency is one of the biggest challenges for which there is a need to adopt innovative ways like use of seed-cum-fertilizer drill, adopting effective use of soil testing/soil health cards and decision-support systems for soil-/plant test-based use of nutrients, use of *neem* coated urea for slow release and better uptake, use of customized fertilizers, fertigation etc.

Agricultural Diversification Including Secondary and Specialty Agriculture

New options

It must be understood that unless smallholder farmers adopt diversified agriculture in a farming systems mode, including both secondary and speciality agriculture, the expected doubling of their income will not be possible. Fortunately, India has made great strides in sectors like horticulture (now the second-largest producer in the world in fruit and vegetable production with more than 304 million t), livestock (the White Revolution achieving the highest milk production in the world, at 155 million t) and fisheries (the Blue Revolution achieving 11 million t of total fish production). All these sectors have shown much faster growth (5–7%) compared to foodgrains over the last two decades. Also, considerable scope exists to increase the income of farmers by adopting agroforestry; rural based, low-cost primary processing for value addition; cool chain; secondary and speci-ality agriculture such as protected cultivation; mushroom production; bee-keeping; sericulture; growing low-volume, high-value crops like nuts, spices, medicinal plants and nutri-crops; seed production of vegetable hybrids; nursery raising to provide disease-free saplings; fish-seed pro-duction; growing of flowers; vegetable seedlings to promote peri-urban agriculture; use of plastic culture; post-harvest processing; rural-based, low-cost value addition etc.

These new options would certainly provide opportunities to enhance farmers' incomes sub-stantially, and attract youth (including women) to agriculture, provided the right knowledge is disseminated, competent human resources are built and enabling policy support and incentives are provided. Youth can also play an important role as technology providers and input suppliers, besides being rural entrepreneurs. For increas-ing income, farmers would need a change in their attitude/perception towards adoption of diversified agriculture.

Innovations in extension

In fact, enlightened farmers of India are more interested today in getting the right knowledge rather than to have subsidies. In this context, ag-ricultural extension needs transformation. The public extension system played a key role during the Green Revolution phase, but it remained confined to irrigated areas. The success was also due to a holy alliance among researchers, exten-sion specialists, farmers and policy makers. At that time, the technology-dissemination approach remained top-down, focusing on demonstrations on individual farmers' fields. As already mentioned, the current scenario of Indi-an agriculture is confronted with multifaceted challenges arising out of inefficient manage-ment of natural resources (soil, water, agrobio-diversity). All these have led to considerable deceleration of factor productivity and decline in farm profitability. Apparently, this complexity cannot be overcome by routine transfer of tech-nologies. Rather, more serious efforts are now needed towards translational research requiring outscaling of innovations through 'out-of-box' extension systems. Also, conscious deployment of rural youth, women and progressive farmers

would help in speedy transfer of technology and the needed impact on the livelihood of small-holder farmers.

Moreover, farmers' welfare needs to be en-sured through a 'farmer first' approach to bene-fit equally producers and consumers. In view of the diverse demand for new innovations, new products, new information and new extension services, there is a need to shift from top-down to a bottom-up approach, involving farmers' par-ticipation at grassroots level, while ensuring confidence-building among farming communi-ties to take risks and adopt more scientific and resilient agriculture. In the process, knowledge-sharing on good agricultural practices (GAP), without dissemination loss, and incentives for timely supply of inputs become highly critical to double farmers' income. At the same time, part-nerships among key stakeholders, especially the private sector, become vital for promoting agri-cultural growth. In the process, care is also needed to overcome complacency that has crept into the public extension system, and greater vibrancy in the National Agricultural Research and Extension System (NARES) is required with active involvement of stakeholders (especially the private sector, NGOs and farmers) and a pol-icy shift in the extension approach towards farming communities rather than individual farmers.

Attracting youth to agriculture

Empowering youth through vocational training and building a cadre of technology agents to provide technical backstopping as well as cus-tom-hire services to smallholder farmers would go a long way in linking research with exten-sion, thereby accelerating agricultural growth (TAAS, 2015). There is also a need to link 'land with lab', 'village with institute' and 'scientists with society' to ensure faster adoption of efficient resource-utilization technologies that would benefit both producers and consumers. In the suggested transformation process, the agricul-ture technology agents will need to become job creators and not job seekers, and provide best technologies as well as quality inputs on farm-ers' doorsteps. Another important action that can change the game is to promote the establish-ment of 'agri-clinics' where technology agents are able to join hands in providing a single-window system of advisory services to farmers.

Another helpful approach would be to involve innovative young farmers as knowledge providers. Their own innovations, once recognized, could help in outscaling economically efficient farming practices. The concept of a demand-driven extension approach around integrated farming systems should henceforth be pursued.

Policy Support and Linking Farmers to Market

National Mission on 'Farmer First'

As stated earlier, a large number of initiatives and new schemes have been started by the government to support farmers, but there appears to be a need to have better coordination and convergence mechanisms to ensure effective outcomes and impact. Accordingly, concerns for collaboration, convergence and synergy need to be addressed along with issues of optimizing institutional arrangements of prevailing pluralistic agricultural extension and farm advisory. Agricultural extension systems urgently need a radical change. For this, a policy reorientation towards farmers' welfare through innovative and efficient technology-delivery systems, remunerative rural-based, low-cost value chains and assured market linkages would help in achieving the 'farmer first' objective. For this, a 'National Mission on Farmer First', by additional funding support and integrating different interrelated ongoing programmes under the Ministry of Agriculture and Farmers' Welfare and other ministries should be established to meet the objective of doubling farmers' incomes. The proposed national mission can oversee the coordination and convergence of various inter-ministries' programmes and have a key role to promote innovations through Krishi Vigyan Kendras (KVKs), the Agriculture Technology Management Agency (ATMA), agri-clinics, Agriculture Technology Information Centres (ATIC) and active involvement of private sector institutions. Hence, a mission on farmer first, with an initial allocation of Rs 100 billion should be mandated to promote the establishment of agri-clinics by encouraging well-trained groups of young individuals as small-scale private entrepreneurs. At least one agri-clinic per district could be targeted to begin with, linked to performance-based incentives and funding support in a phased manner. Also, under this mission, a farmers' innovation fund could be established for the validation and refinement of cost-saving/efficient technologies for outscaling. This mission should also be mandated to support the self-help groups/associations of progressive farmers/cooperatives or even farmers' producer companies to link them with markets. In addition, it must oversee and support the initiatives related to knowledge-/technology-sharing and capacity-building by private entrepreneurs using ICT, media, TV, smart phones and market advisory services. As the information needs of the farmers are exploding, and presently accessible to only 45% of farmers, innovative ways need to be found with the greater involvement of youth in agriculture. The initiative of DD Kisan, a dedicated TV channel for farmers, is indeed a good beginning, but its programmes need to be made more innovative and attractive, especially to attract youth around new options by which they can enhance income while adopting sustainable and diversified agriculture. Penetration of mobile phones and the use of the internet in rural areas can be another goal under the proposed mission on farmer first.

It is a fact that despite being the custodians of the country's food security, Indian farmers, especially smallholders (around 86%) are stuck in a low-income rut. As already stated, their per capita income (Rs 15,000 p.a.) is just one fifth of the national average. Only around 7% of marginal farmers earn more than Rs 50,000 per capita p.a. In their case, 60% of the income comes from non-farming sources. Also, they are engaged in diversified agriculture like animal husbandry, horticulture and growing cash crops. Unfortunately, allocation of R&D resources to these allied sectors like livestock, fishery and agro-forestry are not proportionate to their actual contribution to agricultural GDP, which, as a matter of policy, needs urgent attention (Government of India, 2018).

Increasing funding support

As already emphasized, in the long run, the boost to farmers' incomes must come from technological breakthroughs that raise yields and resource-use efficiency, reduce cost of production and ensure resilience in agriculture (Government of India, 2018). It is also a fact that those developing nations that have supported their agricultural research for development (AR4D) have made faster progress. China currently spends almost twice that spent by India on agricultural R&D, whereas challenges before Indian agriculture are equally daunting (Lele, 2017). Current funding of 0.4% of its agricultural GDP on AR4D is indeed much less than many developed and developing countries. This, therefore, calls for an immediate increase in resource allocation (almost tripled) to address the emerging challenges in agriculture. India would do much better if the government allocated a minimum of 1% of its agricultural GDP on R&D.

It is also clear that for successful scaling of innovations there is a need to enable the following: (i) institutional policies for facilitation of farmers' collectives like self-help groups, cooperatives, FPOs (commensurate with a legal framework), establishment of a cadre of agri-business professionals at the village level, creation of agri-clinics, provision of credit at low interest rates (<4%) to the farmers across the value chain, machine rental services etc.; (ii) promotion of ecoregional research, marketing and trade policy, agro-processing, value-chain development, sustainable livelihood, new funding models for translational research by the state governments etc.; (iii) price policies like a minimum support price (MSP) for most crops/commodities, incentive support around efficiency, avoidance of risk through provision of insurance, compensation for ecosystem/environmental services etc; (iv) investment policies to ensure higher capital investment (around 15–20%) in the states needing critical infrastructure like roads, irrigation, power, markets etc.; gradual reduction in subsidies but linked to incentives that are performance-orientated, promoting the private sector; and (v) policies on land and water use that encourage more efficient use of these natural resources. There is also considerable scope for attracting the private sector and youth for developing wholesale markets, warehouses, cold-storage facilities, rural-based agro-processing infrastructure, promoting micro-irrigation systems, sale of quality inputs, and providing agricultural extension services.

Market reforms

It is urgent that perishable commodities like fruit and vegetables are immediately delinked from centralized sales through Mandis, as at present, by revisiting and amending the Agriculture Produce Marketing Committee (APMC) Act. The initiative to implement the new Model Agricultural Produce and Livestock Marketing (APLM) Act 2017 is a right step but its implementation by all states is to be facilitated and monitored by NITI Aayog. Also, for the proposed electronic network for agricultural marketing (e-NAM), it is necessary that movement of agricultural produce is not restricted by the state governments. We need bold export-import (EXIM) policy, keeping in view long-term goals to take advantage of globalization of agriculture. Present short-term policies of allowing exports sometimes and putting restrictions on them is counter-productive. This has happened in the recent past by imposing restrictions on export of cotton, meat and foodgrains. Even creating positions of agricultural attachées in the embassies of selected countries would be a great help in boosting agricultural exports, thus benefitting indirectly the farmers.

Land laws for tenancy, contract/collective farming, long leases (so that farmers/tenants are encouraged to invest in land development), consolidation of holdings with no more fragmentation below 1 ha, being uneconomical, must be revised and put into implementation at the soonest. Also, the implementation of the Model Land Leasing Act (2016) should be a high priority for state governments. Similarly, for better value and efficient use of precious water resources, both pricing of water and banning of flood irrigation systems must be considered, and incentives for micro-irrigation for greater area coverage must become a national priority. Obviously, bold policy decisions are required and 'business as usual' will not suffice.

Given the limits on land holdings, income growth has to be by raising cropping intensity, improving resource-use efficiency and agricultural diversification. Expansion in agriculture needs to exploit intensive cultivation, as only 40% of crop land is cultivated more than once. This can be enhanced by improving farmers' access to quality seeds of short-duration, high-yielding crop varieties/hybrids and by adopting efficient cropping systems that are more sustainable. More area coverage under quality seeds of improved varieties and hybrids would need reforms, as proposed under the Seed Bill 2004, which has been pending for a long time in Parliament. The needed incentives and handholding of the private seed sector, especially for making available seeds of promising hybrids of different crops, would go a long way in bridging the existing yield gaps and for increasing farmers' income.

The focus should also be on diversification towards high-value crops/commodities, especially horticulture, by bringing a minimum 10% area in each of the states into play. Also, increased support for the animal husbandry and fishery sectors will be of great benefit. Demand for these commodities is growing fast and there is considerable potential for their value addition, including export. These enterprises have, however, not received much policy support, except horticulture. For example, animal husbandry receives just 5% of total public investment and institutional credit to the agricultural sector, though it contributes more than 30% to agricultural GDP. Higher allocation of resources would thus be justified to accelerate the growth of these highly potential sectors. Further, there is a need to create required infrastructure, focusing on improving complementarities, since lack of any of these may restrict farmers in capturing the benefits of investment in others. A typical case is that of Bihar and north-eastern states, where despite some improvement in road networks, farmers have not benefitted much owing to poor electricity supply, irrigation infrastructure and marketing facilities.

Linking farmers to market

There is no doubt that linking farmers to markets is critical for improved livelihood of smallholder farmers and beneficial for consumers. Smallholders are more efficient in production, yet they face serious disadvantages, mainly on account of marketing their produce. As a result, smallholders are often bypassed in the process of transformation of agriculture, agri-food and marketing systems. Although, it is relatively easy for smallholders to diversify towards high-value crops owing to their higher resource flexibility

and better family labour availability, they face disadvantages in terms of scale in production and market. Moreover, they have small marketable surpluses that are costlier to trade in the distant urban markets due to higher transportation and transaction costs. Hence, efforts to improve productivity on small farms may not directly result in higher income unless these are appropriately linked with markets. Their integration into markets or value chains would thus require pro-smallholder policies that create an enabling environment for attracting various stakeholders to act together in processing, marketing and sharing the benefits on account of emerging market opportunities. As stated, these include innovative institutional mechanisms, better infrastructure, greater involvement of the private sector, easy access to agricultural and market-related information and risk-management mechanisms, and, above all, a favourable business environment through stable marketing and trade policies (TAAS, 2013).

The Way Forward

To make agriculture both remunerative and attractive as a profession, and especially to double farmers' income, an action plan for implementing the three-pronged strategy proposed above is described below.

Policy interventions

• A national mission on Farmer First, with an annual allocation of Rs 100 billion to begin with, by merging/clubbing of various central schemes as well as new initiatives to empower farmers, needs to be initiated. This will help to catalyse the activities/programmes specifically designed for scaling innovations that will increase farmers' income and have direct impact on smallholder farmers through adoption of a three-pronged strategy defined earlier.

• Needed regulatory reforms in the existing acts, especially pertaining to land, water, seed, fertilizer, energy and market must be brought about as a matter of national priority by the government. Also, an effective coordination and convergence mechanism for various schemes, programmes and activities by different ministries would help in achieving the desired outcomes much faster. For this, a high-level, inter-ministerial committee to be chaired by the prime minister and co-chaired by the vice-chairman of NITI Aayog and the agriculture minister will help ensure effective monitoring of the outcomes of various programmes aimed at Farmer First. This coordination committee will be assisted by a standing advisory panel of agricultural experts.

- A remunerative MSP for most of the commodities needs to be fixed and announced well in advance of planting season by the Ministry of Agriculture and Farmers' Welfare (MOA&FW) with assurance for either procurement or compensation directly to the producers for prevailing price differences in the market, so that the farmer does not lose out. Also, reforms in the methodology for fixing the MSP by the Commission for Agricultural Costs and Prices (CACP) is needed, for which a high-level external review committee of experts should be established immediately.

- For accelerating agricultural growth, needed incentives and rewards must be put in place quickly to attract youth (including women) to diversified, secondary and speciality agriculture as individual producers, SHGs, cooperatives, farmers-producer organizations/companies or as knowledge/service providers. In the process, farmer-led innovations should be scaled out through required validation, refinement and incentives in the form of credit at low interest rates (<4%), bank support for required commercialization, insurance to avoid any initial risks, and practically no or very low tax on rural-based value additions and marketing of produce/value-added products. Incentives to innovators/entrepreneurs could be in the form of state/national recognition and awards.

- The right policy support for an accelerated role of the private sector will change the game much faster. Hence, an enabling environment to embrace the private sector is the most critical need. In this context, support for hybrid seed production; fabrication of equipment/implements/tools for scaling CA and small-farm mechanization; micro-irrigation (drip and sprinkler); protected cultivation, including fertigation; agro-processing and value addition; fertilizers, including customized and biofertilizers; pesticides, including biopesticides etc. would help accelerate agricultural growth.

Research and development

- Besides the focus on productivity and production growth, we need increased R&D emphasis on post-production, value addition and market linkages (both domestic and foreign).

- There is an urgent need to improve the empowerment of targeted smallholder farmers and ensure delivery of last-mile services. Hence, technology dissemination-related programmes will have to be tailored and re-orientated according to present-day needs. In fact, a paradigm shift from public to private innovation extension systems is the need of the hour to provide much-needed knowledge, quality inputs and much-needed custom-hire services on the farmer's doorstep.

- Ensure that smallholder farmers, especially youth and female farmers, get their entitlements and are not sidelined.

- Identification of agencies/institutions responsible to take specific actions at the local, state and central level, and their effective coordination, will be very helpful. Also, an independent monitoring and evaluation process for much-needed impact will be extremely useful.

Capacity development

- Knowledge-sharing and capacity development (especially women and youth) need to be considered a top priority to bridge yield gaps, achieve diversification, scale innovations that can save on production costs and help in rational use of natural resources, ensure value addition and link farmers to market.

- Greater emphasis must be given to skill (on-farm as well as off-farm activities) development at all levels. This will greatly help the farmers, especially the smallholders, to raise their income.

Financial support

- There is an urgent need to triple the annual budget allocation for the Indian Council of Agricultural Research (ICAR), an apex AR4D organization with a proven track record, in order to continue meeting emerging challenges while providing national public goods for the betterment of farmers as well as Indian agriculture.
- Capital investment in agriculture for much-required infrastructure in the states that were left behind during the Green Revolution (especially the eastern region) must immediately be enhanced (at least to a minimum level of 15–20% from the current less than 10%) to create much-needed infrastructure to help farmers increase their production as well as their income. Such an effort will also help in achieving SDGs much faster.
- The state governments (as they have major responsibility, agriculture being a state subject) must provide necessary financial support and commitment for implementation of the above three-pronged strategy to double farmers' income. The role of NITI Aayog is thus very critical in this context.

Conclusion

In India, while farmers are the major producers, they also constitute the largest proportion of consumers. Hence, improving small-farm production and productivity, as a major development strategy, can make significant contribution towards elimination of hunger and poverty, provided farming is made efficient and remunerative. Experience of countries that have succeeded in reducing hunger and malnutrition shows that growth originating in agriculture, through smallholder farmers is at least twice as effective in benefitting the poorest as is the growth from non-agriculture sectors. The World Development Report of the World Bank (World Bank, 2008) has clearly emphasized that 'Using agriculture as the basis for economic growth in agriculture-based countries requires a productivity revolution in smallholder farming.' As stated earlier, higher productivity requires higher investment in agriculture and agricultural research – a fact that needs to be heeded by policy makers to make sure that 1% of agricultural GDP is invested in AR4D, as against the present level of just 0.4%. Hence, a three-fold increase in resource allocation for the National Agricultural Research System (NARS) must be considered a prerequisite to doubling farmers' income.

It is also a fact that India will remain predominantly an agricultural country during most of the 21st century. Therefore, we must have both vision and a national strategy for shaping the destiny of agriculture by making it highly productive, efficient and economically attractive for the smallholder farming community. The target of doubling farmers' income by 2022, not an easy, yet laudable, goal, augurs well for the government's intention to help farmers. It is also clear that if concerted efforts, as per the suggested action plan above, are made, the prospects of making agriculture the engine of national economic growth and a respectable profession for smallholder farmers are much brighter.

References

Birthal, P.S., Negi, D.S. and Roy, D. (2017) Enhancing Farmers' Income: Who to Target and How? Policy Paper No. 30. ICAR-National Institute of Agricultural Economics and Policy Research, New Delhi.

Chand, R. (2017) Doubling Farmers' Income: Rationale, Strategy, Prospects and Action Plan. NITI Policy Paper No. 1/2017. NITI Aayog, Government of India, New Delhi.

Chand, R., Saxena, R. and Rana, S. (2015) Estimates and analysis of farm income in India: 1983–84 to 2011–12. *Economic and Political Weekly* 50(22), 139–145.

Government of India (2006) *Report of National Commission on Farmers on 'Faster and More Inclusive Growth'*, headed by M.S. Swaminathan. Government of India.

Government of India (2018) Report of the DFI Committee: *Strategy for Doubling Farmers' Income by 2022*. Chaired by Ashok Dalwai, Ministry of Agriculture and Farmers' Welfare, Government of India.

Gulati, A. and Saini, S. (2016) Farm incomes: dreaming to double. *The Indian Express*, 28 July. Available at: http://epaper.indianexpress.com/c/12056965 (accessed 18 June 2018).

Hooda Committee Report (2010) *Report of Working Group on Agriculture Production (WGAP)*, headed by B.S. Hooda. Government of India.

Jat, M.L., Dagar, J.C., Sapkota, T.B., Singh, Y., Govaerts, B. *et al.* (2016) Climate change and agriculture: adaptation strategies and mitigation opportunities for food security in South Asia and Latin America. *Advances in Agronomy* 137, 127–235.

Lele, U. (2017) Climate Change and Doubling Farmers' Income Experience Around the World. Paper presented in XIII Agricultural Science Congress, UAS, Bengaluru, 21–24 February.

Paroda, R.S. (2017) Indian Agriculture for Achieving Sustainable Development Goals. Strategy paper. Trust for Advancement of Agricultural Sciences, New Delhi.

Pratt, A.N. and Fan, S. (2010) R&D Investment in National and International Agricultural Research: an Ex-ante Analysis of Productivity and Poverty Impact. Discussion paper 986. IFPRI, Washington, DC.

TAAS (2013) *Proceedings of Brainstorming on Achieving Inclusive Growth by Linking Farmers to Markets*. IARI, New Delhi, 24 June.

TAAS (2015) Proceedings of *National Dialogue on Innovative Extension Systems for Farmers' Empowerment and Welfare*. NASC complex, New Delhi, 17–19 December.

TAAS (2017) *Scaling Conservation Agriculture for Sustainable Intensification in South Asia*. TAAS, New Delhi.

World Bank (2008) *World Development Report on Agriculture for Development*. The International Bank for Reconstruction and Development, Washington DC.

31

Future Challenges and Opportunities in Agriculture

Since the Green Revolution, a paradigm shift has been noticed from food scarcity to self-sufficiency, monocropping to crop diversification, flood irrigation to drip irrigation, conventional varieties to hybrid seeds, saplings to tissue-culture plants and traditional to secondary and speciality agriculture. The pressure on land and water is continuously increasing, and it is a daunting challenge to feed the growing population, which is currently 1.34 billion. Along with these, an unprecedented increase has been observed in consumer demand for more diversified and nutritious foods – fruits, vegetables, meat, fish etc. Above 6% growth over the last decade in the fishery and horticultural sectors is indeed remarkable. Through R&D initiatives, farmers harvested a record 277.49 million t in 2017–18. The average agricultural sector growth over the last three years has remained at around 4.7%.

India will need 70% more foodgrains by 2030; that, too, from declining natural resources. Thus, to produce more from less is an enormous challenge, especially when the farmers are facing second-generation problems of the Green Revolution as well as the adverse impact of climate change. These are: factor productivity decline, poor soil health, loss of soil organic carbon, ground and surface water pollution, water-related stress, increased incidence of pests and disease, increased cost of inputs and decline in farm profits. The major concerns in agriculture are the declining total factor productivity, diminishing and degrading natural resources, increased incidence of diseases and pests, and stagnating farm income. The impact of trade liberalization on agriculture and global climate change are also new challenges. Other challenges are: (i) weakening of input delivery and local agri-governance systems; (ii) increasing risk in agriculture due to weather, prices and trade policies, including the impact of globalization; (iii) small, declining and fragmented holdings; (iv) growing marketing inefficiencies and increasing agri-waste; and (v) limited employment opportunities in non-farm sectors. These challenges have serious implications for farm income and the future of Indian agriculture. In many ways, these can neutralize even the contributions of many technological breakthroughs. If not addressed immediately, these challenges may adversely affect national food and livelihood security.

In spite of the enormous challenges, Indian agriculture continues to remain at the forefront of development and providing livelihood to half of India's population. Despite liberalization and fast growth in services and manufacturing sectors, the contribution of agriculture is still around 17.4% of national GDP, which compares fairly well with the contribution of the industrial sector, which is currently 18%. In the present scenario, increasing productivity and farmers' income are two major challenges when land holdings are diminishing among the majority of farmers. Other critical areas needing priority are access to good knowledge and required appropriate infrastructure in rural areas. Problems related to infrastructure for irrigation, power, markets and roads affect the farming sector adversely, mainly in eastern India. Unlike other business enterprises, agriculture is prone to risks on account of factors beyond the control of farmers. At the same time, the number of initiatives undertaken for agricultural development does not translate into effective delivery mechanisms at ground level in terms of increasing productivity, decreasing cost and increased income by linking farmers to markets. For effective delivery of products and knowledge, we need reforms in agriculture to encourage participation of the private sector through the creation of an enabling environment, which is crucial in this context.

How we meet the emerging challenges is a question before all. Will technology-led agriculture succeed in producing more from less? Historical experiences of producing more through various revolutions are a testimony to inspire farmers to take up new challenges to be successful. The emergence of new sciences like biotechnology, information technology, nanotechnology, bioinformatics etc. provides new hope. The need of the hour is to embrace climate-smart agriculture, precision agriculture and good agronomic practices. Innovations like CA, micro-irrigation, protected cultivation, tissue culture, GM crops, hybrid technology, aeroponics, precision nutrient management and IPM offer greater opportunities for outscaling and for greater impacts, provided they are supported well by the right policies, development-related activities and higher investments (at least 1% of agricultural GDP from the current level of 0.4%) (ACIAR, 2016; Tangermann, 2016).

On the greener side, new opportunities are unfolding in the form of increased demand for agricultural commodities in both domestic and global markets as a result of higher economic growth and rising consumer income level. The growing international demand for rice, wheat and maize, besides cotton, soymeal, fruits, vegetables, fish, meat and poultry, have opened up enormous opportunities for boosting exports. In addition, the increasing demand for high-value commodities such as fruits, vegetables, milk, meat, flowers and agri-processed products in the domestic market points towards potential prosperity that can be brought about in the farm sector. The entry of the corporate sector in developing and delivering market-driven technologies,

contract farming, processing agri-products, developing organized retailing and exploring markets for exports is providing a new dimension to Indian agriculture. Some of these encouraging developments are taking place around the value chain from farm to plate. But the main question still remains as to how to involve the farming community, especially small-scale farmers, in capitalizing markets and sharing benefits arising from new opportunities. Failing to address these issues can lead to further exploitation of the farming community, culminating in distress to smallholders. Innovative policies, appropriate institutional arrangements and market-driven initiatives can, on the contrary, harness untapped opportunities and provide much-needed benefits to smallholder farmers, representing 80% of the 141 million farming households (Government of Telangana, 2015).

Moreover, only agriculture can liberate India from the triple burden of poverty, hunger and malnutrition while ensuring conservation of natural resources and sustainability of environment. It can address effectively concerns of poor health and nutrition of children and empowerment of women, being important SDGs. Thus agriculture should be seen as an important sector of the national economy, sustaining as it does around 55% of the population.

Agri-business is currently the single largest sector in India, worth around Rs 20,000 billion. Hence, India needs to focus more on agri-business, a generic term for many businesses involving agriculture and food production, including cooperative/contract farming, seed supply, agrochemicals, farm machinery/equipment, wholesale distribution, food processing and marketing. In future, agri-business may contribute to approximately 27% of India's GDP, involving both production and processing components. The agri-business segment may nearly double in future years driven largely by growth in per capita income related to higher consumption and changes in consumer preferences towards value-added and processed foods. Hence, a greater focus on post-production-related activities through processing, value addition and efficient marketing, including export, would go a long way in accelerating agricultural growth as well as farmers' income. We must, therefore, promote low-cost, rural-based agri-processing and value-chain-related technologies/approaches.

A few scattered successful models have taken advantage of new options and have addressed key challenges. It is a daunting task to upscale such successful models, and to reform the agricultural sector, which urgently requires an enabling policy environment. Considering past successes of the Green, White and Blue Revolutions, based on policy support, including higher capital investment, it is now evident that appropriate policies, institutions and technologies must play a key role in facing the challenges. The following are key suggestions for the way forward to ensure higher and inclusive growth in Indian agriculture (MoA and FW, 2015; MoE and CC, 2015; CHAI, 2016; FAO, 2017):

1. Increasing agricultural productivity is the key challenge for ensuring national food security. To increase production, exploitation of the potential existing yield gaps offers tremendous opportunities. Hence, a mission-mode programme on Bridging the Productivity Gap, employing real missionary zeal and effective monitoring, is required to be launched with meticulous planning for each state as a matter of priority. For this, attention to agriculture in science policy is needed, and the existing technology dissemination and input supply system needs to be revitalized and tuned to meet emerging needs of smallholder farmers. Special emphasis on the seed sector, input-use efficiency, financial and insurance institutions and a paradigm shift in technology-transfer mechanisms involving both private sector and NGOs are critical in achieving desired goals.

2. Rainfed areas have a huge potential to raise production and increase farm income. These 'grey' areas can be made 'green' to harness a second Green Revolution. The role of technologies, policies and infrastructure would be very important in realizing the potential of rainfed agriculture. In this context, it has to be ensured that public policies and technologies have the appropriate synergies to move forward. The initiative of the Government of India to establish a National Rainfed Authority of India was a welcome step. However, this authority needs a proper policy framework, legal and funding support as well as empowerment for effective coordination and monitoring of all rainfed-related programmes run by various ministries/departments. The earlier it is ensured, the better it will be for the national interest.

3. Linking farmers to markets is a prerequisite for augmenting farm production and farmers' income. The role of innovative institutions like e-NAM will be critical in this context to reap the benefits of emerging opportunities. A silent revolution of innovative institutions is already taking place in the Indian agricultural production and marketing system (farm-to-plate continuum), encompassing effective functioning of value chains and marketing efficiencies. Therefore, our current need is to replicate such best practices through the formation of producers' associations, self-help groups, cooperatives or farmers-producer companies. Krishi Vigyan Kendras (KVKs), being an existing institutional mechanism at district level, could play an important role in the entire supply chain through access to best practices in the production-to-marketing continuum. ICT also offers new opportunities to support this.

4. Agricultural extension in India and elsewhere requires constant transformation. The current transitional phase also needs a renewed interest and policy attention. The public extension system, therefore, needs revamping towards 'translational research', requiring outscaling of innovations through an 'out-of-the-box' extension system. Also, conscious deployment of rural youth, women and progressive farmers may help in much speedier transfer of technology for needed impact on the livelihood of smallholder farmers. For this, farmers' participatory approach for testing, refinement and adoption of farmer-led innovations is to be ensured. Also, empowering youth (both men and women) through vocational training and building a cadre of technology agents to provide technical backstopping as well as custom-hire services to smallholder farmers would go a long way in linking research with extension, and thereby accelerating agricultural growth. Linking 'land with lab', 'village with institute' and 'scientists with society' is essential to ensure faster adoption of efficient resource-utilization technologies, benefitting both producers and consumers. In the transformation process, the agricultural technology agents need to be job creators and not job seekers and provide best technologies as well as quality inputs on farmers' doorsteps. Another strategy could be to create agri-clinics, where technology agents can join hands to ensure a single-window system of advisory services for

farmers, so that they need not run from 'pillar to post' to obtain best technical inputs.

5. To ensure inclusive growth in agriculture through innovative and synergistic approaches for achieving sustainable food and nutrition security, AR4D would require a paradigm shift to ARI4D, with increased (at least double) resource allocations, accountability and monitoring. In the process, complacency that has crept into public research, education and the extension system has to be overcome. This necessitates greater vibrancy in the NARES, requiring active involvement of stakeholders (farmers, NGOs, private sector, scientists and policy makers) to remain technology-wise and globally competitive.

6. There is an urgent need for agricultural diversification by identifying key crops/commodities that can help small-farm holders to raise their income. Incremental gains in income through diversification would help capital formation, which would be instrumental in attaining higher productivity and profitability. In this context, agro-ecological zone-wise planning; adoption of scientific land-use planning, such as new-areas new-crops approaches using GIS; land-use planning; and effective district-level implementation of the strategies by involving grassroots organizations and stakeholders would be the best options to move forward. Towards agricultural diversification, many horticultural crops, especially perennial fruit trees, spices and plantation crops and agro-forestry species have an important role and would help in carbon sequestration as well as mitigating climate change. Promoting agro-horticulture and agro-forestry would ensure sustainable agriculture. In addition, we shall have to promote both urban and peri-urban agriculture, and adoption of post-harvest practices including grading and packaging, processing, value addition, and cool-chain marketing and export. All these would ensure higher economic returns to farmers. Also, emphasis now should be on secondary and speciality agriculture as well as on peri-urban agriculture for higher productivity and income. In this context, promoting precision farming and protected cultivation on a larger scale would need major policy support in an aggressive mode.

7. Water is the most critical natural resource for future agricultural growth. Currently, the water sector for irrigation is invariably neglected both at central and state level. High inefficiencies in water delivery, distribution and on-farm use adversely affect agricultural production. Irrigated areas can be expanded easily, up to 30%, with improved micro-irrigation techniques and by discouraging flood-irrigation practices. Innovations in governance and pricing of surface and ground water for desired water-use efficiency through an integrated approach among irrigation and agriculture departments, private sector and farmers' water-user associations are urgent issues for coordinated action.

8. Precision nutrient management using decision-support systems, aimed at targeted yields, keeping in view site-specific nutrient availability in the soil, would help achieve much-needed resilience in agriculture. Nutrient-use efficiency needs to be improved, which is invariably quite low (30% for N fertilizers). Also, use of biofertilizers, organic matter recycling, CA and organic farming would help achieve sustainable/evergreen agriculture.

9. To address biotic stress of diseases and pests, outscaling of available IPM technologies, while keeping pests below the economic threshold level (ETL), is an emerging option to be harnessed. Also, increased use of biopesticides, at least up to 10% from the present 3% of total pesticides used (60,000 t of active ingredient), would help greatly in reducing environmental load due to pesticides.

10. Biotech crops hold considerable promise for smallholder farmers. For crops that are proven to work for both consumers and producers, regulatory uncertainties and excessive restrictions surrounding biotech crops must be removed in order to widen the technology options and provide both private and public sectors with the confidence to invest.

11. The food-processing and food-distribution sectors need to be strengthened further by proper policies for greater private sector or farmers' cooperatives'/self-help groups'/producer-companies' participation in the entire value chain. Incentives, through appropriate tax structure and exclusive rights, such as agro-processing, especially in rural areas, becomes a lucrative option for farmers as well as the private sector. Current post-harvest losses in foodgrains are also to be minimized, for which construction of modern silos for foodgrain storage is a matter of national priority. Also, primary processing and

value addition in rural areas would need a different tax structure and support for building infrastructure.

12. There is a dire need to significantly enhance capital investment in agriculture by both public and private institutions in non-Green Revolution regions, particularly in eastern and north-eastern India, where there is a great potential for agricultural growth. Hence, investment priorities should now be orientated towards the realistic growth of agriculture to meet the emerging needs of the people. Therefore, public policies should be such that these trigger much-needed private sector investment for infra-structure development. Unfortunately, this has not happened.

References

ACIAR (2016) The Future of Agriculture in Development: Challenges and Opportunities for Australia in the Asia-Pacific. Organized by the Australian Centre for International Agricultural Research, Canberra, December.

CHAI (2016) Future Challenges and Options in Agriculture. National Agricultural Conference 2016 organized by the Amit Singh Memorial Foundation, Confederation of Horticultural Association of India and Jain Irrigation Systems Ltd, Jalgaon, 28–30 May.

FAO (2017) *The Future of Food and Agriculture – Trends and Challenges*. Food and Agriculture Organization of the United Nations, Rome.

Government of Telangana (2015) Task Force Report: Agricultural Challenges and the Way Forward. Submitted to NITI Aayog by the Government of Telangana, Agriculture and Cooperation Department.

MoA and FW (2015) State of Indian agriculture. In: *Indian Agriculture: Performance and Challenges*. Ministry of Agriculture & Farmers' Welfare, Government of India, New Delhi, pp. 1–20.

MoE and CC (2015) *Development Goals in India: A Study of Financial Requirements and Gaps*. Ministry of Environment, Forest and Climate Change, Government of India.

Tangermann, S. (2016) Agriculture and Food Security: New Challenges and Options for International Policy. E15 Expert Group on Agriculture, Trade and Food Security – Policy Options Paper. E15 Initiative. International Centre for Trade and Sustainable Development (ICTSD) and World Economic Forum, Geneva.

32

Change We Must – But Change is Difficult

Introduction

Change your thoughts and you can change your world. Small changes are hard and big changes are even more difficult. Having made impressive progress by any standards, India is presently faced with numerous challenges to be urgently addressed so as to achieve its ultimate goal of being a 'developed nation through progress in agriculture'. It was emphasized long ago that although change is difficult, we must bring about those needed changes in order to meet the emerging challenges and harness the opportunities for faster and sustainable growth of agriculture (Paroda, 2014). In this endeavour, as we move forward, the immediate task before us is to address the following issues as priorities:

- to ensure both economic and ecological access to food and nutrition security, particularly for those living below the poverty line;
- to secure higher productivity combined with profitability through minimum input use and improved efficiency of production systems;
- to address the second-generation problems of the Green Revolution followed by other revolutions in agriculture, such as White, Yellow and Blue;
- to remain competitive and take full advantage of globalization of agriculture through advanced preparedness for the new World Trade Organization regime;

- to generate resources in the wake of dwindling donor support for agricultural research and human resource development;
- to improve preparedness to meet effectively the economic and technological sanctions presently imposed or likely to be imposed in future as we demonstrate our scientific excellence and capabilities.

All these require a strong NARS committed to a paradigm shift from the present 'productive and purposeful' to a 'responsive and responsible' organization. In order to accomplish this, we have to introduce major changes, however difficult they may be, to revamp the institutional system for agricultural research in India. Our NARS, despite being one of the largest in the world, has its own strengths and weaknesses, which must be clearly understood. As a matter of fact, we are still functioning as a National Agricultural Research Institute (NARI). Hence, we must move fast to become, in a true sense, the NARS involving, besides the ICAR institutes and the SAUs, all other stakeholders such as traditional universities and institutions, NGOs, private sector institutions, farmers and agri-business entrepreneurs. Obviously, the change from NARI to NARS is not a simple task, and it would require appropriate policy initiatives, change of mindset and, above all, commitment of all those involved in the process. We have also to guard against the possible danger of complacency creeping into the system.

This will require self-examination, reorganization and revamping of the system for its 'renewal', thus demanding the fullest involvement of all concerned. Similarly, we need to revisit our Land Grant System of education on which India had built the foundations of its SAUs and the Indian Council of Agricultural Research (ICAR) national institutes, having 'deemed-to-be university status'. In other words, we have to reinvigorate the system faster and bring in the required change for the better. A new research agenda will have to be drafted for the competent cadre of our young scientists, around ICT, GIS and crop modelling, agribusiness management, post-harvest technology etc. Also, the institutional mechanisms for effective governance will have to be put in place through requisite organization and management reforms.

The new work culture linked with incentives and accountability would first demand a change in the mindset of senior research managers. This, in itself, is a major challenge. Those organizations that have changed in time have survived and prospered, whereas those that have not have lagged behind. Despite these formidable challenges, Indian agriculture offers tremendous 'uncommon opportunities' that can be harnessed to take full advantage in the near future. Some of these are:

- a vast institutional and human resource base that can be further strengthened and made more efficient and effective;
- a threshold of low productivity that can be further enhanced substantially through increased input use and production efficiency;
- a reservoir of proven technologies that have yet to reach the farmers/stakeholders;
- a vibrant private sector whose potential is yet to be tapped for R&D in agriculture;
- a strong network of public and private sector institutions, well organized to provide needed technical backstopping for agricultural advancement;
- opening up of world markets for Indian agri-products, particularly new crops, commodities and value-added products, as well as health foods;
- the present low-input-use efficiency, which can be enhanced considerably through adoption of available technological options as well as policy interventions;
- availability of vast arable land, all kinds of climate, cheap labour and, above all, hardworking farmers;
- future possibilities of resource generation by bringing a new corporate culture into the existing research organizations.

There is no doubt that these 'uncommon opportunities' can be harnessed to our advantage, provided we bring in the needed change despite stiff resistance from within. This paradigm shift is a must now in order to make our NARS both responsive and responsible.

The following are the four major areas where change has become imminent and must be accomplished as priority.

Institutional Change

Institutions are the foundations of required social change and advancement of any society. Most of our research institutions are 40–50 years old. Also, their equipment has become old and obsolete. They need immediate renovation and replacements. The process of 'mushrooming' of institutions needs to be curbed. Rather than horizontal expansion, we need to consolidate and revamp existing institutions and bring in inter-institutional partnership in order to maximize the returns from our investments in agricultural research. The doubling of plan allocation from Rs 125 billion to Rs 250 billion during the current 12th plan, as agreed to by the Planning Commission (now NITI Aayog), would indeed be a timely step in the right direction. We hope this justified demand is met by the Government.

Another area of institutional concern is to remove the imbalance from difficult agro-ecologies, especially the remote and difficult ecoregions. This change is warranted out of a concern for equity and the required institutional support for those areas that have been denied the benefits of new technologies in the past. This decision of the government to spend 10% of the allocations of each department for activities in the north-eastern region reinforces this concern. Support of this kind is critical for the faster growth of the hitherto bypassed regions as well as the social sector of the country. This calls for a major change in policies and programmes.

Organizational Change

As stated earlier, there is an urgent need to move from the NARI sytem to NARS, through an effective involvement of all the stakeholders. The need for organization and management (O&M) reforms in areas of human resource development, incentives and rewards for the performers, impact assessment and evaluation with in-built transparency, project-based budgeting, and decentralization linked with accountability are some of the critical elements associated with the future growth of the system. Hence, the enforcement of required change in the O&M system is more justified now than ever before. Public–private sector linkages are also to be built and institutionalized faster. Similarly, institutional collaboration with the advanced research institutions and international agricultural research centres will have to be strengthened for required excellence in science as well as for human resource development. ICT networking at global level will provide access to value-added information and knowledge, critical for the advancement of science. This demands a massive change in the existing IT culture.

Globally, donor support for agricultural research and training is declining. At the same time, we have to have the human resources that are globally competitive. To obviate this paradox, the best option is to generate resources internally and to build the required facilities for excellence in science. In the past, our scientists did not face this challenge, mainly on account of unstinted support from the government and policy makers. However, as this pressure is now building up, it is critical that our scientists and the system start to respond to this paradigm shift and start to mobilize internal resources fast. Many international organizations are already adjusting to this change. In future, a system's sustainability will have to be addressed more seriously. We must, therefore, respond favourably to this 'wake-up call'. Areas of contract research, consultancy, training, generation of technology-linked inputs in institutional laboratories/farms/workshops, patenting and corporatization are some of the options that need to be explored through appropriate change in our policies and procedures.

Change in the Research Portfolio

Radical changes are also called for in our method of conducting research. We have to continuously prioritize as well as re-prioritize our research portfolio to be in tune with the fast-changing global, regional and national needs. The top-down approach adopted in the past will have to be changed to a 'bottom-up' approach. A shift from 'project' to 'programme' mode and also from commodity/crop to a systems approach is now warranted. This would require a matrix mode of research management necessitating interdisciplinary teamwork among scientists. We can no longer afford individual scientist-orientated research agendas. Research must address institutional priorities and open-ended research will have to be made time-bound and targeted. A matrix mode of management would demand effective partnership between both the divisional head and the programme leader, besides sharing of responsibilities among the scientists involved. This change is most critical for the future success of our system and would demand commitment and a positive mindset from all the partners involved.

Excellence in science will have to be recognized through needed change in our incentive and reward system. In future, centres of excellence will have to be built around scientists and not around institutions. These centres of excellence will have to take added responsibilities for human resource development in their field of expertise. Also, institutions will have to undertake an ambitious programme for human resource development through careful planning and separate allocation of resources. As stated earlier, the research portfolio will have to be carefully balanced to meet the concern of different ecologies, conservation of natural resources and the protection of our environment. Globalization would also demand preparedness in areas of ICT, intellectual property rights, sanitary and phytosanitary systems, possible impact of removal of quantitative restrictions and likely imposition of non-tariff barriers. Those NARS prepared to change fast to address these concerns would be ahead of others. Hence, the need for urgency cannot be over-emphasized.

Change in Technology Dissemination

We have run out of soft options in the area of technology dissemination. Also, it is recognized more now than ever before that with available technologies, significant advancements in agriculture can be made provided they are effectively disseminated to farmers. The training-and-visit system has outlived its usefulness. It had mainly relied on the 'technology generation, technology transfer' model and presumed that all technologies would have wide acceptance and adoption, whereas it is well understood now that a continuum between 'technology generation-assessment-refinement and transfer' is critical for the success of new technologies. Hence, there is a need to change the front-line extension approach for the assessment and refinement of research information by establishing links between scientists and farmers and between institutions and laboratories. To ensure suitability of new technologies, scientists will have to adopt now the farmers' participatory approach and move out to use farmers' fields for revalidation and refinement of technologies. Also, the existing gap between the scientists and the farmers will have to be bridged. The tested ICAR model of Institute Village Linkage Programme (IVLP) is a bottom-up initiative in this direction focusing on farmers' specific needs rather than providing input-related package technology that has been found to be unsustainable in the long run. Scientist–farmer linkages also ensure a reduction in technology dissemination losses, critical for the success of any new technology.

In an information age, the role of appropriate information packages and their dissemination is equally important. It is not enough just to generate information; we must see that the required information is delivered to the end-user at the earliest opportunity and with the least dissemination loss. Thus, there is a need to have a single-window system of delivery for the farmers/end-users at the institute level. The establishment of an Agricultural Technology Information Centre (ATIC) would provide such a mechanism and contribute towards dissemination of information with the objective of helping farmers and other stakeholders to provide solutions to their problems and make available all technological information along with technological products for their testing and use.

To meet the changing needs, it is essential to create a cadre of technology agents from among the unemployed youth who are better-trained, -equipped and -committed to serve our farming community, while generating employment for themselves. Also, it is being felt that a publicly supported system of technology transfer may not be the best model in the future. We may, therefore, have to generate a new breed of competent technology agents who are well-trained and committed to provide specialized services on a custom-hire basis. In this process, not only are technology dissemination losses avoided but also appropriate technologies are disseminated faster. Another advantage of this approach is that these technology agents will become job creators not job seekers. Obviously, this would demand the institutions and the SAUs to undertake greater responsibility in future for vocational training programmes, thus requiring a change from the existing formal degree system to a more informal education system catering to different areas of agriculture.

An effective transfer of technology approach would demand quick delivery of technology-related inputs. For this purpose, provision of a revolving fund to the institutes/scientists to generate more of the technology-related inputs for effective dissemination would be a welcome development and would put pressure on the system to be more accountable in future.

The Krishi Vigyan Kendras are emerging as an effective institutional mechanism at rural district level for technology assessment, refinement and dissemination of the latest technologies. Their growing utility and demand has raised their number to almost 700, thus ensuring at least one KVK in each of the districts of the country. Such a vast network of KVKs raises the question of their performance and financial sustainability as well as their effective governance. To make them more effective and useful, joint ownership of these institutions, besides the ICAR, by the departments of agriculture of the central and the state governments, Panchayati Raj institutions, NGOs, farmers etc. has become necessary. All the stakeholders involved will have to own these KVKs and provide required backstopping. Such an approach would provide appropriate reinforcement of the programmes as well as the required interface at grassroots level, so critical for reaping the benefits from available new technologies. These KVKs would also have to serve as agricultural technology information centres in future and also as information centres for distance education and public awareness programmes using mass media and better communication mechanisms. Also, these institutions will have to make a paradigm shift from farmers' training at individual level to that of a group- or community-training approach, so that a larger section of society is benefitted.

All these initiatives require a strong interface between the research organizations and development departments at central, state and regional levels. While the Department of Agriculture and Cooperation (DAC)-ICAR interface provides such an opportunity at the centre, a mechanism needs to be worked out and institutionalized at state and district levels.

Epilogue

Change is a sign of growth. No organization that shows resistance to change can grow. Change is also a difficult process and requires commitment of not only the leaders but also the entire organization and system. Often the process of change in mindset meets with stiff internal resistance; yet the most dynamic institutions have grown through needed reforms to meet new challenges. The ICAR, as an apex organization for research and education in the field of agriculture, has grown with time. In the process, it achieved recognition and visibility, as evidenced by various revolutions (Green, White, Yellow and Blue), which many developing countries are still unable to achieve.

Today, the NARS, comprising the ICAR and the SAUs, has emerged as a strong organization through timely policy and structural reforms. The system must now gear to meet future challenges that are daunting. This will demand yet another critical self-examination coupled with mandatory change in the system. Change must always be welcomed despite difficulties that may be encountered. Dynamic change will require the commitment of the entire scientific community and all those associated with the system to make India's agriculture strong and resilient as we move through the new millennium.

Reference

Paroda, R.S. (2014) Change we must, but change is difficult. *A Compendium of Addresses by Past Presidents*. National Academy of Agricultural Sciences, NASC Complex, New Delhi.

Index

CABI – who we are and what we do

This book is published by **CABI**, an international not-for-profit organisation that improves people's lives worldwide by providing information and applying scientific expertise to solve problems in agriculture and the environment.

CABI is also a global publisher producing key scientific publications, including world renowned databases, as well as compendia, books, ebooks and full text electronic resources. We publish content in a wide range of subject areas including: agriculture and crop science / animal and veterinary sciences / ecology and conservation / environmental science / horticulture and plant sciences / human health, food science and nutrition / international development / leisure and tourism.

The profits from CABI's publishing activities enable us to work with farming communities around the world, supporting them as they battle with poor soil, invasive species and pests and diseases, to improve their livelihoods and help provide food for an ever growing population.

CABI is an international intergovernmental organisation, and we gratefully acknowledge the core financial support from our member countries (and lead agencies) including:

Ministry of Agriculture
People's Republic of China

Agriculture and
Agri-Food Canada

Ministry of Foreign Affairs of the
Netherlands

Discover more

To read more about CABI's work, please visit: **www.cabi.org**

Browse our books at: **www.cabi.org/bookshop**,
or explore our online products at: **www.cabi.org/publishing-products**

Interested in writing for CABI? Find our author guidelines here:
www.cabi.org/publishing-products/information-for-authors/